Steroid Hormone Action

EDITED BY

Malcolm G. Parker
*Imperial Cancer Research Fund,
Lincoln's Inn Fields, London*

OXFORD UNIVERSITY PRESS
Oxford New York Tokyo

Oxford University Press, Walton Street, Oxford OX2 6DP

Oxford New York Toronto
Delhi Bombay Calcutta Madras Karachi
Kuala Lumpur Singapore Hong Kong Tokyo
Nairobi Dar es Salaam Cape Town
Melbourne Auckland Madrid
and associated companies in
Berlin Ibadan

Oxford is a trade mark of Oxford University Press

Published in the United States
by Oxford University Press Inc., New York

A catalogue record for this book is available from the British Library

Library of Congress Cataloging in Publication Data
Steroid hormone action/edited by Malcolm G. Parker.
(Frontiers in molecular biology)
Includes bibliographical references and index.
1. Steroid hormones–Physiological effect. 2. Genetic regulation.
3. Steroid-binding proteins. I. Parker, Malcolm G. II. Series.
[DNLM: 1. Steroids–physiology. 2. Gene Expression Regulation.
3. Receptors, Steroid–physiology. WK 150 S83566 1993]
QP572.S7S742 1993 612.4'05–dc20 93–1165
ISBN 0–19–963393–2 (h/b)
ISBN 0–19–963392–4 (p/b)

Typeset by
Footnote Graphics, Warminster, Wilts.
Printed in Great Britain by
The Bath Press, Avon

Preface

In the past decade there has been dramatic progress in our understanding of the molecular mechanism of action of steroid hormones. Much of this progress has resulted from technical advances in molecular biology which have allowed the analysis of hormonally regulated genes, steroid hormone receptors and, more recently, steroid hormone binding proteins and steroidogenic enzymes. The first breakthrough came with the characterization of specific target genes and the identification of DNA binding sites that were shown to have the properties of transcriptional enhancers. Since that time transcriptional activation has been an extremely active area of research and this subject is reviewed in the chapter by Hinrich Gronemeyer. Transcriptional repression on the other hand was often overlooked but this is no longer the case and a number of potential mechanisms are described by Jacques Drouin.

Another major breakthrough came with the isolation of DNA clones for steroid hormone receptors. Analysis of these clones demonstrated that these receptors are not only very similar to one another but also to many other proteins that form members of a large superfamily of nuclear receptors. The availability of DNA clones for receptors has made it possible to study their structure and function at the molecular level. The superfamily of nuclear receptors is reviewed by Bert O'Malley. In addition, Bill Pratt discusses the role of heat shock proteins in receptor function, and Sophie Dauvois and Malcolm Parker discuss current views on the mechanism of action of hormone antagonists. With the availability of purified receptors and isolated domains there is now a considerable effort to determine their 3-D structure. The first success has been the elucidation of the structure of the DNA binding domain for the glucocorticoid and oestrogen receptor and Len Freedman summarizes this work. Naturally occurring mutations in receptors are the subject of Michael McPhaul's chapter on defective receptors associated with a number of clinical diseases.

Finally, two areas which are often not covered in books on steroid hormone action are the extracellular steroid binding proteins and steroidogenic enzymes. Although the importance of metabolic changes in many of the steroid hormones has been recognized for a long time, their significance is only just beginning to emerge. These two topics are reviewed in chapters by Geoffrey Hammond and John Funder.

In view of a role for steroid hormones in so many developmental pathways, I feel sure the book will be of interest to readers from a wide range of fields including molecular biologists, endocrinologists, pharmacologists, and clinicians. I would like to thank the authors for their hard work and patience during the editorial process and hope their efforts will be well-received.

London M. G. P.
April 1993

Contents

List of contributors xiii
Abbreviations xiv

1 Extracellular steroid-binding proteins 1

GEOFFREY L. HAMMOND

1. Introduction 1
2. Sex steroid-binding proteins 1
2.1 Structure 1
2.2 Gene expression 7
2.3 Function 9
3. Corticosteroid-binding globulin 10
3.1 Structure 10
3.2 Gene expression 12
3.3 Function 15
4. Conclusions 17
Acknowledgements 17
References 17

2 Importance of steroidogenesis in specific hormone action 26

JOHN W. FUNDER

1. Signals 26
2. Receptors 26
2.1 Specificity 26
3. Mineralocorticoid receptors 28
3.1 Tissue specificity of aldosterone binding 28
3.2 Clinical studies: apparent mineralocorticoid excess 29
3.3 Licorice and carberoxolone 30
3.4 11β Hydroxylation and 18 oxidation 30

4. Aldosterone synthesis and mineralocorticoid specificity 32
4.1 Glucocorticoid remediable hyperaldosteronism 33
4.2 11β Hydroxysteroid dehydrogenase: a family of enzymes? 33
4.3 Steroidogenesis: summary 34
5. Hormone response elements 35
5.1 Experimental mineralocorticoid specificity 35
5.2 Mineralocorticoid response elements 38
5.3 Other possible sources of MR/GR selectivity 38
6. Biosynthesis: secretion and activation 39
6.1 Vitamin D 39
6.2 Androgens 39
6.3 Oestrogens 39
7. Metabolism: activation and inactivation 40
7.1 Aldosterone 40
7.2 Antagonists 40
7.3 Non-selectivity *in vitro* 41
7.4 Further directions 41
References 41

3 Overview of the steroid receptor superfamily of gene regulatory proteins 45

BERT W. O'MALLEY and MING-JER TSAI

1. Introduction 45
2. Conserved sequences and function 45
3. Regulation of gene expression 48
4. 'Orphan' receptors 54
5. Activation of receptors 56
Summary 57
References 58

4 Role of heat-shock proteins in steroid receptor function 64

WILLIAM B. PRATT

1. Introduction 64

2. Receptor-associated proteins 65

2.1 Hsp90 66

2.2 Hsp70 71

2.3 Hsp56 73

2.4 Other receptor-associated proteins 75

2.5 Stoichiometry of the heat-shock protein heterocomplex 76

3. Cell-free receptor heterocomplex assembly 78

3.1 Binding of hsp90 to *in vitro* translated steroid receptors 78

3.2 Reconstitution of receptor heterocomplexes with reticulocyte lysate 80

3.3 General properties of heterocomplex reconstitution 80

3.4 Is protein unfoldase activity of hsp70 required to form the receptor–hsp90 complex? 81

3.5 A model of heterocomplex assembly 82

4. The heat-shock protein complex and steroid hormone action 83

4.1 Model of receptor transformation driven by the trapped folding energy of the HBD 83

4.2 Relevance of the heat-shock protein complex to hormone action in the cell 83

References 85

5 Nuclear hormone receptors as transcriptional activators 94

HINRICH GRONEMEYER

1. Introduction 94

2. Nuclear receptors are composed of structural elements that contain redundant and overlapping functions 96

2.1 Co-operation of proto-signals is responsible for nuclear localization 97

2.2 DNA binding, dimerization, and response element selection 98

3. Transcriptional activation 105

3.1 Nuclear receptors contain two transcription activation functions that act in a cell- and promoter-specific fashion 105

3.2 Evidence for transcription intermediary factors which mediate the enhancer function of nuclear receptors 106

3.3 Isoform-specific activation of transcription 108

4. Perspectives 108

Acknowledgements 109

References 109

6 Repression of transcription by nuclear receptors 118

JACQUES DROUIN

1. Introduction 118

2. Mechanisms of transcriptional repression 119

2.1 Competition for overlapping binding sites 120

2.2 Null binding site 120

2.3 Silencer or direct repressor element 120

2.4 Repression by interference 120

2.5 Repression by quenching 121

2.6 Repression by squelching 121

2.7 Repression by tethering 122

3. Negative hormone response elements 122

3.1 Negative glucocorticoid response elements 122

3.2 Negative thyroid hormone response elements 125

4. Dominant negative forms of nuclear receptors 127

5. Receptor heterodimers as repressors 128

6. Silencer or direct repressor activity 129

7. Competition for overlapping binding sites 129

7.1 Competition between receptors 130

7.2 Competition for binding sites of other factors 130

8. Steroid repression by quenching mechanism 133

9. Repression without direct DNA binding of receptor 134

10. Conclusion 135

References 135

7 Structure and function of the steroid receptor zinc finger region 141

LEONARD P. FREEDMAN

1. Introduction 141

2. Hormone response elements 143

3. Functional dissection of the steroid receptor DNA-binding domain 145

3.1 Mutagenesis of the DNA-binding domain 145

3.2 Role of dimerization in DNA binding 148

4. Structure of the steroid receptor zinc finger region 150
4.1 Biochemistry of zinc co-ordination 150
4.2 Solution structure by two-dimensional nuclear magnetic resonance 151
4.3 Crystallographic analysis of the GR zinc finger–DNA complex 153
5. Stereochemical basis of nuclear receptor dysfunction in disease 158
6. Conclusions 159
References 160

8 Mechanism of action of hormone antagonists 161

S. DAUVOIS and MALCOLM G. PARKER

1. Introduction 166
2. Antioestrogens 166
2.1 General classification and structure 166
2.2 Clinical uses 168
2.3 *In vitro* activity 169
3. Antiandrogens 170
4. Antiprogestins 171
5. Mechanism of action of steroid hormones 171
6. Anti-hormone binding 174
7. Mechanism of action of antagonists with partial agonist activity 175
8. Mechanism of action of 'pure' oestrogen antagonists 177
9. Mechanism of action of other types of antagonist 178
10. Concluding remarks 179
References 180

9 Mutations in steroid hormone receptors causing clinical disease 186

MICHAEL J. McPHAUL

1. Introduction 186
2. Androgen resistance 186
2.1 Molecular basis of androgen resistance 188
2.2 Distribution and type of mutations causing androgen resistance 190
2.3 Patient phenotype and genetic mutations 191

3. Thyroid hormone resistance 193

3.1 Receptor-binding abnormalities in thyroid hormone resistance 193

3.2 Generalized resistance to thyroid hormone 194

3.3 Genetic mutation and phenotype 196

4. Vitamin D resistance 197

4.1 Receptor binding in fibroblasts in patients with vitamin D resistance 197

4.2 Genetic defects causing vitamin D resistance 198

5. Glucocorticoid resistance 198

6. Conclusions 200

References 201

Index 209

Contributors

S. DAUVOIS
Molecular Endocrinology Laboratory, Imperial Cancer Research Fund, Lincoln's Inn Fields, London WC2A 3PX, UK.

JACQUES DROUIN
Institut de Recherches Cliniques de Montréal, 110 avenue des Pins Ouest, Montréal, Québec H2W 1R7, Canada.

LEONARD P. FREEDMAN
Cell Biology and Genetics Program, Sloan-Kettering Institute, 1275 York Avenue, New York, NY 10021, USA.

JOHN W. FUNDER
Baker Medical Research Institute, PO Box 348, Prahran 3181, Victoria, Australia.

HINRICH GRONEMEYER
Laboratoire de Génétique Moléculaire des Eucayotes, CNRS Unité 184 de Biologie Moléculaire et de Génie Génétique de l'INSERM, Institut de Chimie Biologique, Faculté de Médecine, 11 rue Humann, 67085 Strasbourg Cedex, France.

GEOFFREY L. HAMMOND
Cancer Research Laboratories, London Regional Cancer Centre, 790 Commissioners Road East, London, Ontario N6A 4L6, Canada.

MICHAEL J. McPHAUL
University of Texas Southwestern Medical Centre, 5323 Harry Hines Boulevard, Dallas, TX 75235, USA.

BERT W. O'MALLEY
Department of Cell Biology, Baylor College of Medicine, One Baylor Plaza, Houston, TX 77030, USA.

MALCOLM G. PARKER
Molecular Endocrinology Laboratory. Imperial Cancer Research Fund, Lincoln's Inn Fields, London WC2A 3PX, UK.

WILLIAM B. PRATT
Department of Pharmacology, University of Michigan Medical School, Ann Arbor, MI 48109-0010, USA.

MING-JER TSAI
Department of Cell Biology, Baylor College of Medicine, One Baylor Plaza, Houston, TX 77030, USA.

Abbreviations

AAT	α_1-antitrypsin
ABP	androgen-binding protein
ACT	α_1-antichymotrypsin
ACTH	adrenocorticotrophic hormone
ARE	androgen response element
ARP1	apolipoprotein AI regulatory protein 1
bp	base pair
cAMP	cyclic AMP
CAT	chloramphenicol acetyl transferase
CBG	corticosteroid-binding globulin
cDNA	complementary DNA
CEF	chick embryo fibroblast (cells)
α-CG	α-chorionic gonadotropin
CHO	Chinese hamster ovary (cells)
COUP-TF	chicken ovalbumin upstream promoter transcription factor
CRABP	cellular retinoic acid binding protein
CREB	cyclic AMP response element binding protein
DBD	DNA-binding domain
DEX	dexamethasone
DMS	dimethylsulfate
DR1	direct repeat of PuG (G/T) TCA spaced by 1 base pair
EAR	v-erbA-related factor
EDTA	ethylenediamine tetra-acetic acid
EGF	epidermal growth factor
ER	oestrogen receptor
ERE	oestrogen response element
FSH	follicle-stimulating hormone
GR	glucocorticoid receptor
GRE	glucocorticoid response element
GRTH	generalized resistance to thyroid hormone
HBD	hormone-binding domain
HEGO	expression vector for hER
HEO	expression vector for hER Gly400-Val
hER	human oestrogen receptor
HRE	hormone response (enhancer) element
11 βHSD	11β-hydroxysteroid dehydrogenase
hsp	heat-shock protein
IGF	insulin-like growth factor

LTR	long terminal repeat
MCM1	mini chromosome maintenance
MMTV	mouse mammary tumour virus
MR	mineralocorticoid receptor
MRE	mineralocorticoid response element
mRNA	messenger RNA
NAD	nicotinamide adenine dinucleotide
NF-KB	nuclear factor kappa B
NGF	nerve growth factor
NLS	nuclear localization signal
NMR	nuclear magnetic resonance
pNLS	proto nuclear localization signal
POMC	pro-opiomelanocortin
PPAR	peroxisome proliferator activated receptor
PR	progestin/progesterone receptor
PRE	progesterone response element
RAR	retinoic acid receptor
RARE	retinoic acid response element
RXR	retinoid X receptor
SEM	standard error of the mean
SERPIN	serine proteinase inhibitor
SHBG	sex hormone-binding globulin
SRE	steroid response element
SWI	class of yeast genes involved in transcription regulation
TAF	transcription activation function
TAT	tyrosine aminotransferase
TBG	thyroxine-binding globulin
TBP	TATA box binding protein; component of TFIID
TF	transcription factor
TGF	transforming growth factor
TIF	transcription intermediary factor
TK	thymidine kinase
TPA	tetradecanoylphorbolacetate

1 | Extracellular steroid-binding proteins

GEOFFREY L. HAMMOND

1. Introduction

Steroid hormones in the blood and extracellular compartments of most vertebrate species interact with two high-affinity plasma proteins that are generally known as sex hormone-binding globulin (SHBG) and corticosteroid-binding globulin (CBG). These proteins are structurally unrelated and do no resemble other plasma proteins that bind steroids with low affinity and limited specificity, such as albumin and orosomucoid, or any other steroid-binding proteins, including the intracellular steroid hormone receptors and enzymes responsible for steroid biosynthesis or metabolism. It is generally assumed that plasma SHBG and CBG function primarily as transport proteins and limit the metabolic clearance of biologically active steroids, but recent advances in our understanding of their structures indicate that they may act in much more sophisticated ways than previously considered (1, 2). It is the purpose of this review to highlight this information, and to provide a perspective on how these proteins may regulate the activities of steroid hormones.

2. Sex steroid-binding proteins

Plasma SHBG is produced by hepatocytes (3), and its testicular homologue, androgen-binding protein (ABP), is synthesized by Sertoli cells (4–6). Their steroid-binding properties have been studied extensively in numerous species and are reviewed in an comprehensive monograph on steroid–protein interactions (7). The sex-steroid binding proteins in primates (8) and bats (9) have relatively high affinities for both androgens and oestrogens compared with their counterparts in other species, but they invariably bind androgens with greater affinity than oestrogens (8). Our understanding of the composition of these proteins has been advanced by the resolution of their primary structures (10–12), and this information will be reviewed with respect to their possible functions.

2.1 Structure

Physicochemical analyses of SHBG and ABP from numerous species (13–16) have indicated that they exist in biological fluids as a homodimeric molecule of approximately 90 kDa. It has not been possible to demonstrate steroid-binding activity using dissociated subunits, and it is generally believed that the two subunits must somehow unite to form a single steroid-binding site. Several models have been proposed to explain this (17), but these are highly speculative in the absence of any tertiary structural information.

These proteins are also glycosylated and the average carbohydrate composition of human SHBG has been defined (18), although variants with additional carbohydrate structures exist (19–21). Variations in glycosylation probably account for the electrophoretic heterogeneity of SHBG and ABP isolated from different tissues (14, 22–24), but little is known about the impact of this on the structure or biological activity of these proteins.

2.1.1 Protein structure

The primary structure of human SHBG was first resolved by direct sequence analyses of the purified protein (10), and confirmed by the isolation and characterization of a complementary DNA (cDNA) encoding the mature polypeptide (11). This cDNA-deduced sequence differed from the chemically derived sequence at residue 259, and reflects an editing error in the cDNA sequence. Furthermore, it is clear that the codon for Leu-259 is TTG, because this has been identified in two genomic fragments that contain the SHBG gene (25, 26) and in two partial cDNA sequences (27, 28). When considered together, these studies indicate that the SHBG subunits are produced as a 402-residue polypeptide, which includes a typical hydrophobic leader sequence of at least 29 residues. The fact that the amino-terminus of purified human SHBG is heterogeneous (13, 19) may reflect limited proteolytic degradation, but it is also possible that cleavage of the leader sequence does not always occur in exactly the same position. Thus, the mature human SHBG subunit comprises maximally 373 amino acids with a polypeptide molecular weight of 40.5 K.

The amino acid sequence of rat ABP has also been deduced from several testicular cDNAs (12, 29), and sequence comparisons of human SHBG and rat ABP have clearly established that these proteins are related at the primary structural level (12, 30). They also share regions of significant sequence similarity with protein S (31) and two basement membrane proteins, laminin and merosin (32). The primary structures of sex steroid-binding proteins in several other species have since been determined (33, 34), and a phylogenetic comparison of these sequences is presented in Figure 1. This reveals several conserved features among these molecules that are likely to be important. These include the two pairs of cysteine residues that form disulphide bridges in human (10) and rabbit (34) SHBG and an unusual pattern of alternating leucyl residues between Leu-267 and Leu-281 in the human sequence. This region has been implicated in steroid binding (10), but there is no evidence for this. Recently, Lys-134 (35), Met-139 (36), and His-235 (37) have been identified as interacting with various steroid-affinity ligands, but only Met-139 is conserved in all species (Fig. 1). Moreover, Lys-134 is located in a poorly conserved region of the molecule (Fig. 1), and the suggestion that this residue is involved in

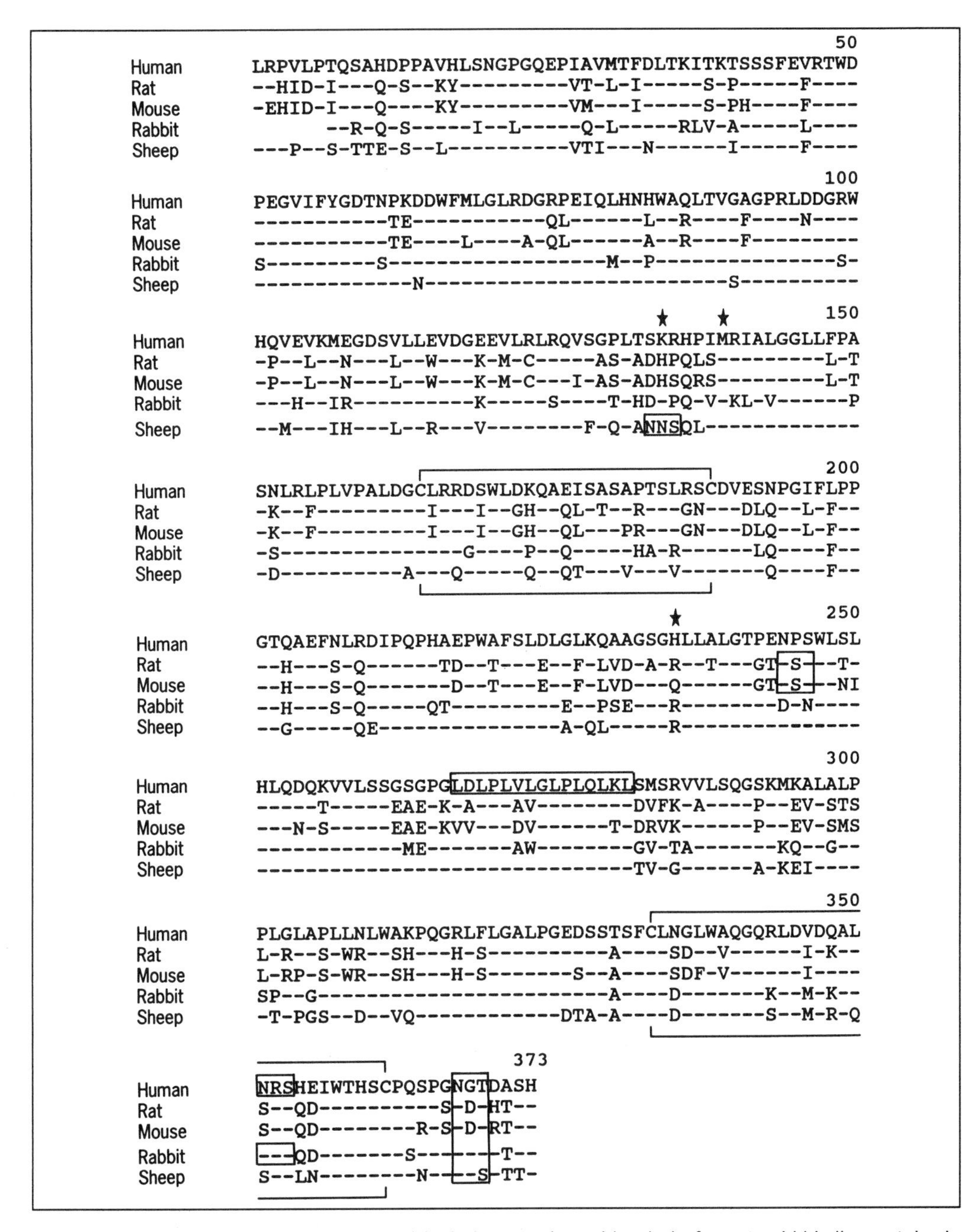

Fig. 1 Comparison of the primary structures (single-letter amino acid codes) of sex steroid-binding proteins in different mammalian species. A dash indicates an identical amino acid in the human SHBG sequence. Stars indicate amino acids that have been affinity labelled or otherwise identified as being involved in steroid binding. The sequence of leucine repeats that has been suggested to play a role in formation of the steroid-binding site is boxed, as are the consensus sites for *N*-glycosylation. The cysteines that form disulphide bridges are indicated in their respective pairs

steroid binding remains speculative. Regions of low sequence homology may represent immunogenic domains, one of which may include His-136 because this residue appears to constitute part of the epitope recognized by a monoclonal antibody that only recognizes SHBG in Old World primates (38, 39).

Recently, we have adopted a molecular approach for the identification of important residues and domains within the sex steroid-binding proteins, and accomplished this by inserting a cDNA for the human SHBG precursor polypeptide into a suitable vector for constitutive expression in mammalian cells (40). The product is secreted as a dimeric glycoprotein that exhibits the same subunit size heterogeneity as SHBG in serum samples, and binds steroids with appropriate afinity and specificity (40). We have also performed experiments to locate the steroid-binding domain within the human SHBG molecule by creating a human SHBG–rat ABP chimera, as well as several mutants in which specific amino acids were altered by site-directed mutagenesis of the human SHBG cDNA (39). The chimeric molecule comprised the first 205 amino-terminal residues of human SHBG fused to the 168 carboxy-terminal portion of rat ABP, and it retained exactly the same steroid-binding characteristics as human SHBG (39). This is remarkable because the dissociation of androgens from rat ABP is significantly more rapid than from human SHBG, and we therefore concluded that the residues responsible for its high-affinity steroid binding are probably located within the amino-terminal half of the protein. We consequently focused attention on the region surrounding a residue (Met-139) in human SHBG that interacts with the photoaffinity ligand 17β-hydroxy-androst-4,6-diene-3-one (36). When this residue was converted to tryptophan, it resulted in a marked reduction in steroid-binding affinity without any detectable perturbation in the immunochemical or physicochemical properties of the recombinant protein (39). This confirms that this residue is probably located within the steroid-binding site, and additional mutations within this region may also be informative.

The resolution of the primary structure of human SHBG has helped to locate a series of amino acids within the amino-terminal portion of the molecule that inhibit SHBG binding to a protein isolated from human prostate plasma membranes (41). This region, between residues 48 and 57 in human SHBG, is highly conserved between species (Fig. 1), and mutations within it may also provide information about its interaction with the plasma membranes of sex-steroid responsive tissues and cell types (2).

2.1.2 Associated carbohydrate structures

Analyses of SHBG from human pregnancy serum have indicated that carbohydrates constitute approximately 12% of the total mass of the glycoprotein (13, 18). More detailed studies of carbohydrate composition have shown that each subunit comprises one *O*-linked oligosaccharide and two *N*-linked biantennary carbohydrate structures of the *N*-acetyl-lactosamine type (18), and sites of attachment of these sugar chains were located by sequence analyses of human SHBG (10). The carbohydrate composition of sex steroid-binding proteins in other species is less well defined, but it would appear from their primary structures that at least one of the two sites for *N*-glycosylation in human SHBG is conserved despite evolutionary drift in the consensus sequence (Fig. 1). It is interesting that the other site in the human sequence is present in rabbit SHBG, but not in the mouse, rat, or sheep sequences (Fig. 1). The rodent sequences contain a second consensus site for *N*-glycosylation at residues 224–226. This is absent in sheep SHBG, but in this species there is an unusual consensus site for *N*-glycosylation within a hypervariable region close to Met-139 (Fig. 1), and whether this is utilized remains to be established.

The *O*-linked carbohydrate chain in the human protein has been assigned to Thr-7 (10). This residue could not be identified during amino-terminal sequence analysis, and it was originally suggested that an *N*-linked carbohydrate chain may be attached to it because adjacent residues conform to a consensus sequence for *N*-glycosylation (13). However, based on observations that an *O*-linked carbohydrate chain was associated with a proline-rich peptide fragment that also contained a threonine residue (18), it was assumed that this corresponded to the unidentifiable residue at position 7 (10). Furthermore, the cDNA-deduced sequence of human SHBG predicts a threonine at this position (11), and the oligosaccharide attached at this position must therefore be *O*-linked. Much less is known about the location of *O*-linked oligosaccharide chains in sex steroid-binding proteins of other species. The only other protein that has been sequenced directly is rabbit SHBG, and this does not appear to contain an *O*-linked oligosaccharide (34). Furthermore, the rodent ABP sequences have an isoleucine at position 7 (Fig. 1) and it is unlikely that this carbohydrate chain is biologically very important.

Attempts to assess the significance of carbohydrates associated with sex steroid-binding proteins have generally relied on chemical and enzymatic methods for deglycosylation (42), but it is difficult to ensure that carbohydrates are completely removed, or that protein structures are unaffected by these treatments. It has, however, been demonstrated that the biological half-life of bovine SHBG in the blood circulation of rats is decreased by desialylation, presumably by a more active clearance of the protein by the asialoglycoprotein receptor (43). These and other questions may now be resolved by producing mutant proteins lacking specific carbohydrate structures. Our studies using this approach (44) have demonstrated that carbohydrates do not influence the overall structure of the molecules, their ability to dimerise or their affinity and specificity for steroid hormones. However, the production and/or secretion of human SHBG mutants lacking both *N*-linked carbohydrate chains from Chinese hamster ovary cells is impaired, although these cells will produce unglycosylated human SHBG with appropriate steroid-binding characteristics (44). This is important because it suggests that biologically active forms of this protein may be produced in bacteria, and this would greatly facilitate further structural analyses of this protein.

In addition to cell-specific differences in glycosylation (40), it has been reported that the carbohydrate composition of SHBG may be influenced by hormonal treatment (45). We have examined this by culturing a BW-1 mouse hepatoma clone containing a human SHBG cDNA expression vector (40) in the presence of oestrogen, thyroid hormone, or insulin, but failed to observe any major differences in the concanavalin A-binding properties of the recombinant products (Fig. 2). This suggests that these hormones do not influence the post-translational modification of SHBG to any great extent, but we cannot exclude the possibility that the BW-1 cell line is inappropriate for this kind of study or that we were unable to resolve subtle changes in carbohydrate composition by a single-step elution from concanavalin A. Nevertheless, it is obvious from isoelectric focusing studies of human SHBG that several molecular isoforms exist, and these are almost certainly due to

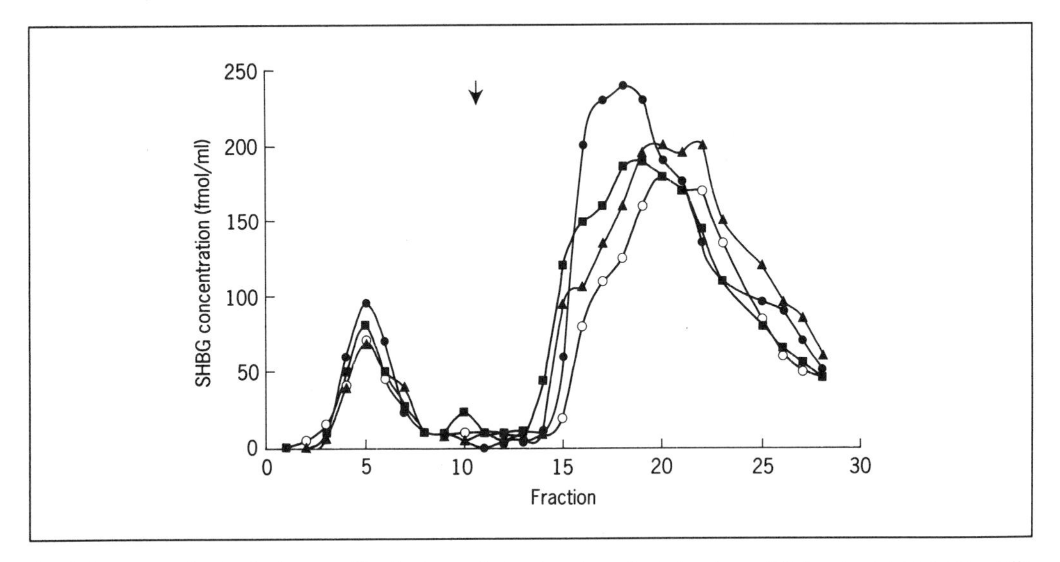

Fig. 2 Concanavalin A–Sepharose (5 ml) chromatography of media taken from BW-1 mouse hepatoma cells containing a human SHBG cDNA expression construct (40), after culture in phenol red-free Modified Eagle Medium containing 5% charcoal-stripped fetal bovine serum (●) or the same medium containing 10^{-8} ethinyl estradiol (○), 10^{-8} thyroxine (▲), and 10^{-8} insulin (■) for 3 days. The SHBG content of 1-ml fractions was measured by a radioimmunometric assay (40), and the arrow represents the point at which buffer containing 5% methyl-α-D-mannopyranoside was added to elute lectin-bound proteins

variations in carbohydrate composition because they can be eliminated by treatment with neuraminidase to remove terminal sialic acid residues (13). It is unclear whether this heterogeneity is biologically significant, but similar variability in the carbohydrate composition of the glycoprotein hormones may influence their ability to interact with their receptors and initiate signal transduction (46, 47). It has also been claimed that sex steroid-binding proteins may increase intracellular level of cyclic AMP when incubated with prostate cancer cells (48). This remains to be verified, but the diversity in the content and composition of carbohydrates associated with individual molecules may also provide an additional level of specificity to any interaction between sex steroid-binding proteins and membrane receptors on specific cell types (49).

2.2 Gene expression

Our understanding of the sex steroid-binding protein genes in various species has been influenced to a great extent by the cloning of the cDNAs and genes for human SHBG (11, 25, 26) and rat ABP (12, 29, 50). In essence, this work has demonstrated that plasma SHBG and testicular ABP in a given species share the same primary structures and are the products of a single gene (25). Attempts have been made to define the rat ABP promoter (50), but the molecular basis of the hormonal regulation and tissue specificity of expression of this gene has not yet been defined; this may be attributed to the fact that the expression of the human and rat genes is complicated by variations in the processing of primary transcripts in different tissues.

2.2.1 Gene structure and chromosomal location

The structural organization of the human SHBG and rat ABP genes is similar, and the coding regions for these proteins are distributed over eight exons that span 3–5 kb of genomic DNA (25, 50). The rat gene has been assigned to chromosome 10 (51) and the human gene is located on the short arm of chromosome 17 (p12–13) in a region that contains the p53 tumour suppressor gene (52). A cDNA for mouse ABP has also been characterized (33) and its gene has been mapped to a region on mouse chromosome 11 syntenic with the short arm of human chromosome 17 (53).

The human and rat genes both contain several repetitive Alu-like elements within intervening sequences adjacent to exons that are differentially utilized (25, 50). This may be significant because repetitive sequences may mediate chromosomal recombination events that result in gene rearrangements (54). It is therefore possible that they are somehow responsible for the differential exon utilization (25, 50) and trans-splicing events (51) associated with the expression of these genes. Indeed, cDNAs for numerous SHBG-related messenger RNAs (mRNAs) have been identified in different tissues from several species (25, 26, 51, 55), and multiple transcripts that hybridize with SHBG cDNAs have been identified in tumour cell extracts (56, 57), but little is known about their structure and whether they encode proteins, or have some other function. However, the SHBG gene lies within a particularly fragile region of the human genome (52), and it is possible that rearrangement or transposition of the gene in transformed cells results in abnormal transcripts of questionable significance.

The most consistent difference in the cDNA sequences observed to date indicates that the exon containing the translation initiation codon for the SHBG precursor polypeptide is alternatively utilized. It is not known whether they represent the first exon within a particular transcript and are under different transcriptional control, but it has been suggested that the promoter responsible for ABP production in the testis is located immediately 5′ to this exon in the rat gene (50). However, there is no direct evidence that this region functions as a promoter, and we have recently identified a sheep SHBG cDNA that includes a 63-bp 5′ sequence with 75% sequence identity to the proposed proximal promoter of the rat ABP gene.

2.2.2 Spatial and temporal expression

Although the liver and testes are major sites of sex steroid-binding protein gene expression, immunoreactive ABP and transcripts that resemble rat ABP mRNA have been identified throughout the adult rat brain (55). Moreover, analyses of several cDNAs from a brain library indicate that their sequences diverge at the 3′ junction of the exon containing the translation initiation codon for the ABP precursor (55). In this context, they resemble the SHBG-related cDNAs from a human testis library (25, 26) and a fetal rat liver ABP-related cDNA (51), but their 5′ regions are unique and are represented by distinct exons located upstream from the proposed promoter region for the testicular ABP transcription unit (50). These findings further illustrate the complexity of sex steroid-binding protein gene expression, and it remains to be determined whether homologues of these tissue-specific gene products are present in other species and are biologically important.

Sex steroids influence the differentiation and maturation of accessory sex tissues and developing fetal organs (58, 59), and the production of SHBG by the fetus undoubtedly modulates the activities of sex steroids during critical periods of development. It is therefore interesting that the gene is expressed transiently in the fetal rat liver between days 15 and 17 of gestation (51). This is remarkable because adult rats function normally without producing a plasma sex steroid-binding protein, and it was generally assumed that the gene was never expressed in the rat liver. During fetal life the presence of plasma SHBG is therefore probably very important, and this may explain why there have been no reliable reports of a total lack of this protein in humans. Apart from studies of the ontogeny of ABP gene expression in the rat testis (29) and liver (51), very little is known about the underlying mechanisms responsible for temporal fluctuations in sex steroid-binding protein gene expression, although there is indirect evidence that the ontogeny of SHBG production during infancy and adolescence is under hormonal control (60).

2.2.3 Hormonal regulation

Several hormones alter the plasma levels of SHBG, including sex steroids (61), thyroxine (61), insulin, and insulin-like growth factor 1 (62), but little is known about how they may influence SHBG gene expression. Studies of SHBG biosynthesis in hepatoma cell lines may help address this issue (56. 63), but may not necessarily reflect the expression of the gene in normal hepatocytes. The expression of the ABP gene in the rat testis has been studied in intact animals (29) and by use of primary Sertoli cell cultures (50). Collectively these studies indicate that follicle-stimulating hormone (FSH) acts on Sertoli cells directly to increase the accumulation of ABP mRNA while testosterone increases testicular ABP mRNA levels in intact animals but not in Sertoli cells in culture. This suggests that testosterone acts indirectly via some other mediator of Sertoli cell function.

2.3 Function

2.3.1 Steroid transport and bioavailability

In human blood, SHBG levels influence the plasma distribution and metabolic clearance of the major sex steroid hormones, testosterone and oestradiol (64), while ABP is thought to play an important role in maintaining the high androgen levels between the testis and epididymis that are considered important for sperm maturation (65). Abnormally low plasma concentrations of SHBG are often found in patients with androgen (61) and oestrogen (66) dependent diseases that are generally associated with a concomitant increase in the amount of non-protein-bound steroid in the blood (64). Many synthetic progestins also bind to SHBG with high affinity, and their biological activities may be influenced by SHBG plasma levels (67). Moreover, synthetic steroids may displace natural steroids from the SHBG steroid-binding site and alter the normal plasma distribution of natural steroid hormones (68).

These clinical observations support the hypothesis that only the non-protein-bound or 'free' steroid fraction in the blood is biologically active. While this may be true for many tissues in intimate contact with plasma, it may not apply in steroid-sensitive tissues that are separated from the blood circulation, such as the brain, or exposed to other biological fluids rich in sex steroid-binding proteins, such as those of the male reproductive tract. There is also a tendency to assume that steroids enter cells in precisely the same way, and to ignore the many factors that influence this process, including tissue temperature, blood flow rates and capillary transit time, the lipid composition of different target cell membranes, the lipophilic characteristics of different steroids, and the intracellular location of their respective receptors. Furthermore, evidence that the sex steroid-binding protein gene is expressed in the rat brain (55) casts doubt on the validity of a large body of experimental data obtained using the rat brain as a model of steroid bioavailability (69). This is a controversial subject, but there is now considerable evidence that steroid-binding proteins play a more active role in determining the entry or exit of steroid hormones from certain cells. However, these alternative mechanisms are often dismissed simply because they do not conform to the prevailing dogma that 'the only active steroid is a free steroid'.

2.3.2 Plasma membrane interactions

The immunocytochemical localization of human SHBG (70) and rat ABP (71) in cells of male reproductive tract tissues provided the first evidence that these proteins interact directly with sex steroid-responsive cells. The specificity of these and other similar studies involving the use of polyclonal antisera (72) may always be questioned, but they have been complemented by parallel investigations that have demonstrated the cellular uptake of protein (73, 74) and characterized plasma membrane binding sites with high affinity and specificity for SHBG in several tissues (75, 76). The affinity of these interactions appears to vary depending on the tissues or cell types examined (75, 77), and may be influenced by the type and amount of ligand in the SHBG steroid-binding site (78, 79). It is therefore possible that SHBG interacts with more than one class of binding site on the surface of different cells. In examining the specificity of these interactions, closely related proteins, such as protein S, laminin, or merosin, have not yet been used as competitors. This is obviously important because laminin interacts with several binding proteins on the surface of cells (80), and the primary structures of these proteins are particularly similar within a region of the SHBG molecule implicated in binding to the prostate plasma membrane (41).

3. Corticosteroid-binding globulin

3.1 Structure

A plasma protein with high affinity for glucocorticoids has been identified in a wide range of vertebrate species (81). The early literature often referred to this protein as transcortin, because it is undoubtedly the major transport protein for glucocorticoids in the blood (82), but it is now generally known as corticosteroid-binding globulin. It is, however, not generally appreciated that this protein also binds progesterone with relatively high affinity, and this may also be physiologically important. In most species the protein circulates in the blood as a monomeric glycoprotein of about 60 kDa (83), but a notable exception is squirrel monkey CBG which appears to circulate as a dimer (84). Nevertheless, there appears to be only one steroid-binding site per CBG molecule and its affinity and specificity has been studied in numerous species; this information is summarized by Westphal (7). These properties are species specific and, when coupled with phylogenetic comparisons of the primary structure of CBG (85), they provide important clues about specific domains and amino acids that may be structurally or functionally important.

3.1.1 Protein structure

Although the physicochemical properties of CBG from several species have been studied in detail (7), little was known about its primary structure, apart from a limited amount of amino-terminal sequence data (86), until cDNAs for human CBG were isolated and their sequences indicated that the mature polypeptide of 383 amino acids was derived from a 405-residue precursor polypeptide (87). This information also revealed that CBG is structurally related to members of the serine proteinase inhibitor (SERPIN) superfamily that includes α_1-antitrypsin (AAT), α_1-antichymotrypsin (ACT), and the major plasma transport protein for thyroxine, thyroxine-binding globulin (TBG). This totally unanticipated finding has provided important additional insight into the structural properties of the molecule because the tertiary structures of several related proteins have been resolved in their native

(88) or proteolytically cleaved (89) states. Indeed, predictions of the tertiary structure of human CBG have already been attempted (90) and it is now possible to test these models by creating CBG mutants with specific amino acid substitutions (91). These studies will be facilitated by the increasing body of information derived from phylogenetic comparisons of CBG primary structures, as well as the identification of natural mutations that result in steroid-binding defects (91, 92).

It is reasonable to assume that the tertiary structure of CBG resembles the highly stressed configuration of AAT, and this is obviously important for the formation of a steroid-binding site because elastase cleavage of CBG at a single location close to its carboxy-terminus relaxes its conformational state and disrupts its steroid-binding properties (93). The location and structure of the steroid-binding site remain speculative (1, 94), but there are considerable physicochemical data demonstrating the importance of various amino acids in steroid binding. In particular, there is evidence that the following residues are located in the steroid-binding pocket: cysteine (95), methionine, histidine (94), tryptophan (96), and tyrosine (97). The location of these residues has not been clearly defined, but indirect evidence, based on phylogenetic sequence comparisons, suggests that at least Cys-228 in the human sequence may represent the residue that can be affinity labelled with 6-bromo-progesterone (95). In addition, we have recently identified mutations responsible for reduced steroid-binding affinity in natural variants of human (92) and rat (91) CBG, and these implicate residues corresponding to Leu-93 and Ile-284 in the human sequence as being involved in steroid binding.

3.1.2 Associated carbohydrate structures

Direct analyses of the carbohydrate composition of human CBG have shown that it contains an average of five *N*-acetyl-lactosamine type biantennary and triantennary oligosaccharides in a ratio of 3:2, respectively (98). The cDNA-deduced sequence of human CBG indicates that there are six consensus sites for *N*-glycosylation, and their locations are shown in Figure 3. It would therefore appear that one of these sites is not available for glycosylation or that several sites are only partially utilized. In support of the latter, the consensus site within the carboxy-terminal fragment of CBG, generated by elastase cleavage, appears to be only partially glycosylated (99). The composition of the oligosaccharides attached to it may also vary (99), and this heterogeneity of carbohydrate additions undoubtedly accounts in part for the complex isoelectrofocusing profiles and diffuse electrophoretic mobility of the protein (100). During pregnancy and specific disease states in humans, this heterogeneity is further complicated by the appearance of molecular variants that contain only triantennary carbohydrate chains (101). Carbohydrate chains that are normally associated with CBG influence the biological half-life of the protein (102), and these variants may have an extended half-life in the blood (103). They also interact very differently with the plasma membranes of syncytiotrophoblasts than with normally glycosylated CBG (104).

Enzymatic deglycosylation of purified human CBG has been reported not to affect its steroid-binding properties (100), but more recent studies of the expression

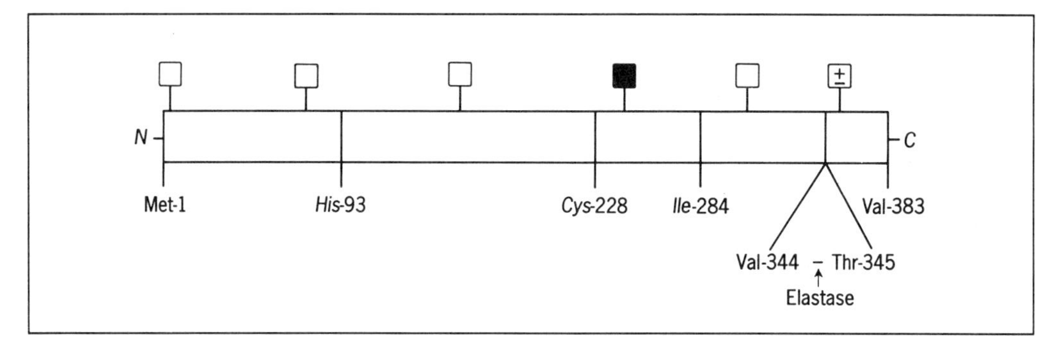

Fig. 3 Schematic representation of human CBG showing the positions of consensus sites for *N*-linked carbohydrate chains (□); the evolutionarily highly conserved site (■) and the site that appears to be partially utilized (□) are indicated. Also shown are the positions of the elastase cleavage site (↑) and residues that have been implicated in steroid binding (in italics)

of a human CBG cDNA in insect cells indicate that inhibition of glycosylation with tunicamycin results in a loss of CBG production (105). This was attributed to a lack of proper folding and destabilization of the protein within the endoplasmic reticulum, and supports earlier data obtained using tunicamycin-treated hepatoma cells (106). Since chemical and enzymatic deglycosylation is almost never completely effective, and the use of tunicamycin is non-specific, it remains to be determined whether individual carbohydrate chains are involved in protein folding and/or the creation of the steroid-binding site. In this regard, our phylogenetic comparisons of CBG primary structures (85) (E. T. M. Berdusco *et al.*, *Endocrinology*, in press) have indicated that only one consensus site for *N*-glycosylation is perfectly conserved between mammalian species, and this corresponds to residues 238–240 in the human sequence (Fig. 3). This site has been maintained throughout evolution despite drift in both codon usage and the amino acids that comprise the consensus sequences; this implies that the addition of a carbohydrate has been conserved at this position for an important reason. Furthermore, when the primary structures of CBG and AAT are compared (Fig. 4), it is apparent that the consensus sites at Asn-74 and Asn-238 in the human CBG sequence are also conserved in the AAT sequence, and this suggests that they may be structurally important.

3.2 Gene expression

The availability of specific cDNAs for CBG have recently allowed more detailed studies of CBG gene expression in several mammalian species. In this context, it should be noted that SERPIN genes are generally not well conserved between species, and the CBG gene is no exception with only 56–79% sequence identity between the coding regions for this protein in several mammalian species. Moreover, given that the human CBG and AAT coding sequences are approximately 43% identical, it is important to use homologous cDNA probes at high stringency

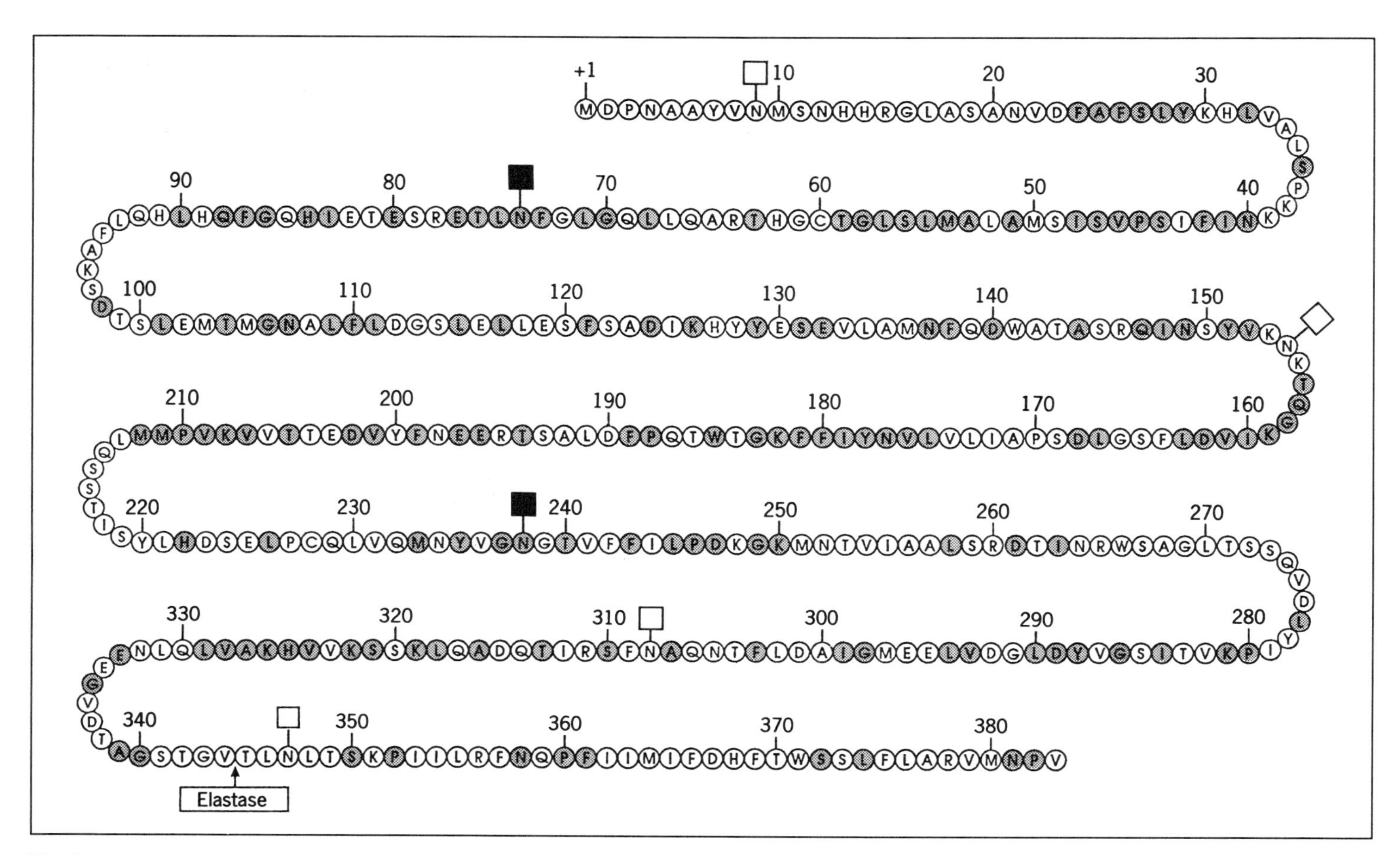

Fig. 4 Illustration of the sequence identity between CBG and α_1-antitrypsin (AAT). The primary structure of human CBG is shown and the location of identical amino acids in the human AAT sequence are shaded. The positions of carbohydrate consensus sites that are conserved between these proteins are indicated (■)

and to ascertain the specificity of any hybridization signals. The isolation of the human CBG gene has also allowed the transcription unit for hepatic CBG mRNA to be defined (107), as well as the structures of variant genes encoding CBG mutants with reduced affinity for steroids (92).

3.2.1 Gene structure and chromosomal location

The close relationship between CBG and certain members of the SERPIN superfamily also extends to the structural organization of their genes. For example, there is a perfect conservation in the intron–exon boundaries throughout the coding regions for the CBG and AAT genes (107). To illustrate this point further, the structural organization of the CBG and AAT genes are even more similar than the AAT and ACT genes (107). It is therefore not surprising that the CBG gene is located in a region (q32.1) of human chromosome 14 that contains a cluster of SERPIN genes (108), supporting the concept that these genes have evolved relatively recently in concert with the acquisition of important biological processes, such as those involved in blood coagulation, tissue damage and repair, and the immune system.

The initiation site for transcription in the liver has been mapped in the human CBG gene; this region contains a series of *cis*-acting elements frequently associated with transcriptionally active promoters in hepatocytes (107). It is not known whether expression of the CBG gene is controlled by the same promoter in other tissues, such as the kidney (109, 110), but by analogy with the AAT gene it might be anticipated that more than one transcriptional start site may exist (111).

3.2.2 Spatial and temporal expression

There is now little doubt that hepatocytes are the major site of CBG biosynthesis in adult mammals (109, 112), but CBG is probably also produced in several other tissues. Most notably, CBG mRNA has been detected in the kidneys of several species (87, 109, 110), and polysomes from rat kidneys have been shown to produce CBG *in vitro* (113). This extrahepatic production of CBG may not influence the plasma concentrations of CBG to any great extent, but it could alter the local bioavailability of steroids that bind to CBG in the extravascular compartments of certain tissues. Nothing is known about the regulation of CBG gene expression in these sites, and the properties of CBG produced outside the liver have never been examined. This may be important because different cells will almost certainly glycosylate CBG in different ways, and may influence the extravascular compartmentalization of CBG or its relocation to other cell types within a given tissue. If so, it is possible that CBG produced in these tissues may influence the bioavailability of steroids in ways that have never been explored.

In mammals, the biosynthesis of CBG undergoes enormous changes during development, and fetal hepatic CBG mRNA and plasma CBG levels are very high in most species during mid to late gestation, but are very low or undetectable at birth (112, 114). These changes in fetal CBG gene expression occur independently of contemporary changes in maternal hepatic CBG biosynthesis, and the gene appears to be under different control during fetal life (109). In rats, postnatal increases in plasma CBG levels lag behind the increase in CBG gene expression in the liver, as measured by th accumulation of CBG mRNA. This may be due to a more rapid plasma clearance of the protein during the first few weeks of life (114).

3.2.3 Hormonal regulation

Plasma levels of CBG are influenced by various hormone treatments, but these effects may vary considerably between species. For instance, administration of synthetic oestrogens results in modest increases in plasma CBG levels in humans (115, 116) and several laboratory animals (7), but causes increases of several orders of magnitude in squirrel monkeys (117). Similarly, thyroid hormone increases plasma CBG levels in rats (118) but has virtually no effect in humans (116). The thyroxine-induced increases in hepatic CBG mRNA levels in the rat appear to be post-transcriptional and may be attributed to a reduction in CBG mRNA turnover (119).

In contrast, glucocorticoids appear invariably to decrease plasma CBG levels in adult mammals (7), and synthetic glucocorticoids, such as dexamethasone, are particularly potent in this regard (119). In rats, this correlates with a marked decrease in hepatic CBG mRNA levels and reduction in the transcription rate of the gene (119). The way in which glucocorticoids exert this negative effect on CBG gene transcription remains to be defined, but it is interesting that the sheep CBG gene appears to respond quite differently to glucocorticoids during fetal life (E. T. M. Berdusco *et al.*, *Endocrinology*, in press). In rodents, there is also a sexual dimorphism in the plasma CBG levels that is directly related to the amounts of CBG mRNA in the liver (114), and this has been attributed to an androgen-dependent imprinting of pulsatile growth hormone secretion patterns (120).

Plasma CBG levels also decrease promptly during acute inflammation in rats (121) and humans (122). This may be attributed to an increase in the proteolytic cleavage and clearance of the protein, as well as a reduction in the relative amount of CBG mRNA in the liver (123). Plasma glucocorticoid levels also increase during this time, but it is unlikely that they are responsible for the marked decrease in hepatic CBG mRNA observed because exogenous glucocorticoid treatment takes several days to reduce plasma CBG levels (119). It is therefore possible that other modulators of the immune response are involved in this process, and it has recently been shown that the accumulation of CBG MRNA is reduced by interleukin-6 in a human hepatoma cell line (L. Bartalena *et al.*, *Endocrinology*, in press).

3.3 Function

3.3.1 Steroid transport and bioavailability

Although approximately 90% of the plasma glucorticoids in most mammalian species are transported in association with CBG, a relatively large fraction of the progesterone in female blood is also bound by this protein (64). In tissues that contain appreciable amounts of progesterone as a result of local synthesis, such as the ovary and placenta, this steroid may competitively displace ligands from CBG and thereby liberate relatively large amounts of biologically active glucocorticoids at these sites. In addition, the protein may then assume a role as a major carrier of progesterone and act to compartmentalize this steroid or target it to specific cells.

Alterations in the plasma concentrations of CBG will also determine the plasma distribution of steroids that interact with its binding site. These changes probably have widespread effects on the bioavailability of various steroid ligands at several levels. For instance, they impact on glucocorticoid-dependent differentiation and maturation of several organ systems during normal development, homoeostatic processes during pregnancy, and diseases involving acute stress and/or activation of the immune system.

3.3.2 Plasma membrane interactions

There have been several descriptions of interactions between CBG and binding sites on the plasma membranes of cells (124) and reports that the localization of CBG in their cytoplasm (125) may involve a process of cellular internalization (64). It has also been reported that CBG binding to cells may increase intracellular cyclic AMP levels (126), although this response is small and remains to be confirmed. The plasma membrane-binding site within the CBG molecule has not been identified but carbohydrates have been shown to be involved in this process (104). Furthermore, a pregnancy-specific CBG variant interacts differently with syncytiotrophoblast plasma membranes than with normal CBG (104). This suggests that the type of carbohydrates associated with CBG may influence this. The biological significance of these observations is obscure, but these interactions may somehow influence the distribution of CBG between various compartments of the organism. In support of this, CBG has not only been shown to translocate from maternal to fetal compartments in the rabbit, but the fetal kidney is also able to discriminate between the maternal and fetal forms of CBG that presumably differ only with respect to their carbohydrate composition (127).

3.3.3 Interactions with proteinases

The close relationship between CBG and other SERPINs (Fig. 4) initially directed our attention to the possibility that CBG is a specific substrate for a serine proteinase (87), and we have since developed this hypothesis to suggest a potentially important mechanism by which CBG may actively promote steroid delivery to certain cells (1). In essence, an interaction between CBG and specific proteinases may cause a local release of bound ligand at sites of inflammation, or other areas where cellular interactions result in a breakdown of basement membranes, such as trophoblast invasion and sites of tissue remodelling during development. One proteinase that appears to be involved in this process is the serine proteinase, neutrophil elastase (93, 99). Cleavage of CBG by this enzyme at a specific location (see Fig. 3) results in a conformational change in the protein and a disruption of steroid binding (93, 99). Furthermore, this occurs on the surface of neutrophils activated either artificially (128) or spontaneously during the course of an inflammatory reaction (99, 128). Since approximately 90% of the cortisol in human plasma is bound to CBG, this event may result in the immediate and local delivery of much larger amounts of steroid than those predicted to be available by the 'free hormone hypothesis', and may have far-reaching implications in terms of the impact of glucocorticoids and progesterone on development, and of the disease processes that respond to treatment with these steroids.

4. Conclusions

The primary structures of the major extracellular steroid-binding proteins have been resolved, their genes have been cloned, and their chromosomal locations defined in several mammalian species. This information has provided novel insights into their functions, and a combination of molecular biological, biochemical, and physiological approaches have been used to assess their biological significance. Extracellular steroid-binding protein genes are not only regulated by the hormones they interact with, but are influenced by other hormonal and nutritional factors.

Acknowledgements

I would like to thank Dr. George Avvakumov, Wayne Bocchinfuso, and Mette Pallesen for their input and review of the manuscript, and Gail Howard for secretarial assistance.

References

1. Hammond, G. L. (1990) Molecular properties of corticosteroid binding globulin and the sex-steroid binding proteins. *Endocr. Rev.*, **11,** 65.
2. Rosner, W. (1990) The functions of corticosteroid-binding globulin and sex hormone-binding globulin: recent advances. *Endocri. Rev.*, **11,** 80.
3. Khan, M. S., Knowles, B. B., Aden, D. P., and Rosner, W. (1981) Secretion of testosterone–estradiol-binding globulin by a human hepatoma-derived cell line. *J. Clin. Endocrinol. Metab.*, **53,** 448.
4. Cheng, C. Y., Musto, N. A., Gunsalus, G. L., Frick, J., and Bardin, C. W. (1985) There are two forms of androgen binding protein in human testes. *J. Biol. Chem.*, **260,** 5631.
5. Hagenas, L., Ritzen, E. M., Ploen, L., Hansson, V., French, F. S., and Nayfeh, S. N. (1975) Sertoli cell origin of testicular androgen-binding protein (ABP). *Mol. Cell. Endocrinol.*, **2,** 339.
6. Cheng, C. Y., Frick, J., Gunsalus, G. L., Musto, N. A., and Bardin, C. W. (1984) Human testicular androgen-binding protein shares immunodeterminants with serum testosterone–estradiol-binding globulin. *Endocrinology*, **114,** 1395.

7. Westphal, U. (1986) *Steroid–Protein Interactions II. Monographs on Endocrinology*, Springer-Verlag: New York, Heidelberg, Berlin.
8. Renoir, J.-M., Mercier-Bodard, C., and Baulieu, E-E. (1980) Hormonal and immunological aspects of the phylogeny of sex steroid binding plasma protein. *Proc. Natl Acad. Sci. USA*, **77**, 4578.
9. Damassa, D. A., Gustafson, A. W., and King, J. C. (1982) Identification of a specific binding protein for sex steroids in the plasma of the male little brown bat, *Myotis lucifugus lucifugus*. *Gen. Comp. Endocrinol.*, **47**, 288.
10. Walsh, K. A., Titani, K., Takio, K., Kumar, S., Hayes, R., and Petra, P. H. (1986) Amino acid sequence of the sex steroid binding protein of human blood plasma. *Biochemistry*, **25**, 7584.
11. Hammond, G. L., Underhill, D. A., Smith, C. L., Goping, I. S., Harley, M. J., Musto, N. A., Cheng, C. Y., and Bardin, C. W. (1987) The cDNA-deduced primary structure of human sex hormone binding globulin and location of its steroid binding domain. *FEBS Lett.*, **215**, 100.
12. Joseph, D. R., Hall, S. H., and French, F. S. (1987) Rat androgen-binding protein: evidence for identical subunits and amino acid sequence homology with human sex hormone-binding globulin. *Proc. Natl Acad. Sci. USA*, **84**, 339.
13. Hammond, G. L., Robinson, P. A., Sugino, H., Ward, D. N., and Finne, J. (1986) Physicochemical characteristics of human sex hormone binding globulin: evidence for two identical subunits. *J. Steroid Biochem.*, **24**, 815.
14. Cheng, C. Y., Gunsalus, G. L., Musto, N. A., and Bardin, C. W. (1984) The heterogeneity of rat androgen-binding protein in serum differs from that in testis and epididymis. *Endocrinology*, **114**, 1386.
15. Danzo, B. J., Taylor, C. A., Jr., and Eller, B. C. (1982) Some physicochemical characteristics of photoaffinity-labeled rabbit testosterone-binding globulin. *Endocrinology*, **111**, 1278.
16. Danzo, B. J., Eller, B. C., and Bell, B. W. (1987) The apparent molecular weight of androgen-binding protein (ABP) in the blood of immature rats differs from that of ABP in the epididymis. *J. Steroid Biochem.*, **28**, 411.
17. Petra, P. H. (1991) The plasma sex steroid binding protein (SBP or SHBG). A critical review of recent developments on the structure, molecular biology and function. *J. Steroid Biochem. Mol. Biol.*, **40**, 735.
18. Avakumov, G. V., Matveentseva, I. V., Akhrem, L. V., Strel'chyonok, O. A., and Akhrem, A. A. (1983) Study of the carbohydrate moiety of human serum sex hormone-binding globulin. *Biochim. Biophys. Acta*, **760**, 104.
19. Gershagen, S., Henningsson, K., and Fernlund, P. (1987) Subunits of human sex hormone binding globulin. *J. Biol. Chem.*, **262**, 8430.
20. Van Baelen, H., Convents, R., Calleau, J., and Heyns, W. (1992) Genetic variation of human sex hormone-binding globulin: evidence for a worldwide bi-allelic gene. *J. Clin. Endocrinol. Metab.*, **75**, 135.
21. Power, S. G. A., Bocchinfuso, W. P., Pallesen, M., Warmels-Rodenhiser, S., Van Baelen, H., and Hammond, G. L. (1992) Molecular analyses of a human sex hormone-binding globulin variant: evidence for an additional carbohydrate chain. *J. Clin. Endocrinol. Metab.*, **75**, 1066.
22. Cheng, C. Y., Musto, N. A., Gunsalus, G. L., and Bardin, C. W. (1983) Demonstration of heavy and light promoters of human testosterone–estradiol-binding globulin. *J. Steroid Biochem.*, **19**, 1379.

23. Khan, M. S., Ehrlich, P., Birken, S., and Rosner, W. (1985) Size isomers of testosterone–estradiol-binding globulin exist in the plasma of individual men and women. *Steroids*, **45,** 463.
24. Danzo, B. J., and Bell, B. W. (1988) The microheterogeneity of androgen-binding protein in rat serum and epididymis is due to differences in glycosylation of their subunits. *J. Biol. Chem.*, **263,** 2402.
25. Hammond, G. L., Underhill, D. A., Rykse, H. M., and Smith, C. L. (1989) The human sex hormone binding globulin gene contains exons for androgen binding protein and two other testicular messenger RNAs. *Mol. Endocrinol.*, **3,** 1869.
26. Gershagen, S., Lundwall, Å., and Fernlund, P. (1989) Characterization of the human sex hormone binding globulin (SHBG) gene and demonstration of two transcripts in both liver and testis. *Nucleic Acids Res.*, **17,** 9245.
27. Gershagen, S., Fernlund, P., and Lundwall, A. (1987) A cDNA coding for human sex hormone binding globulin: homology to vitamin K-dependent protein S. *FEBS Lett.*, **220,** 129.
28. Que, B. G. and Petra, P. H. (1987) Characterization of a cDNA coding for sex steroid-binding protein of human plasma. *FEBS Lett.*, **219,** 405.
29. Reventos, J., Hammond, G. L., Crozat, A., Brooks, D. E., Gunsalus, G. L., Bardin, C. W., and Musto, N. A. (1988) Hormonal regulation of rat androgen-binding protein (ABP) messenger ribonucleic acid and homology of human testosterone–estradiol-binding globulin and ABP complementary deoxyribonucleic acids. *Mol. Endrocrinol.*, **2,** 125.
30. Bardin, C. W., Gunsalus, G. L., Musto, N. A., Cheng, C. Y., Reventos, J., Smith, C., Underhill, C. A., and Hammond, G. (1988) Corticosteroid binding globulin, testosterone–estradiol-binding globulin and androgen binding protein belong to protein families distinct from steroid receptors. *J. Steroid Biochem.*, **30,** 131.
31. Baker, M. E., French, F. S., and Joseph, D. R. (1987) Vitamin K-dependent protein S is similar to rat androgen binding proteins. *Biochem. J.*, **243,** 293.
32. Joseph, D. R. and Baker, M. E. (1992) Sex hormone-binding globulin, androgen-binding protein, and vitamin K-dependent protein S are homologous to laminin A, merosin, and *Drosophila crumbs* protein. *FASEB J.*, **6,** 2477.
33. Wang, Y-M., Sullivan, P. M., Petrusz, P., Yarbrough, W., and Joseph, D. R. (1989) The androgen-binding protein gene is expresed in CD1 mouse testis. *Mol. Cell. Endocrinol.*, **63,** 85.
34. Griffin, P. R., Kumar, S., Shabanowitz, J., Charbonneau, H., Namkung, P. C., Walsh, K. A., Hunt, D. F., and Petra, P. H. (1989) The amino acid sequence of the sex steroid-binding protein of rabbit serum. *J. Biol. Chem.*, **264,** 19066.
35. Namkung, P. C., Kumar, S., Walsh, K. A., and Petra, P. H. (1990) Identification of lysine 134 in the steroid-binding site of the sex steroid-binding protein of human plasma. *J. Biol. Chem.*, **265,** 18345.
36. Grenot, C., De Montard, A., Blachere, T., Mappus, E., and Cuilleron, C.-Y. (1988) Identification d'un site de photomarquage de la protéine plasmatique de liaison de la testostérone et de l'oestradiol (SBP) par l'hydroxy-17β oxo-3 androstadiéne-4,6 tritié. *C. R. Acad. Sci.*, **307,** 391.
37. Khan, M. S. and Rosner, W. (1990) Histidine 235 of human sex hormone-binding globulin is the covalent site of attachment of the nucleophilic steroid derivative, 17β-bromoacetoxydihydrotestosterone. *J. Biol. Chem.*, **265,** 8431.
38. Hammond, G. L. and Robinson, P. A. (1984) Characterization of a monoclonal antibody to human sex-hormone binding globulin. *FEBS Lett.*, **168,** 307.

39. Bocchinfuso, W. P., Warmels-Rodenhiser, S., and Hammond, G. L. (1992) Structure/function analyses of human sex hormone-binding globulin by site-directed mutagenesis. *FEBS Lett.*, **301,** 227.
40. Bocchinfuso, W. P., Warmels-Rodenhiser, S., and Hammond, G. L. (1991) Expression and differential glycosylation of human sex hormone-binding globulin by mammalian cell lines. *Mol. Endocrinol.*, **5,** 1723.
41. Khan, M. S., Hryb, D. J., Hashim, G. A., Romas, N. A., and Rosner, W. (1990) Delineation and synthesis of the membrane receptor-binding domain of sex hormone-binding globulin. *J. Biol. Chem.*, **265,** 18362.
42. Danzo, B. J., Black, J. H., and Bell, B. W. (1991) Analysis of the oligosaccharides on androgen-binding proteins: implications concerning their role in structure/function relationships. *J. Steroid Biochem. Mol. Biol.*, **40,** 821.
43. Suzuki, Y. and Sinohara, H. (1979) Hepatic uptake of desialylated testosterone–estradiol-binding globulin in the rat. *Acta Endocrinol.*, **90,** 669.
44. Bocchinfuso, W. P., Ma, K.-L., Warmels-Rodenhiser, S., Lee, W. M., and Hammond, G. L. (1992) Selective removal of glycosylation sites from sex hormone-binding globulin by site-directed mutagenesis. *Endocrinology*, **131,** 2331.
45. Tardivel-Lacombe, J. and Degrelle, H. (1991) Hormone-associated variation of the glycan microheterogeneity pattern of human sex steroid-binding protein (hSBP). *J. Steroid Biochem. Mol. Biol.*, **39,** 449.
46. Keene, J. L., Matzuk, M. M., and Boime, I. (1989) Expression of recombinant human choriogonadotropin in Chinese hamster ovary glycosylation mutants. *Mol. Endocrinol.*, **3,** 2011.
47. Sairam, M. R. and Bhargavi, G. N. (1985) A role for glycosylation of the α subunit in transduction of biological signal in glycoprotein hormones. *Science*, **229,** 65.
48. Nakhla, A. M., Khan, M. S., and Rosner, W. (1990) Biologically active steroids activate receptor-bound human sex hormone-binding globulin to cause LNCaP cells to accumulate adenosine 3′,5′-monophosphate. *J. Clin. Endocrinol. Metab.*, **71,** 398.
49. Avvakumov, G. V., Zhuk, N. I., and Strel'chyonok, O. A. (1988) Biological function of the carbohydrate component of the human sex steroid-binding globulin. *Biochem. USSR*, **53,** 726.
50. Joseph, D. R., Hall, S. H., Conti, M., and French, F. S. (1988) The gene structure of rat androgen-binding protein: identification of potential regulatory deoxyribonucleic acid elements of a follicle-stimulating hormone-regulated protein. *Mol. Endocrinol.*, **2,** 3.
51. Sullivan, P. M., Petrusz, P., Szpirer, C., and Joseph, D. R. (1991) Alternative processing of androgen-binding protein RNA transcripts in fetal liver. *J. Biol. Chem.*, **266,** 143.
52. Bérubé, D., Seralini, G.-E., Gagné, R., and Hammond, G. L. (1990) Localization of the human sex hormone-binding globulin gene (SHBG) to the short arm of chromosome 17 (17p12–p13). *Cytogenet. Cell Genet.*, **54,** 65.
53. Joseph, D. R., Adamson, M. C., and Kozak, C. A. (1991) Genetic mapping of the gene for androgen-binding protein/sex hormone-binding globulin to mouse chromosome 11. *Cytogenet. Cell Genet.*, **56,** 122.
54. Lehrman, M. A., Goldstein, J. L., Russell, D. W., and Brown, M. S. (1987) Duplication of seven exons in LDL receptor gene caused by Alu–Alu recombination in a subject with familial hypercholesterolemia. *Cell*, **43,** 827.
55. Wang, Y.-M., Bayliss, D. A., Millhorn, D. E., Petrusz, P., and Joseph, D. R. (1990) The androgen-binding protein gene is expressed in male and female rat brain. *Endocrinology*, **127,** 3124.

56. Mercier-Bodard, C., Nivet, V., and Baulieu, E.-E. (1991) Effects of hormones on SBP mRNA levels in human cancer cells. *J. Steroid Biochem. Mol. Biol.*, **40,** 777.
57. Plymate, S. R., Loop, S. M., Hoop, R. C., Wiren, K. M., Ostenson, R., Hryb, D. J., and Rosner, W. (1991) Effects of sex hormone binding globulin (SHBG) on human prostatic carcinoma. *J. Steroid Biochem. Mol. Biol.*, **40,** 833.
58. Wilson, J. D. (1978) Sexual differentiation. *Annu. Rev. Physiol.*, **40,** 279.
59. Ballard, P. L. (1989) Hormonal regulation of pulmonary surfactant. *Endocr. Rev.*, **10,** 165.
60. Forest, M. G., Bonneton, A., Lecoq, A., Brébant, C., and Pugeat, M. (1986) Ontogenèse de la protéine de liaison des hormones sexuelles (SBP) et de la transcortine (CBG) chez les primates: variations physiologiques et étude dans différents milieux biologiques. In: *Binding Proteins of Steroid Hormones,* Vol. 149. Forest, M. G. and Pugeat, M. (eds). INSERM/John Libbey: London, p. 263.
61. Anderson, D. C. (1974) Sex-hormone-binding globulin. *Clin. Endocrinol. (Oxf.)*, **3,** 69.
62. Pugeat, M., Crave, J. C., Elmidani, M., Nicolas, M. H., Garoscio-Cholet, M., Leueune, H., Dechaud, H., and Tourniaire, J. (1991) Pathophysiology of sex hormone binding globulin (SHBG): relation to insulin. *J. Steroid Biochem. Mol. Biol.*, **40,** 841.
63. Rosner, W., Aden, D. P., and Khan, M. S. (1984) Hormonal influences on the secretion of steroid-binding proteins by a human hepatoma-derived cell line. *J. Clin. Endocrinol. Metab.*, **59,** 806.
64. Siiteri, P. K., Murai, J. T., Hammond, G. L., Nisker, J. A., Raymoure, W. J., and Kuhn, R. W. (1982) The serum transport of steroid hormones. *Recent Prog. Horm. Res.*, **38,** 457.
65. Tindall, D. J. and Means, A. R. (1980) Properties and hormonal regulation of androgen binding proteins. In: *Advances in Sex Hormone Research,* Vol. 4. Thomas, J. A. and Singhal, R. L. (eds). Urban and Schwarzenberg: Baltimore, p. 295.
66. Nisker, J. A., Hammond, G. L., Davidson, B. J., Frumar, A. M., Takaki, N. K., Judd, H. L., and Siiteri, P. K. (1980) Serum sex hormone-binding globulin capacity and the percentage of free estradiol in postmenopausal women with and without endometrial cancer. *Am. J. Obstet. Gynecol.*, **138,** 637.
67. Hammond, G. L., Lähteenmäki, P. L. A., Lähteenmäki, P., and Luukkainen, T. (1982) Distribution and percentages of non-protein bound contraceptive steroids in human serum. *J. Steroid Biochem.*, **17,** 375.
68. Fotherby, K. (1988) Interactions of contraceptive steroids with binding proteins and the clinical implications. *Ann. N.Y., Acad. Sci.*, **538,** 313.
69. Pardridge, W. M. (1981) Transport of protein-bound hormones into tissues *in vivo. Endocr. Rev.*, **2,** 103.
70. Bordin, S. and Petra, P. H. (1980) Immunocytochemical localization of the sex steroid-binding protein of plasma in tissues of the adult monkey *Macaca nemestrina. Proc. Natl Acad. Sci. USA,* **77,** 5678.
71. Pelliniemi, L. J., Dym, M., Gunsalus, G. L., Musto, N. A., Bardin, C. W., and Fawcett, D. W. (1981) Immunocytochemical localization of androgen-binding protein in the male rat reproductive tract. *Endocrinology,* **108,** 925.
72. Egloff, M., Vendrely, E., Tardivel-Lacombe, J., Dadoune, J.-P., and Degrelle, H. (1982) Etude immunohistochimique du testicule et de l'epididyme humains à l'aide d'un antisérum monospécifique dirigé contre la protéine plasmatique liant les hormones sexuelles. *C.R. Acad. Sci.*, **295,** 107.
73. Bordin, S., Torres, R., and Petra, P. H. (1982) An enzyme-immunoassay (ELISA) for the sex steroid-binding protein (SBP) of human serum. *J. Steroid Biochem.*, **17,** 453.

74. Porto, C. S., Musto, N. A., Bardin, C. W., and Gunsalus, G. L. (1992) Binding of an extracellular steroid-binding globulin to membranes and soluble receptors from human breast cancer cells (MCF-7 cells). *Endocrinology,* **130,** 2931.
75. Strel'chyonok, O. A., Avvakumov, G. V., and Survilo, L. I. (1984) A recognition system for sex-hormone-binding protein–estradiol complex in human decidual endometrium plasma membranes. *Biochim. Biophys. Acta,* **802,** 459.
76. Hryb, D. J., Khan, M. S., and Rosner, W. (1985) Testosterone–estradiol-binding globulin binds to human prostatic cell membranes. *Biochem. Biophys. Res. Commun.,* **128,** 432.
77. Rosner, W., Hryb, D. J., Khan, M. S., Nakhla, A. M., and Romas, N. A. (1991) Sex hormone-binding globulin: anatomy and physiology of a new regulatory system. *J. Steroid Biochem. Mol. Biol.,* **40,** 813.
78. Avvakumov, G. V., Zhuk, N. I., and Strel'chyonok, O. A. (1986) Subcellular distribution and selectivity of the protein-binding component of the recognition system for sex-hormone-binding protein–estradiol complex in human decidual endometrium. *Biochim. Biophys. Acta,* **881,** 489.
79. Hryb, D. J., Khan, M. S., Romas, N. A., and Rosner, W. (1990) The control of the interaction of sex hormone-binding globulin with its receptor by steroid hormones. *J. Biol. Chem.,* **265,** 6048.
80. Mecham, R. P. (1991) Laminin receptors. *Annu. Rev. Cell Biol.,* **7,** 71.
81. Seal, U. S. and Doe, R. P. (1965) Vertebrate distribution of corticosteroid-binding globulin and some endocrine effects on concentration. *Steroids,* **5,** 827.
82. Slaunwhite, W. R., Jr. and Sandberg, A. A. (1959) Transcortin: a corticosteroid-binding protein of plasma. *J. Clin. Invest.,* **38,** 384.
83. Kato, E. A., Hsu, B. R.-S., and Kuhn, R. W. (1988) Comparative structural analyses of corticosteroid binding globulin. *J. Steroid Biochem.,* **29,** 213.
84. Kuhn, R. W., VestWeber, C., and Siiteri, P. K. (1988) Purification and properties of squirrel monkey (*Saimiri sciureus*) corticosteroid binding globulin. *Biochemistry,* **27,** 2579.
85. Hammond, G. L., Smith, C. L., and Underhill, D. A. (1991) Molecular studies of corticosteroid binding globulin structure, biosynthesis and function. *J. Steroid Biochem. Mol. Biol.,* **40,** 755.
86. Le Gaillard, F., Han, K.-K., and Dautrevaux, M. (1975) Caractérisation et propriétés physico-chimiques de la transcortine humaine. *Biochimie,* **57,** 559.
87. Hammond, G. L., Smith, C. L., Goping, I. S., Underhill, D. A., Harley, M. J., Reventos, J., Musto, N. A., Gunsalus, G. L., and Bardin, C. W. (1987) Primary structure of human corticosteroid binding globulin, deduced from hepatic and pulmonary cDNAs, exhibits homology with serine protease inhibitors. *Proc. Natl Acad. Sci. USA,* **84,** 5153.
88. Stein, P. E., Leslie, A. G. W., Finch, J. T., Turnell, W. G., McLaughlin, P. J., and Carrell, R. W. (1990) Crystal structure of ovalbumin as a model for the reactive centre of serpins. *Nature,* **347,** 99.
89. Loebermann, H., Tokuoka, R., Deisenhofer, J., and Huber, R. (1984) Human α_1-proteinase inhibitor. Crystal structure analysis of two crystal modifications, molecular model and preliminary analysis of the implications for function. *J. Mol. Biol.,* **177,** 531.
90. Mornon, J.-P., Bissery, V., Gaboriaud, C., Thomas, A., Ojasso, T., and Raynaud, J-P. (1989) Hydrophobic cluster analysis (HCA) of the hormone-binding domain of receptor proteins. *J. Steroid Biochem.,* **34,** 355.

91. Smith, C. L. and Hammond, G. L. (1991) An amino acid substitution in BioBreeding rat corticosteroid binding globulin results in reduced steroid binding affinity. *J. Biol. Chem.*, **266,** 18555.
92. Smith, C. L., Power, S. G. A., and Hammond, G. L. (1992) A Leu → His substitution at residue 93 in human corticosteroid binding globulin results in reduced affinity for cortisol. *J. Steroid Biochem. Mol. Biol.*, **42,** 671.
93. Pemberton, P. A., Stein, P. E., Pepys, M. B., Potter, J. M., and Carrell, R. W. (1988) Hormone binding globulins undergo serpin conformational change in inflammation. *Nature,* **336,** 257.
94. Le Gaillard, F. and Dautrevaux, M. (1977) Affinity labelling of human transcortin. *Biochim. Biophys. Acta,* **495,** 312.
95. Khan, M. S. and Rosner, W. (1977) Investigation of the binding site of human corticosteroid-binding globulin by affinity labeling. *J. Biol. Chem.*, **252,** 1895.
96. Akhrem, A. A., Sviridov, O. V., Strel'chyonok, O. A., and Prischchepov, A. S. (1981) Study of steroid-binding site in human transcortin using tryptophan-specific reagents. *Bioorg. Khim.*, **7,** 662.
97. Le Gaillard, F., Racadot, A., Aubert, J. P., and Dautrevaux, M. (1982) Modification of human transcortin by tetranitromethane. Evidence for the implication of a tyrosine residue in cortisol binding. *Biochimie,* **64,** 153.
98. Akhrem, A. A., Avvakumov, G. V., Akhrem, L. V., Sidorova, I. V., and Strel'chyonok, O. A. (1982) Structural organization of the carbohydrate moiety of human transcortin as determined by methylation analysis of the whole glycoprotein. *Biochim. Biophys. Acta,* **714,** 177.
99. Hammond, G. L., Smith, C. L., Paterson, N. A. M., and Sibbald, W. J. (1990) A role for corticosteroid-binding globulin in delivery of cortisol to activated neutrophils. *J. Clin. Endocrinol. Metab.*, **71,** 34.
100. Mickelson, K. E., Harding, G. B., Forsthoefel, M., and Westphal, U. (1982) Steroid–protein interactions. Human corticosteroid-binding globulin: characterization of dimer and electrophoretic variants. *Biochemistry,* **21,** 654.
101. Strel'chyonok, O. A., Avvakumov, G. V., and Akhrem, A. A. (1984) Pregnancy-associated molecular variants of human serum transcortin and thyroxine-binding globulin. *Carbohydr. Res.*, **134,** 133.
102. Hossner, K. L. and Billiar, R. B. (1981) Plasma clearance and organ distribution of native and desialylated rat and human transcortin: species specificity. *Endocrinology,* **108,** 1780.
103. Avvakumov, G. V. and Strel'chyonok, O. A. (1987) Properties and serum levels of pregnancy-associated variant of human transcortin. *Biochim. Biophys. Acta,* **925,** 11.
104. Avvakumov, G. V. and Strel'chyonok, O. A. (1988) Evidence for the involvement of the transcortin carbohydrate moiety in the glycoprotein interaction with the plasma membrane of human placental syncytiotrophoblast. *Biochim. Biophys. Acta,* **938,** 1.
105. Ghose-Dastidar, J., Ross, J. B. A., and Green, R. (1991) Expression of biologically active human corticosteroid binding globulin by insect cells: acquisition of function requires glycosylation and transport. *Proc. Natl Acad. Sci. USA,* **88,** 6408.
106. Murata, Y., Sueda, K., Seo, H., and Matsui, N. (1989) Studies on the role of glycosylation for human corticosteroid-binding globulin: comparison with that for thyroxine-binding globulin. *Endocrinology,* **125,** 1424.
107. Underhill, D. A. and Hammond, G. L. (1989) Organization of the human corticosteroid binding globulin gene and analysis of its 5′-flanking region. *Mol. Endocrinol.*, **3,** 1448.

108. Seralini, G.-E., Bérubé, D., Gagné, R., and Hammond, G. L. (1990) The human corticosteroid binding globulin gene is located on chromosome 14q31–q32.1 near two other serine protease inhibitor genes. *Hum. Genet.*, **86,** 73.
109. Seralini, G.-E., Smith, C. L., and Hammond, G. L. (1990) Rabbit corticosteroid-binding globulin: primary structure and biosynthesis during pregnancy. *Mol. Endocrinol.*, **4,** 1166.
110. Scrocchi, L. A., Hearn, S. A., Han, V. K. M., and Hammond, G. L. (1993) Corticosteroid-binding globulin biosynthesis in the mouse liver and kidney during postnatal development. *Endocrinology*, **132,** 910.
111. Ruther, U., Tripodi, M., Cortese, R., and Wagner, E. F. (1987) The human alpha-1-antitrypsin gene is efficiently expressed from two tissue-specific promoters in transgenic mice. *Nucleic Acids Res.*, **15,** 7519.
112. Smith, C. L. and Hammond, G. L. (1989) Rat corticosteroid binding globulin: primary structure and messenger ribonucleic acid levels in the liver under different physiological conditions. *Mol. Endocrinol.*, **3,** 420.
113. Kraujelis, K., Ulinskaite, A., and Meilus, V. (1991) Transcortin in rat kidney: subcellular distribution of transcortin-synthesizing polyribosomes. *J. Steroid Biochem. Mol. Biol.*, **38,** 43.
114. Smith, C. L. and Hammond, G. L. (1991) Ontogeny of corticosteroid-binding globulin biosynthesis in the rat. *Endocrinology*, **128,** 983.
115. Sandberg, A. A., Slaunwhite, W. R., and Carter, A. C. (1960) Transcortin: a corticosteroid-binding protein of plasma. III. The effects of various steroids. *J. Clin. Invest.*, **39,** 1914.
116. Brien, T. G. (1981) Human corticosteroid binding globulin. *Clin. Endocrinol. (Oxf.)*, **14,** 193.
117. Coe, C. L., Murai, J. T., Wiener, S. G., Levine, S., and Siiteri, P. K. (1986) Rapid cortisol and corticosteroid-binding globulin responses during pregnancy and after estrogen administration in the squirrel monkey. *Endocrinology*, **118,** 435.
118. Labrie, F., Pelletier, G., Labrie, R., Ho-Kim, M. A., Delgado, A., MacIntosh, B., and Fortier, C. (1968) Liaison transcortine–corticostérone et contrôle de l'activité hypophyso-surrénalienne chez le rat. Interactions hypophyse–thyroide–surrénales-gonades. *Ann. Endocrinol (Paris)*, **29,** 29.
119. Smith, C. L. and Hammond, G. L. (1992) Hormonal regulation of corticosteroid binding globulin biosynthesis in the male rat. *Endocrinology*, **130,** 2245.
120. Jansson, J.-O., Oscarsson, J., Mode, A., and Ritzen, E. M. (1989) Plasma growth hormone pattern and androgens influence the levels of corticosteroid-binding globulin in rat serum. *J. Endocrinol.*, **122,** 725.
121. Savu, L., Lombart, C., and Nunez, E. A. (1980) Corticosterone binding globulin: an acute phase 'negative' protein in the rat. *FEBS Lett.*, **113,** 102.
122. Savu, L., Zouaghi, H., Carli, A., and Nunez, E. A. (1981) Serum depletion of corticosteroid binding activities, an early marker of human septic shock. *Biochem. Biophys. Res. Commun.*, **102,** 411.
123. Hammond, G. L. and Smith, C. L. (1991) Bioavailability of glucocorticoids during inflammation. In *The New Biology of Steroid Hormones*, Vol. 74. Hochberg, R. B. and Naftolin, F. (eds). Raven Press: New York, p. 177.
124. Strel'chyonok, O. A. and Avvakumov, G. V. (1991) Interaction of human CBG with cell membranes. *J. Steroid Biochem. Mol. Biol.*, **40,** 795.

125. Kuhn, R. W., Green, A. L., Raymoure, W. J., and Siiteri, P. K. (1986) Immunocytochemical localization of corticosteroid-binding globulin in rat tissues. *J. Endocrinol.*, **108,** 31.
126. Nakhla, A. M., Khan, M. S., and Rosner, W. (1988) Induction of adenylate cyclase in a mammary carcinoma cell line by human corticosteroid-binding globulin. *Biochem. Biophys. Res. Commun.*, **153,** 1012.
127. Seralini, G.-E., Underhill, C. M., Smith, C. L., Nguyen, V. T. T., and Hammond, G. L. (1989) Biological half-life and transfer of maternal corticosteroid-binding globulin to amniotic fluid in the rabbit. *Endocrinology*, **125,** 1321.
128. Hammond, G. L., Smith, C. L., Underhill, C. M., and Nguyen, V. T. T. (1990) Interaction between corticosteroid binding globulin and activated leukocytes *in vitro. Biochem. Biophys. Res. Commun.*, **172,** 172.

2 | Importance of steroidogenesis in specific hormone action

JOHN W. FUNDER

1. Signals

Six classes of hormonal steroids are currently recognized in mammalian systems—mineralocorticoids, glucocorticoids, androgens, oestrogens, progestins, and vitamin D. This classification was made historically on effector grounds—growth of a cock's comb for androgens, hepatic glycogen deposition for glucocorticoids—long before anything was known of receptors or the subcellular mechanism of steroid action. In invertebrates, ecdysones are clearly an additional class of physiological steroid hormone, and in a variety of species both endocrine and paracrine (neurosteroid) roles are currently being explored for a number of steroids (e.g. dehydroepiandrosterone, pregnenolone). What remains true, however, is that of the hundreds of different steroids in various body compartments a (patho)physiological role as a signal is assigned to fewer than a dozen, with the rest being considered as precursors, biosynthetic intermediates, or degradation products.

2. Receptors

What distinguishes a steroid hormone signal from background circulating steroid noise is the relative configuration of steroid and receptor, so that the likelihood of recognition, interaction, and activation is high for signal and low for noise. A shorthand for the likelihood of a receptor being occupied by a particular ligand is the affinity of that ligand for the receptor. Affinity is commonly given as K_d, or equilibrium dissociation constant, the molar concentration of signal required to occupy half of the receptors; the lower the concentration of ligand needed, the higher the affinity.

2.1 Specificity

One way, therefore, to build specificity into steroid hormone–receptor systems would be to evolve highly selective receptors, which recognize 'their' appropriate physiological ligand with high affinity, but other steroids either minimally or not at all. For some steroid hormones this has more or less happened, so that oestradiol 17β has at least an order of magnitude higher affinity for oestrogen receptors than other ligands (oestrone, Δ5-androstanediol). Vitamin D receptors also have an affinity for the physiological ligand (1,25-dihydroxycholecalciferol) at least an order of magnitude higher than for other vitamin D derivatives, and steroids of other classes have very low affinity for vitamin D receptors.

Receptors for the other steroid hormones, however, are much less specific. This might be predicted from a consideration of their structure (Fig. 1), in that oestrogens (in which the A-ring is aromatized) and vitamin D (in which the B-ring is cut to give a much more flexible secosteroid) are clearly the outliers in structural terms. In contrast, progesterone hydroxylated at C21 becomes deoxycorticosterone, a potent mineralocorticoid, which in turn hydroxylated at C11 becomes corticosterone, the physiological glucocorticoid in the rat. That the mineralocorticoid–glucocorticoid–progestin–androgen receptors (MR–GR–PR–AR) are a subfamily could also be predicted from early binding studies, which showed considerable non-specificity between receptors, most noticeably for mineralocorticoid receptors, which have equivalent affinity for aldosterone and the physiological glucocorticoids corticosterone and cortisol (1–3). With the cloning of steroid receptors, the structural

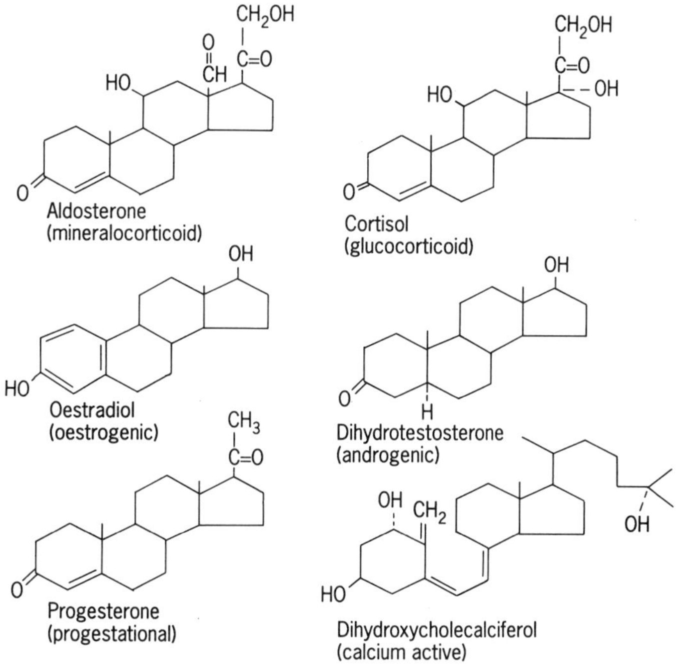

Fig. 1 Two-dimensional representation of physiological ligands for the six recognized classes of steroid hormone

basis for this lack of specificity became apparent; in the ligand-binding region MR–GR–PR–AR show ⩾50% amino acid identity, whereas oestrogen receptors (ER) and vitamin D receptors (VDR) have ⩽15% identity with each other or the MR–GR–PR–AR subfamily (3–8).

3. Mineralocorticoid receptors

To illustrate the importance (and limitations) of steroidogenesis in specific hormone action we will focus on the question of how aldosterone can occupy and act via mineralocorticoid receptors that have equal affinity for physiological glucocorticoids — particularly given that glucocorticoids circulate at concentrations that are orders of magnitude higher than that of aldosterone, even in sodium deficiency. This equivalence in MR affinity has been shown in cytosol preparations from kidney (1), and in whole-cell preparations such as human mononuclear leukocytes (2) and rat vascular smooth muscle cells (9); other non-epithelial sites in which equivalent binding of aldosterone and physiological glucocorticoids has been shown include the hippocampus (1, 10). Although such sites are clearly not 'mineralocorticoid receptors', in that they are not involved in steroid modulation of unidirectional transepithelial sodium transport, it is of some interest that the rat mineralocorticoid receptor was cloned from a hippocampal library (11), on the basis of the sequence of the previously cloned human renal mineralocorticoid receptor (3).

By a variety of criteria, these hippocampal 'MR' — and perhaps those in other non-epithelial tissues — appear to be high-affinity glucocorticoid (type I) receptors, in that they are occupied by cortisol (corticosterone in the rat) under physiological conditions, at circulating concentrations substantially lower than those required to occupy classical (type II) GR. They were first described as 'corticosterone-preferring sites' by McEwen and co-workers, on the basis of the preferential hippocampal uptake and retention of [^{3}H]corticosterone over [^{3}H]dexamethasone, when injected into adrenalectomized rats (12). Subsequent studies from de Kloet's group (13, 14) have shown that such sites are physiologically glucocorticoid rather than mineralocorticoid receptors, in that hippocampal changes following adrenalectomy are reversed by corticosterone but not by aldosterone, which blocks the action of co-administered corticosterone. More recent studies by Joels and de Kloet, comparing corticosterone action in rat hippocampal preparations *in vitro* via type I and II receptors, will be discussed later in the context of the specificity (or otherwise) of response elements for MR and GR.

3.1 Tissue specificity of aldosterone binding

Although MR type I receptors in a variety of tissues and species had been shown in a number of laboratories to be unable to distinguish aldosterone and the physiological glucocorticoids *in vitro*, this was not the case *in vivo* (15). As shown in Figure 2, when [^{3}H]aldosterone or [^{3}H)corticosterone is injected into 10-day-old, 1-day adrenalectomized rats, the uptake and retention in the hippocampus 15 min later is

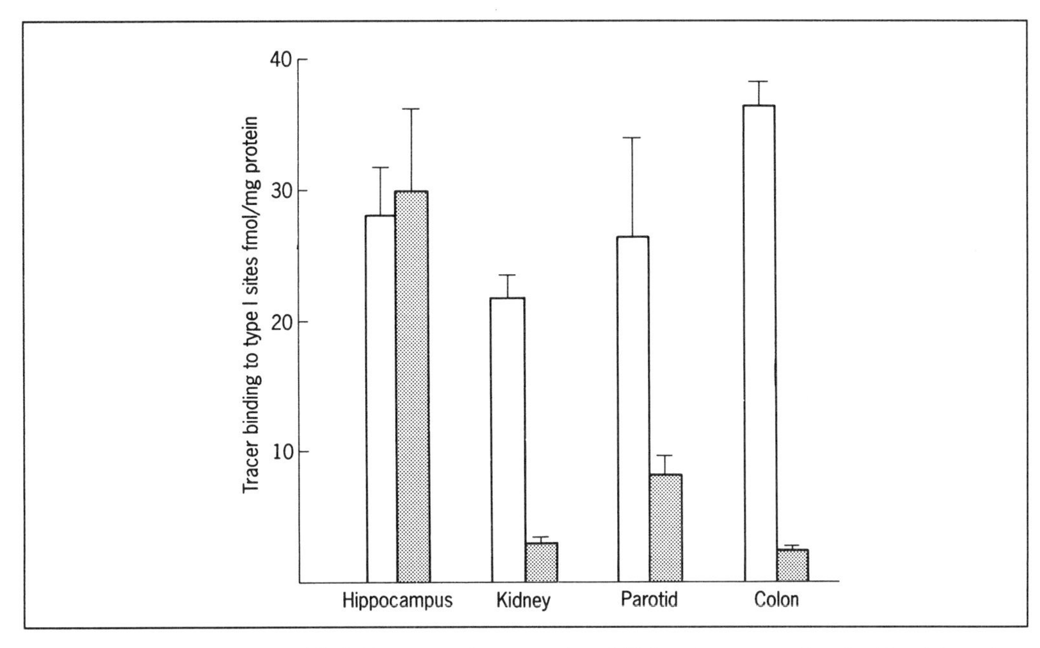

Fig. 2 Uptake and retention of [^{3}H]aldosterone (open bars) and [^{3}H)corticosterone (hatched bars) in various tissues of 10-day-old, 1-day adrenalectomized rats. Comparable doses of either steroid were injected, in the presence of excess RU28362 to exclude tracer from glucocorticoid (type II) receptors, with and without excess unlabelled aldosterone or corticosterone. Tissues were harvested 15 min after injection, and specific type I receptor binding was determined. Values are mean ± SEM (n = 6)

indistinguishable. In contrast, however, there are marked differences in tracer binding in the classical mineralocorticoid target tissues (kidney, parotid, colon). One possibility is that aldosterone may be 'locked in' to the receptor *in vivo*, thus giving it a higher operational affinity, a possibility that is consistent with some otherwise puzzling subsequent co-transfection studies (16).

The principal specificity-conferring mechanism for mineralocorticoid receptors, however, is the exclusion of physiological glucocorticoids from the immediate vicinity of the receptor by the operation of the enzyme 11β-hydroxysteroid dehydrogenase (11βHSD). There have been four stages in establishing this to be the case: astute clinical investigation of a rare syndrome (apparent mineralocorticoid excess), reproduction of the syndrome by licorice administration to normal volunteers, enzyme blockade and binding studies in rats, and the purification and cloning of 11βHSD. The evolution of our current state of knowledge thus draws from studies from molecular biology to clinical medicine, and will be given in some detail to highlight their complementarity and interdependence.

3.2 Clinical studies: apparent mineralocorticoid excess

First, in 1979 Ulick and colleagues thoroughly investigated and documented the index case of apparent mineralocorticoid excess (17). As the name implies, the syndrome was characterized by high levels of sodium retention and blood pressure, with normal or suppressed levels of cortisol and aldosterone in the plasma. The patient's urinary steroid profile, however, was clearly abnormal, with marked elevations in the ratio of cortisol to cortisone metabolites, and in the ratio of 5α to 5β reduced cortisol metabolites. One possible explanation suggested by these authors for the inappropriate activation of renal mineralocorticoid receptors was that they were occupied by the abnormally high intrarenal cortisol levels. This concept was further developed by Edwards *et al.* (18) in their discussion of a subsequent patient as that of abrogation of the normal intrarenal 'cortisol–cortisone shuttle'.

3.3 Licorice and carbenoxolone

Second, Edwards and co-workers made a crucial contribution to the elucidation of the underlying physiological mechanisms by their demonstration that the clinical syndrome of apparent mineralocorticoid excess could be mimicked, to a modest extent, by giving normal volunteers licorice over a period of 10 days (19). It had long been appreciated that licorice abuse could be followed by hypokalaemia and hypertension, effects considered to reflect the ability of the active components of licorice (glycyrrhizic acid, glycyrrhetinic acid) to bind to mineralocorticoid receptors, albeit with very low affinity. What Edwards and colleagues found was that patients given licorice showed elevated urinary cortisol:cortisone metabolite ratios, and prolonged half-times of infused [^{3}H]cortisol, evidence that the effect of licorice was on the enzyme normally responsible for the conversion of cortisol to cortisone, rather than on the mineralocorticoid receptor *per se*.

Third, in parallel studies from Mebourne (20) and from Edinburgh and Utrecht (21), inhibition of 11βHSD by glycyrrhizic acid or carbenoxolone (the hemisuccinate of glycyrrhetinic acid) was shown to allow [^{3}H]corticosterone occupancy of otherwise aldosterone-selective MR in mineralcorticoid target tissues; data for carbenoxolone, in studies similar in design to those shown in Figure 2, are shown in Figure 3. In the hippocampus, binding of [^{3}H]aldosterone and [^{3}H]corticosterone is indistinguishable, and neither is affected by pretreatment of the rats with carbenoxolone. In contrast, pretreatment with carbenoxolone raised [^{3}H]corticosterone binding to levels equal to (kidney, parotid) or approaching (colon) those of aldosterone, evidence for the crucial role of 11βHSD in excluding corticosterone frm MR under normal conditions.

3.4 11β Hydroxylation and 18 oxidation

If conversion of corticosterone (in the rat) to 11-dehydrocorticosterone (compound A) by 11βHSD is the basis for the aldosterone selectivity of MR in aldosterone target tissues, then a number of points may be made; one of these is that compound A should have negligible affinity for MR, and another is that aldosterone is

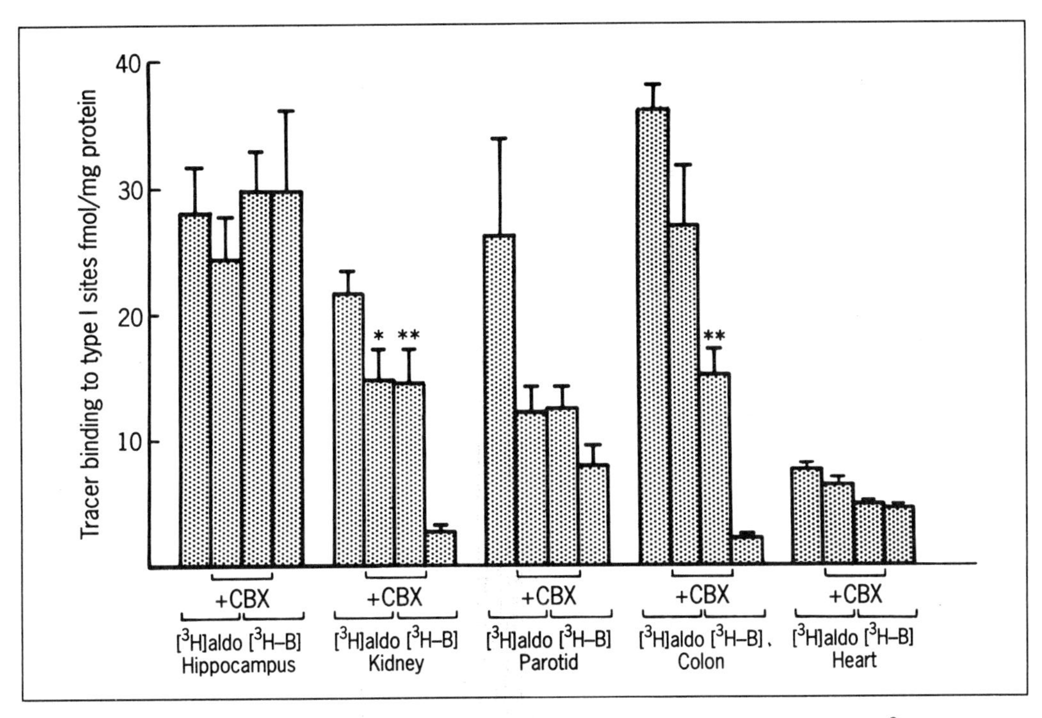

Fig. 3 Effect of pretreatment with carbenoxolone (CBX) on the uptake and retention of [^{3}H]aldosterone ([^{3}H]aldo) or [^{3}H]corticosterone ([^{3}H]B) in various tissues of 10-day-old, 1-day adrenalectomized rats. Values are mean ± SEM (n = 6). Reproduced with permission from Funder *et al.* (20)

either not a substrate for 11βHSD, or that 11-dehydrogenation of aldosterone does not affect its affinity for and activity on MR. Figure 4 shows the relative ability of corticosterone (compound B) and its 11-dehydro metabolite (compound A) to compete for MR and GR. Note that *in vitro*, in a kidney cytosol preparation from adrenalectomized rats, corticosterone and aldosterone have indistinguishable affinity for MR; in contrast, the affinity of compound A for these receptors is only 0.3% that of corticosterone, on the basis of its ability to compete for [^{3}H]aldosterone binding. Note also that the affinity of compound A is <1% that of corticosterone for GR, measured by their relative ability to compete for [^{3}H]dexamethasone binding in thymus cytosol preparations from adrenalectomized rats: for both MR and GR 11-dehydrogenation lowers ligand affinity considerably.

The same is true for aldosterone: in a survey of the effect of various substitutions on steroid affinity for MR, 11-keto-aldosterone was found to have approximately 1% the affinity of native aldosterone for these receptors (S. Ulick and J. W. Funder, unpublished results). On the other hand, aldosterone is essentially a non-substrate for the 11βHSD enzyme, either *in vivo* or *in vitro*, as gauged by trace or absent 11-keto-aldosterone metabolites in urine and the inability of even very high concentrations of aldosterone to lower cortisol to cortisone conversion *in vitro*. What this reflects is the *in vivo* structure of aldosterone, a product of its unique and highly

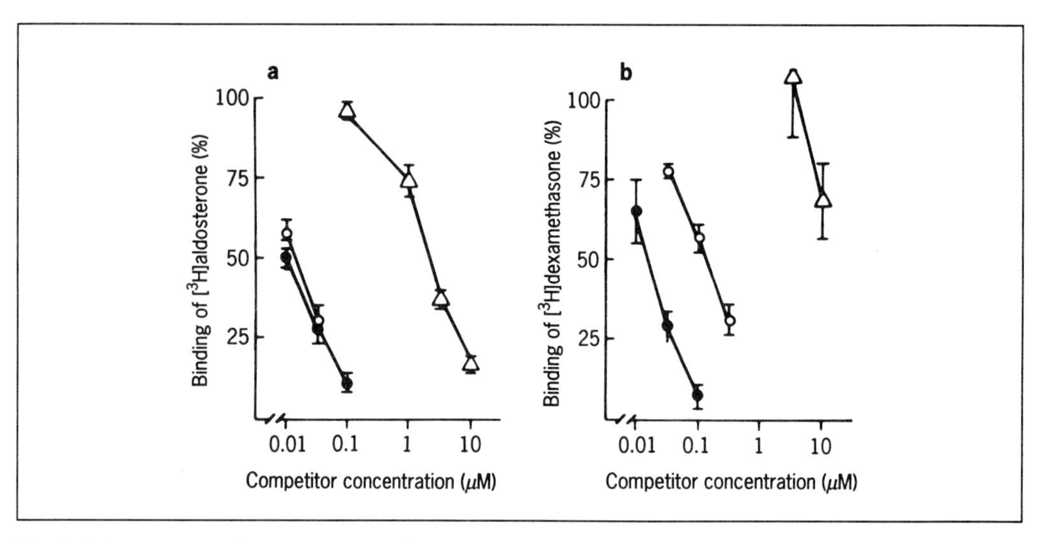

Fig. 4 Relative ablity of corticosterone (○), 11-dihydrocorticosterone (△), aldosterone (●, a), and dexamethasone (●, b) to compete for [^{3}H]aldosterone binding to renal mineralcorticoid receptors (a) or [^{3}H]dexamethasone binding to thymic glucocorticoid (type II) receptors (b). Values are mean ± SEM. Reproduced with permission from Funder *et al.* (20)

reactive aldehyde group at C18. Although aldosterone can be depicted as being 11-hydroxylated (Fig. 5a), at physiological pH this hydroxyl is immediately cyclized with the C18 aldehyde group to the very stable 11,18-hemiketal form (Fig. 5b), rendering the molecule impervious to attack at C11.

4. Aldosterone synthesis and mineralocorticoid specificity

In this regard, then, steroidogenesis is of the utmost importance for the specificity of aldosterone action. Aldosterone is a product uniquely of the adrenal glomerulosa, the outermost zone of the adrenal cortex. For many years it was believed that the

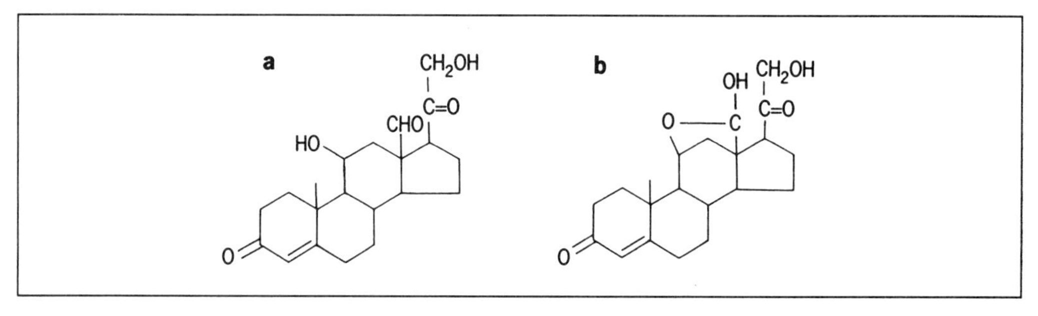

Fig. 5 Two-dimensional representation of (a) non-cyclized aldosterone and (b) aldosterone in its physiological 11,18-hemiketal form

enzyme responsible for 11β-hydroxylation in the adrenal was also responsible for 18-hydroxylation, i.e. the conversion of the methyl group at C18 to an aldehyde group. If this were the case, then what remained unexplained was how the single enzyme was bifunctional in the glomerulosa (11-hydroxylation, 18-hydroxylation) but catalysed only 11-hydroxylation in the zona fasciculata. Recently, the genes coding for both C11 hydroxylase and C18 methyloxidase have been cloned and sequenced, and the apparent enigma explained. CYP11B1 (11-hydroxylase) is expressed generally in the adrenal; CYP11B2 (18-methyloxidase) has 95% amino acid identity, but is confined to the zona glomerulosa, where it can effect both C18 and C11 hydroxylation (22).

4.1 Glucocorticoid remediable hyperaldosteronism

Considerable insight into the physiology of CYP11B1 and CYP11B2 has even more recently come from genetic analysis of the condition of glucocorticoid remediable aldosteronism. This syndrome is characterized by elevated plasma levels of aldosterone; unlike primary aldosteronism, however, levels of aldosterone are lowered following suppression of adrenocorticotrophic hormone (ACTH) with dexamethasone, and the patient's blood pressure and electrolyte status restored to normal. Chromosomal analyses have shown CYP11B1 and CYP11B2 to be adjacent on 8q22; a recent, very elegant, study from Lifton *et al.* (23) has shown that glucocorticoid remediable aldosteronism reflects a recombination event between the two genes, so that affected subjects have a chimæric gene with the 5′ end up to exon 4 derived from CYP11B1, and the remainder of the gene from CYP11B2. Although the site of recombination varies between kindred, it appears that the specific C18 methyl oxidase activity is preserved if the recombination site is 5′ to exon 4. This chimæric gene appears to be expressed in the adrenal fasciculata, as high levels of steroids modified at both C17 and C18 (e.g. 18-oxo-cortisol) are described in this condition; accordingly it would appear that the sequences responsible both for zona glomerulosa localization and ACTH responsivity will be found in the respective upstream 5′ UT regions, as presumably are the regions responsible for the angiotensin sensitivity of CYP11B2 in culture.

4.2 11β Hydroxysteroid dehydrogenase: a family of enzymes?

A similar but currently less well explained complexity surrounds the activity of what appear to be multiple species of 11βHSD. Rat liver 11βHSD has been isolated, purified, cloned, and sequenced by Monder and co-workers (24, 25). On immunohistochemical examination the enzyme is seen to be localized predominantly in the liver, testis, lung, and renal proximal tubule, none of which are classical mineralocorticoid target tissues; specifically, in studies using antisera to 11βHSD and MR, 11βHSD and MR were shown to be confined to distinct and separate populations of cells in rat kidney sections (26). This species is also found in the hippocampus on immunohistochemical examination (27); hippocampal 11βHSD in the rat differs from that in the kidney, however, in that it converts corticosterone to 11-dehydrocorticosterone but not cortisol to cortisone (20), and *in vivo* does not operate to exclude corticosterone from hippocampal type I receptors. The role of the species isolated and cloned by Monder and asociates, then, would appear to be that of modulating glucocorticoid occupancy of GR in various tissues, of fractionating the response of different tissues; all nucleated cells appear to have GR, and essentially all tissues see identical levels of circulating signal.

If this species of 11βHSD (11βHSD1, for precedence) is absent or at very low levels in physiological mineralocorticoid target tissues, the question remains of how the MR are protected. *In vitro*, there is clear evidence for 11βHSD activity in such tissues, either in dissected nephron segments (28), cultured cortical collecting tubules (29), or in sections of rat (30) and pig (31) kidney. Bonvalet and colleagues (28) have studied 11βHSD activity along the nephron, and shown that levels in distal tubules are higher than those in proximal tubules, although the extent of the difference appears to vary between species. Nary-Fejes-Toth and colleagues (29) have reported that cultured cortical collecting tubules show high levels of 11βHSD activity, with a K_m for corticosterone approximately 40-fold lower than that reported for 11βHSD1. Mercer and Krozowski (30) have co-localized 11βHSD activity to cortical-collecting tubules and connecting segments in rat kidney sections by a cytochemical technique, involving the deposition of diformazan blue consequent upon the conversion of 11βOH androstenedione to 11-keto-androstenedione and the reduction of nicotinamide adenine dinucleotide (NAD) to NADH *in vitro*; on the basis of the tubular localization, and the preference for NAD over NADP as co-factor (the reverse is true for 11βHSD1) they have dubbed this species 11βHSD2. Provencher *et al.* (31) have extended these studies in sections of pig kidney, and have defined the K_m and V_{max} of this NAD-preferring species with cortisol as substrate in microsomal preparations *in vitro*. At the time of writing, then, there exists powerful but circumstantial evidence for at least one other member of the 11βHSD subfamily in the family of dehydrogenases (32); as for the 11- and 18-hydroxylases, definitive evidence awaits the purification and/or cloning of 11βHSD2, currently the focus of attention in several laboratories.

4.3 Steroidogenesis: summary

Finally, in summary it is clear that the extent to which steroidogenesis and unique structural features confer steroid specificity varies widely over the class of steroid hormones. Steroidogenesis and specificity of steroid binding are crucial features for the specificity of action of oestradiol and vitamin D; steroidogenesis and specificity of steroid metabolism are crucial in aldosterone action. For the other members of the MR–GR–PR–AR subfamily, steroidogenesis is clearly of considerable importance in specific hormone action, but it does not appear to be sufficient to guarantee high signal-to-noise ratios at the receptor level. The possibility that specificity may be physiologically limited beyond the signal–receptor interaction in this subfamily and mechanisms that might contribute to whatever specificity is present are discussed in the next section.

5. Hormone response elements

For more than two decades, the consensus model of steroid hormone action at the genomic level has included a step of binding of activated steroid–receptor complexes to DNA, which in turn occasions a cascade of DNA-directed, RNA-mediated protein synthesis (33). Given the clear and widely accepted distinction between the six classes of hormonal steroids, it has come as rather a surprise that prima facie there appears to be little in the way of discrimination between classes in terms of the sequences of the nuclear response elements. For the MR–GR–PR–AR receptor subfamily, the canonical sequence is the pentadecamer palindrome GGTACAnnnTGTTCT, and for ER–VDR the sequence is AGGTCAnnnTGACCT; in addition, the consensus sequence for the family of thyroid hormone receptors and
retinoic acid receptors appears to be a half turn less, GGTCATGACC. Given that many operating response elements are approximations to these canonical sequences, it is possible that for a particular subfamily specific variations in sequence favour binding of one steroid–receptor complex over another; on the other hand, there can be no doubt that there is a high degree of overlap, of non-specificity in terms of sequence, and that mechanisms in addition to nucleotide sequence must operate if there is to be specificity at this level.

At this point it is important to leave the question open, and not to asume that there is necessarily specificity at the level of the interaction of activated receptor and DNA, as the following example—again taken from recent work on aldosterone action—illustrates. In their investigation of the index patient with apparent mineralocorticoid excess, New and colleagues noted that the patient appeared more sensitive, in terms of sodium retention and blood pressure elevation, to infused cortisol than to aldosterone (34). Given the usual asumption that aldosterone is the benchmark, the 'ne plus ultra' mineralocorticoid, this finding appears counterintuitive, particularly in the light of the co-transfection studies of Arriza *et al.* cited above (16). One possible explanation of such a finding is that the syndrome represents not only inappropriate occupancy of MR by cortisol, but also inappropriate occupany of GR. If this is the case, then infusion of aldosterone would be predicted to have little effect, as the high-affinity, unprotected MR would already be largely occupied by cortisol, and aldosterone has low affinity for, and activity at, GR. Infused cortisol, on the other hand, might be expected to occupy additional GR to those occupied by the normal circulating levels of cortisol in such patients, and might thus be able to increase the severity of the symptoms.

5.1 Experimental mineralocorticoid specificity

The hypothesis to be tested, therefore, is that under physiological conditions 11βHSD operates to exclude cortisol (in the rat corticosterone) not only from MR but also from GR in the kidney, by conversion to compounds with low affinity for both MR and GR (see Fig. 4). To test this hypothesis, we initially used patients with a congenital absence of type I receptors, the syndrome of pseudohypoaldosteronism. In the first report of the syndrome over 30 years ago (35) an abnormality of aldosterone action was posited as the underlying cause, and in 1985 Armanini *et al.* (36) showed that such patients, in contrast with normal controls, had no type I receptors in their circulating mononuclear leukocytes. In studies to test the hypothesis that 11βHSD normally protects both MR and GR in the kidney (37), a woman and her three surviving children, all with the diagnosis of pseudohypoaldosteronism verified by receptor assay, were given carbenoxolone 50–150 mg/day for 2 weeks after a run-in period of 5 days. If 11βHSD protects only MR, carbenoxolone would be expected to be without significant effects in patients with pseudohypoaldosteronism. If, on the other hand, 11βHSD protects not only MR but also GR in the kidney, then carbenoxolone given to such patients may have effects reflecting cortisol occupancy of normally forbidden GR.

In the event, patients with pseudoaldosteronism show a classical mineralocorticoid response to carbenoxolone administration (Fig. 6), despite the absence of MR. The left-hand bar in each panel represents the basal value, at the end of the run-in period; over the ensuing 2 weeks, urinary sodium excretion progressively fell, as a

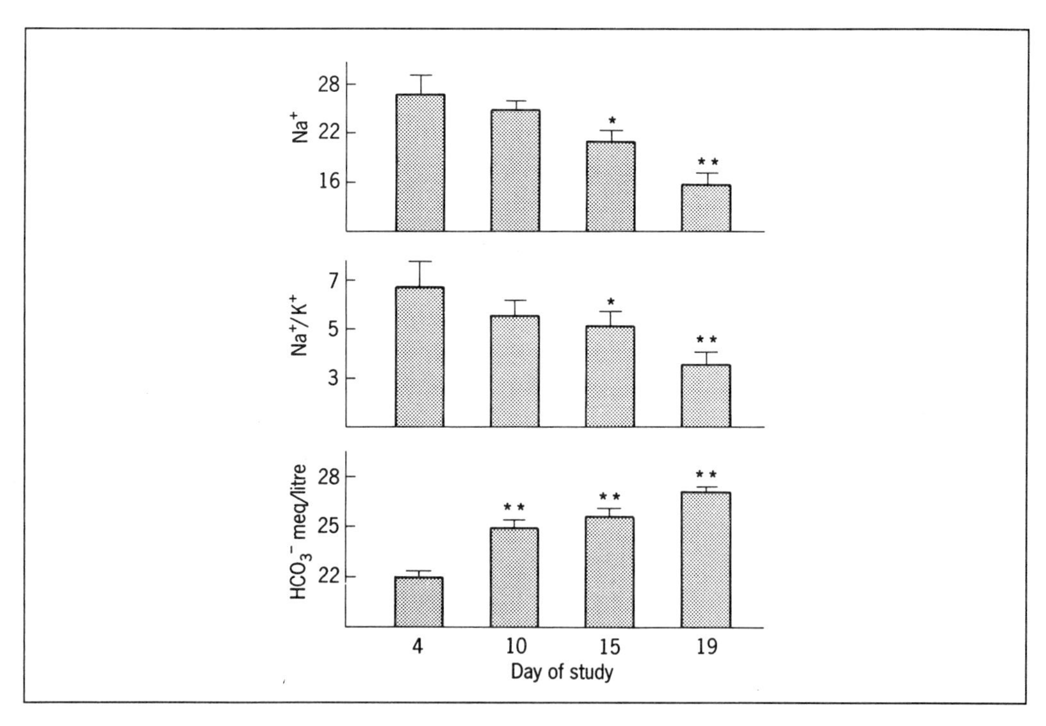

Fig. 6 Response of patients with pseudohypoaldosteronism (n = 4) to administration of carbenoxolone from days 5–19. Mean ± SEM values for urinary Na^+: creatinine excretion, urinary Na^+/K^+ ratio, and plasma bicarbonate concentration. *is $P < 0.05$, and **is $P < 0.01$. From Funder *et al.* (37)

function of urinary creatinine, as did the urinary sodium to potassium ratio; in addition a brisk rise in plasma bicarbonate concentration was seen.

It would thus appear that under circumstances in which cortisol can access distal tubular GR, a response indistinguishable from that normally seen for aldosterone can be produced. To test this hypothesis further — that in the renal distal tubule an activated GR may mimic an activated MR in terms of ion flux — adrenalectomized rats were pretreated with carbenoxolone and/or the highly selective synthetic glucocorticoid RU28362, and the effects of various treatments on urinary sodium to potassium ratio compared. The *y*-axis in Figure 7 expresses the urinary electrolyte ratios as a comparison of concentrations of both ions before and after carbenoxolone and steroid administration; so expressed, the within-group variation is relatively small (as can be seen from the SEM values), and a doubling of response follows a near-maximal dose of aldosterone. As can be seen in Figure 7, carbenoxolone given alone is without effect; RU28362 at a modest dose (10 μg per rat) produces a significant effect on electrolyte ratios when given alone, and a near-maximal effect when given with carbenoxolone (right-hand bar). These rat studies (37) complement those in the patients with pseudohypoaldosteronism shown in Figure 6; together, they suggest very strongly that activation of GR — by cortisol in the patients, or by the very selective glucocorticoid RU28362 in adrenalectomized rats — can be followed by a classical mineralocorticoid effect.

Both of these experiments were carried out *in vivo*, with the power but also the drawbacks of *in vivo* studies. For example, carbenoxolone is a 'dirty' drug, with effects on a variety of enzymes; in addition, RU28362 has cardiovascular effects, and changes renal free water clearance rates when given to adrenalectomized rats. In this context, then, complementary *in vitro* studies are very welcome, and have come from Naray-Fejes-Toth and colleagues. These authors (38) have raised a panel of monoclonal antibodies against epitopes expressed in the kidney, and thus can identify the cells expressing such epitopes by incubating kidney slices with the

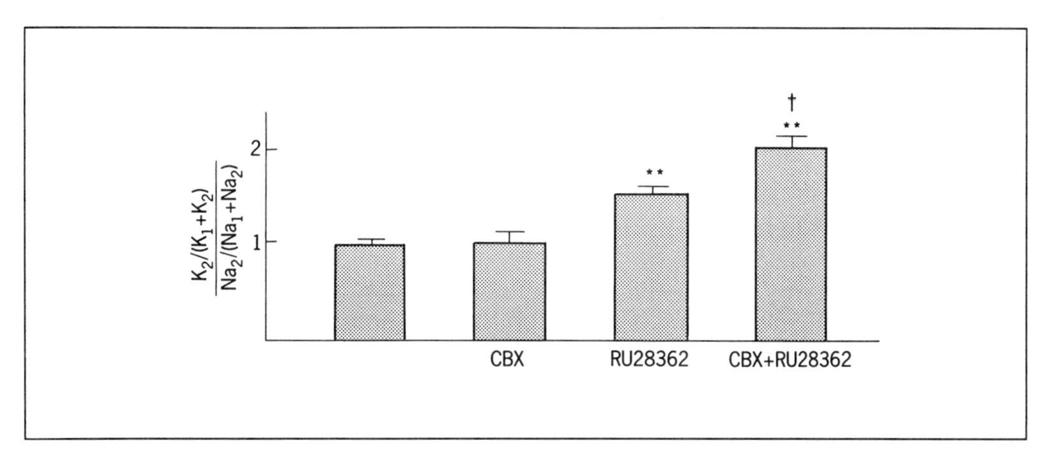

Fig. 7 Urinary electrolyte responses of adrenalectomized rats to carbenoxolone (1 mg) and/or RU28362. Values are mean ± SEM (n = 7–9). From Funder *et al.* (37)

antibody. Having found an antibody that recognizes epitopes expressed uniquely in cortical collecting tubule cells, they used this to immunodissect a collagenase-perfused kidney.

The cortical collecting tubules so harvested are then plated on to a grid between two chambers and grown to confluence, so forming a seal; the cells then start to pump sodium in one direction and potassium in the other. This ion flux can be measured directly, by sampling the chambers; in addition, it sets up a potential difference between the chambers, which can be measured as a short circuit current. When such preparations are exposed to aldosterone, there follows the classical stimulus of unidirectional sodium transport, with potassium secreted in the opposite direction. When similar preparations are exposed either to dexamethasone or to the highly specific GR agonist RU28362, then the changes in ion flux, potential difference, and short circuit current are indistinguishable from those seen with aldosterone.

5.2 Mineralocorticoid response elements

At the time of writing, there have yet to be described uniquely, or even particularly, mineralocorticoid-responsive genes, although there are clearly actions of aldosterone that are physiologically specific. In whatever way these are mediated, the series of *in vivo* and *in vitro* studies cited above provide powerful evidence for the non-specificity, presumably at the genomic level, of activated MR and GR in the cortical collecting tubule cells. Given that the effect on transepithelial sodium transport is the definition of the mineralocorticoid action of aldosterone, it seems possible that in such cells there may be no specific mineralocorticoid response element, inasmuch as the putative MRE responds equally to an activated GR.

5.3 Other possible sources of MR/GR selectivity

What this does *not* mean is that activated MR and GR are equivalent in all—or indeed any—other cell types. In the hippocampus, for example, there is excellent evidence from *in vitro* studies that low and high doses of corticosterone have different effects on neuronal electrical activity, depending upon the receptor(s) occupied (39). This is in clear contradistinction to the kidney, where in studies using specific type I (RU28318) and type II (RU38486) receptor antagonists corticosterone produced identical effects on sodium retention via types I and II receptors in the presence of carbenoxolone (37). Second, a number of laboratories have shown that type II receptors can bind proto-oncogene products (*fos*/c-*jun*), and that their activity can be altered by such binding, which is not the case for type I receptors (40–42). Clearly this allows the possibility of distinction between MR and GR effects in target tissue cells—perhaps even including those of the cortical collecting tubule.

Very recently, Pearce and Yamamoto (43) have produced evidence for differential activity of GR and MR at what they term composite response elements, distinct from the non-selective GRE/MRE/PRE/ARE. One such composite response element (plf–G) is a 25-nucleotide sequence which acts both as a steroid receptor binding site and an AP–1 binding site, via which dimers of the *fos–jun* family of proto-oncogene products can modulate transcription. In the absence of ligand, neither receptor affects AP–1 induced transcription; MR activated by corticosterone are similarly without effect, whereas corticosterone-activated GR very profoundly inhibit AP–1 induced transcription in cotransfection studies. From receptor chimera and domain swap studies this GR-specific activity resides in the N-terminus, between amino acids 105 and 440, and very clearly represents a mechanism whereby GR may have effects at the transcriptional level which cannot be mimicked by MR.

6. Biosynthesis: secretion and activation

A summary overview of the importance of steroidogenesis in specific hormone action might then run as follows. Of the physiological ligands for the six classes of mammalian steroid receptors, three (aldosterone, cortisol/corticosterone, progesterone) are active as secreted, two are converted to active or more active species following secretion (dihydrotestosterone, 1,25-dihydroxycholecalciferol), and one (oestradiol) varies with the reproductive state.

6.1 Vitamin D

1,25-Dihydroxycholecalciferol is synthesized from a skin precursor (7-dihydrocholesterol) by a series of biosynthetic steps. In the skin, the first step in its synthesis requires ultraviolet light, explaining why it was initially termed a vitamin rather than a hormone; many early human studies were carried out on subjects with inadequate exposure to sunlight. The secosteroid formed in the skin is then subjected to a sequence of hydroxylation steps in the liver and kidney to produce the mature product, 1,25-dihydroxycholecalciferol, which is active not only in kidney but also on bone, gut, and cells of the immune system.

6.2 Androgens

The conversion of testosterone to the irreversibly committed androgen dihydrotestosterone is largely peripheral, and reflects the action of the enzyme 5α-reductase. The 5α-dihydrotestosterone product is not only irreversibly committed but also has substantially higher efficacy as an androgen, although this may vary somewhat between species. In the human, 5α-reductase deficiency in phenotypic males produces genital ambiguity in infants and commonly female gender assignation. At puberty, however, testosterone rises to normal adult male levels, sufficient for more complete virilization despite the absence of 5α-reductase.

6.3 Oestrogens

Whereas dihydrotestosterone is very largely a peripheral product of circulating testosterone, aromatization of precursor testosterone to oestradiol is a feature of both ovarian and peripheral oestrogen biosynthesis. When the ovaries are quiescent, i.e. before menarche or after the menopause, the bulk of circulating oestrogens in females derives from peripheral conversion of testosterone by aromatase in a variety of tissues, prominent among which is adipose tissue; in men, circulating oestrogens are similarly derived. In the cycling female, in contrast, the principal source of circulating oestrogens is the ovary, produced by aromatization in the thecal cells of the androgenic precursors from the granulosa cells.

7. Metabolism: activation and inactivation

The first thing to remark, then, is that although steroidogenesis is commonly assumed to be confined to 'steroidogenic' tissues such as adrenals, gonads, and placenta, it seems clear that additional steps can and do occur after secretion, if steroidogenesis is taken to be generation of physiologically active ligands. Second, in all these instances post-secretory steroidogenesis is irreversible. This contrasts with the interconversion of cortisol and cortisone catalysed by 11βHSD, where the balance between oxidation and reduction varies between tissues and with development, by mechanisms yet to be established. This reversible interconversion of more to less active ligand within the same class is not confined to glucocorticoids: 17-hydroxysteroid dehydrogenase is responsible for the interconversion of oestradiol and oestrone, and of testosterone and androstenedione. This lends weight to the notion that such interconversion may be primarily a way of fractionating or modulating target tissue response, dependent on the expression and directionality of expression of dehydrogenase enzymes, for and within a particular class of steroids. The role of 11βHSD in excluding glucocorticoids from MR in aldosterone target tissues may thus represent an adventitious or opportunistic evolutionary step.

7.1 Aldosterone

Third, the unique steroidogenic step in aldosterone biosynthesis appears to be keyed to the generation of a molecule that will escape a particular metabolic fate, rather than one that has inherently higher specificity for the appropriate receptors. This is in sharp contrast to oestradiol and 1,25-dihydroxycholecalciferol, where substantial structural changes to the steroid skeleton appear to be keyed to increased receptor selectivity. In terms of synthetic steroids, similar strategies have been employed, albeit at times unwittingly: 9α-fluorination makes cortisol a very poor substrate for 11βHSD, and thus 9α-fluorocortisol is widely used for mineralocorticoid replacement therapy.

7.2 Antagonists

Fourth, the promiscuity of physiological ligands for the MR–GR–PR–AR subfamily is reflected in the lack of specificity of therapeutic antagonists. For example, the antimineralocorticoid spironolactone is also a potent antiandrogen, and will also compete for GR and PR; the antiprogestin RU38486 is also a potent antiglucocorticoid. In contrast, for ER, non-steroidal agonists (e.g. diethylstilboestrol) and antagonists (e.g. tamoxifen, ICI 164 384) have been synthesized, and are both relatively selective in terms of steroid receptors. Whether non-steroidal ligands can be developed specific for MR–GR–PR, along the lines of flutamide for AR, remains to be determined.

7.3 Non-selectivity *in vitro*

Fifth, in a variety of transfection studies, considerable non-selectivity at the level of the nuclear response element has been shown for steroid hormone receptors; for example, the long terminal repeat of mouse mammary tumour virus responds to MR–GR–PR–AR. This promiscuity is reflected *in vivo*, and in non-transfected cells *in vitro*, in that both GR and MR appear to be able to activate transepithelial sodium transport, provided that glucocorticoids can access the receptor. Although this may be the case in this cell type under the conditions of study chosen, this equivalence may not hold up under other conditions or for other cells and tissues that contain both MR and GR.

7.4 Further directions

Finally, in any consideration of signals and receptor specificity a range of conceptual and mechanistic questions needs to be addressed, of which only a few have been touched on in the present chapter. For example, the possibility of metamessage(s) in the steroid–receptor interaction, conveyed by the episodic rather than constant release of signal, has not been considered. Also, antagonists are commonly considered important in pharmacology and therapeutics: to what extent have physiological circulating antagonists for various steroid receptors simply not been looked for? A wider biology than the laboratory rat, transfected cells, and patients with rare congenital syndromes needs to be drawn on if we are to understand the flexibility and diversity of steroid signal–receptor systems. Above all, we need to keep an open mind: given the recent demonstration of membrane receptors for aldosterone on human monocytes (44) and for corticosterone on Ticarda neurones (45), it may be that the present focus on transactivation and nuclear mechanisms is only half the story, and that a whole new chapter on steroidogenesis in specific hormone action will need to be written.

References

1. Krozowski, Z. S. and Funder, J. W. (1983) Renal mineralcorticoid receptors and hippocampal corticosterone binding species have identical intrinsic steroid specificity. *Proc. Natl Acad. Sci. USA*, **80,** 6056.
2. Armanini, D., Strasser, T. and Weber, P. C. (1985) Characterization of aldosterone binding sites in circulatory human mononuclear leukocytes. *Am. J. Physiol.*, **248,** E388.

3. Arriza, J. L., Weinberger C., Glaser, T., Handelin, B. L., Housman, D. E., and Evans, R. M. (1987) Cloning of human mineralocorticoid receptor complementary DNA; structural and functional kinship with the glucocorticoid receptor. *Science,* **237,** 268.
4. Hollenberg, S. M., Weinberger, C., Ong, E. S., Cerelli, G., Oro, A., Lebo, R., Thompson, E. B., Roenfeld, M. G., and Evans, R. M. (1985) Primary structure and expression of a functional human glucocorticoid receptor cDNA. *Nature,* **318,** 635.
5. Green, S., Walter, P., Kumar, V., Krust, A., Bornert, J. M., Argos, P., and Chambon, P. (1986) Human oestrogen receptor cDNA: sequence, expression and homology to v-*erb*-A. *Nature,* **320,**, 134.
6. Loosfelt, H., Atger, M., Misrahi, M., Guiochon-Mantel, A., Meriel, C., Logeat, F., Benarous, R., and Milgrom, E. (1986) Cloning and sequence analysis of rabbit progesterone–receptor complementary DNA. *Proc. Natl Acad. Sci. USA,* **83,** 9045.
7. Baker, A. R., McDonnell, D. P., Hughes, M., Crisp, T. M., Manglesdorf, D. J., Haussler, M. R., Pike, J. W., Shine, J., and O'Malley, B. W. (1988) Cloning and expression of full-length cDNA encoding vitamin D receptor. *Proc. Natl Acad. Sci. USA,* **85,** 3294.
8. Chang, C., Kokontis, J., and Liao, S. (1986) Molecular cloning of human and rat complementary DNA encoding androgen receptors. *Science,* **240,** 324.
9. Meyer, W. J. and Nichols, N. R. (1981) Mineralocorticoid binding in cultured smooth muscle cells and fibroblasts from rat aorta. *J. Steroid Biochem.,* **14,** 1157.
10. Beaumont, K. and Fanestil, D. D. (1983) Characterization of rat brain aldosterone receptors reveals high affinity for corticosterone. *Endocrinology,* **113,** 2043.
11. Patel, P. D., Sherman, T. G., Goldman, D. J., and Watson, S. J. (1989) Molecular cloning of a mineralocorticoid (type I) receptor complementary DNA for rat hippocampus. *Endocrinology,* **3,** 1877.
12. de Kloet, E. R., Wallach, G., and McEwen, B. S. (1975) Differences in corticosterone and dexamethasone binding to rat brain and pituitary. *Endocrinology,* **96,** 598.
13. de Kloet, E. R., Kovacs, G. L., Szabo, G., Telegdy, D., Bohus, B., and Versteeg, D. H. G. (1982) Decreased serotonin turnover in the dorsal hippocampus of rat brain shortly after adrenalectomy: selective normalization after corticosterone substitution. *Brain Res.,* **239,** 659.
14. de Kloet, E. R., Sybesma, H., and Reul, J. H. M. (1986) Selective control by corticosterone of serotonin-1 receptor capacity raphe–hippocampal system. *Neuroendocrinology,* **42,** 513.
15. Sheppard, K. and Funder, J. W. (1987) Type I receptors in parotid, colon and pituitary are aldosterone-selective *in vivo. Am. J. Physiol.,* **253,** E467.
16. Arriza, J., Simerly, R. B., Swanson, L. W., and Evans, R. M. (1988) Neuronal mineralocorticoid receptor as a mediator of glucocorticoid response. *Neuron,* **1,** 887.
17. Ulick, S., Levine, L. S., Gunczler, P., Zanconato, G., Ramirez, L. C., Rauh, W., Rosler, A., Bradlow, H. L., and New, M. I. (1979) A syndrome of apparent mineralocorticoid excess associated with defects in the peripheral metabolism of cortisol. *J. Clin. Endocrinol. Metab.,* **49,** 757.
18. Stewart, P. M., Shackleton, C. H. L., and Edwards, C. R. W. (1987) The cortisol–cortisone shuttle and the genesis of hypertension. In *Corticosteroids and Peptide Hormones in Hypertension,* Serono Symposia Publications, Vol. 39. Mantero, F., and Vecsei, P. (eds). Raven Press, New York, p. 163.
19. Stewart, P. M., Wallace, A. M., Valentino, R., Burt, D., Shackleton, C. H. L., and Edwards, C. R. W. (1987) Mineralocorticoid activity of liquorice: 11-beta-hydroxysteroid dehydrogenase deficiency comes of age. *Lancet,* **ii,** 821.

20. Funder, J. W., Pearce, P. T., Smith, R., and Smith, A. I. (1988) Mineralocorticoid action: target-tissue specificity is enzyme, not receptor, mediated. *Science*, **242**, 583.
21. Edwards, C. R. W., Stewart, P. M., Burt, D., Brett, L., McIntyre, M. A., Sutanto, W. S., de Kloet, E. R., and Monder, C. (1988) Localization of 11β hydroxysteroid dehydrogenase tissue specific protector of the mineralocorticoid receptor. *Lancet*, **ii**, 986.
22. White, P. C. (1992) Molecular basis of congenital adrenal hyperplasia (CAH). *Proceedings of the 74th Annual Meeting of The Endocrine Society, San Antonio*, p. 32 (abstract).
23. Lifton, R. P., Dluhy, R. G., Powers, M., Rich, G. R., Cook, S., Ulick, S., and Lalouel, J.-M. (1992). A chimæric 11β-hydroxylase–aldosterone synthase gene causes glucocorticoid-remediable aldosteronism and human hypertension. *Nature*, **355**, 262.
24. Lakshmi, V. and Monder, C. (1989) Purification and characterization of the corticosteroid 11β-dehydrogenase component of the rat liver 11β-hydroxysteroid dehydrogenase complex. *Endocrinology*, **23**, 2390.
25. Agarwal, A. K., Monder, C., Eckstein, B., and White, P. C. (1989) Cloning and expression of rat cDNA encoding corticosteroid 11β-dehydrogenase. *J. Biol. Chem.*, **264**, 18 939.
26. Rundle, S. E., Funder, J. W., Lakshmi, V., and Monder, C. (1989) The intrarenal localization of mineralocorticoid receptors and 11β-dehydrogenase: immunocytochemical studies. *Endocrinology*, **125**, 1700.
27. Sakai, R. R., Lakshmi, V., Monder, C., and McEwen, B. S. (1992) Immunocytochemical localization of 11β hydroxysteroid dehydrogenase in hippocampus and other brain regions of the rat. *J. Neuroendocrinol.*, **4**, 101.
28. Bonvalet, J.-P., Kenouch, S., Blot-Chabaud, M., and Farman, N. (1991) Colocalization of mineralocorticoid receptors and 11β-hydroxysteroid dehydrogenase along the nephron. In *Aldosterone: Fundamental Aspects*, Vol. 215. Bonvalet, J.-P., Farman, N., Lombes, M., Rafestin-Oblin, M. E. (eds). John Libbey Eurotext, INSERM, France, p. 85.
29. Rusvai, E., Fejes-Toth, G., and Naray-Fejes-Toth, A. (1992) A novel form of 11-β hydroxysteroid dehydrogenase (11-OHS) in renal cortical collecting duct (CCD) cells. *Proceedings of the 74th Annual Meeting of The Endocrine Society, San Antonio*, p. 205 (abstract).
30. Mercer, W. R. and Krozowski, Z. S. (1990) Localization of an 11β-hydroxysteroid dehydrogenase activity to the distal nephron of the rat kidney. Evidence for the existence of two species of dehydrogenase in the rat kidney. *Endocrinology*, **130**, 540.
31. Provencher, P. H., Mercer, W. R., Funder, J. W., and Krozowski, Z. S. (1992) Identification and characterization of an NAD-dependent 11β-HSD in pig kidney. *Proceedings of the 74th Annual Meeting of The Endocrine Society, San Antonio*, p. 128 (abstract).
32. Krozowski, Z. K. (1992) 11β-hydroxysteroid dehydrogenase and the short-chain alcohol dehydrogenase (SCAD) superfamily. *Mol. Cell. Endocrinol.*, **84**, C25.
33. Feldman, D., Funder, J. W., and Edelman, I. S. (1972) Subcellular mechanisms in the action of adrenal steroids. *Am. J. Med.*, **53**, 545.
34. Oberfield, S. E., Levine, L. S., Carey, R. M., Greig, F., Ulick, S., and New, M. I. (1983) Metabolic and blood pressure responses to hydrocortisone in the syndrome of apparent mineralocorticoid excess. *J. Clin. Endocrinol. Metab.*, **56**, 332.
35. Cheek, D. B. and Perry, J. W. (1958) A salt-wasting syndrome in infancy. *Arch. Dis. Child.*, **33**, 252.
36. Armanini, D., Kuhnle, U., Strasser, T., Dorr, H., Butenandt, I., Weber, P. C., Stockigt, J. R., Pearce, P., and Funder, J. W. (1985) Aldosterone-receptor deficiency in pseudohypoaldosteronism. *N. Engl. J. Med.*, **313**, 1178.
37. Funder, J. W., Pearce, P., Myles, K., and Roy, L. P. (1990) Apparent mineralocorticoid

excess, pseudohypoaldosteronism and urinary electrolyte excretion: towards a redefinition of 'mineralocorticoid' action. *FASEB J.*, **4**, 3234.

38. Naray-Fejes-Toth, A. and Fejes-Toth, G. (1990) Glucocorticoid receptors mediate mineralocorticoid-like effects in cultured collecting duct cells. *Am. J. Physiol.*, **259**, F672.
39. Joels, M. and de Kloet, E. R. (1990) Mineralocorticoid receptor-mediated changes in membrane properties of rat CA1 pyramidal neurons *in vitro*. *Proc. Natl Acad. Sci. USA*, **87**, 4495.
40. Jonat, C., Rahmsdorf, H. J., Park, K.-K., Cato, A. C. B., Gebel, S., Ponta, H., and Herrlich, P. (1990) Antitumor promotion and antiinflammation: down-modulation of AP-1 (*fos/jun*) activity by glucocorticoid hormones. *Cell*, **62**, 1189.
41. Yang-Yen, H.-F., Chambard, J.-C., Sun, Y.-L., Smeal, T., Schmidt, T. J., Drouin, J., and Karin, M. (1990) Transcriptional interference between c-*jun* and the glucocorticoid receptor: mutual inhibition of DNA binding due to direct protein–protein interaction. *Cell*, **62**, 1205.
42. Schule, R., Rangarajan, P., Kliewer, S., Ransone, L. J., Bolado, J., Yang, N., Verma, I. M., and Evans, R. M. (1990) Functional antagonism between oncoprotein c-*jun* and the glucocorticoid receptor. *Cell*, **62**, 1217.
43. Pearce, D. and Yamamoto, K. (1993). Mineralocorticoid and glucocorticoid receptor activities distinguished by nonreceptor factors at a composite response element. *Science*, **259**, 1161.
44. Wehling, M., Christ, M., and Theisen, K. (1991) High affinity binding to plasma membrane rich fractions from mononuclear leukocytes: is there a membrane receptor for mineralocorticoids? *Biochem. Biophys. Res. Commun.*, **181**, 1306.
45. Orchinik, M., Murray, T. F., and Moore, F. L. (1991) A corticosteroid receptor in neuronal membranes. *Science*, **252**, 1848.

3 | Overview of the steroid receptor superfamily of gene regulatory proteins

BERT W. O'MALLEY and MING-JER TSAI

1. Introduction

Over the past two decades, a great deal of evidence has accumulated in favour of the hypothesis that steroid hormones act at the level of nuclear DNA to regulate gene expression (1–4). The earliest studies were qualitative and involved experiments which showed that steroid hormones caused: (1) accumulation of new species of hybridizable RNAs that did not exist before stimulation; (2) stimulation of synthesis of new specific proteins; (3) a corresponding increase in the cellular levels of specific messenger RNAs (mRNAs); and (4) stimulation of the rate of transcription of select endogenous nuclear genes (5). At this point, the early 1970s, the primary pathway for steroid hormone action was defined as follows: steroid → (steroid–receptor) → (steroid–receptor–DNA) → mRNA → protein → functional response (6). Steroid was thought to enter cells by passive diffusion and to activate receptors in either the cytoplasm or nucleus. The activated receptor was predicted to bind at the regulatory region of target genes and to stimulate transcription, and subsequently protein synthesis. Although initially controversial, this pathway has been generally accepted for the past 15 years (5–13).

2. Conserved sequences and function

The complementary DNAs (cDNAs) for all major steroid hormone receptors have been cloned and sequenced (14, 15). As expected, they were related in structure and led to the definition of an even larger family of putative regulatory proteins. This superfamily of regulatory molecules was shown to include the receptors for thyroid hormone (T_3), vitamin D_3, and retinoic acid. Oncogenes such as v-*erb*A were also shown to be members of this family. The latter is one of the two genes that comprise an oncogenic cassette in the avian myeloblastosis virus, and appears to be a mutated form of a cellular gene coding for one of the multiple receptors for thyroid hormone. Finally, a large number of cDNAs have been cloned from various eukaryotic tissues and shown to code for authentic members of the superfamily. Since the regulators (ligands) for these receptors are presently unknown, they have been termed 'orphan receptors'.

A summary of selected members of the steroid receptor superfamily is shown in Figure 1. The schematic is based on primary amino acid sequence and reveals three major regions of conserved amino acids (I–III) set against the glucocorticoid receptor as a reference. Regions II and III are located within the C-terminal or hormone-binding domain of the molecule. Regions II and III are thought to be functionally important but it is unclear whether they participate directly in ligand binding, protein–protein structural interactions, or transcriptional activation. It is likely that the receptor makes multiple contacts with its activating ligand in this domain, but, until recently, our understanding of ligand binding and activation has been unclear.

Region I, a highly conserved 66-amino-acid sequence, is located within the interior of the molecule and comprises the DNA-binding domain. Of special note are nine conserved cysteines, eight of which are thought to form two zinc fingers, each containing one molecule of zinc (14–16). The zinc-finger structure was observed initially in the amphibian transcription factor TFIIA, but was of the type I variety, employing two cysteines and two histidines to co-ordinately bind one zinc ion (type I). The steroid receptor superfamily can be distinguished from other

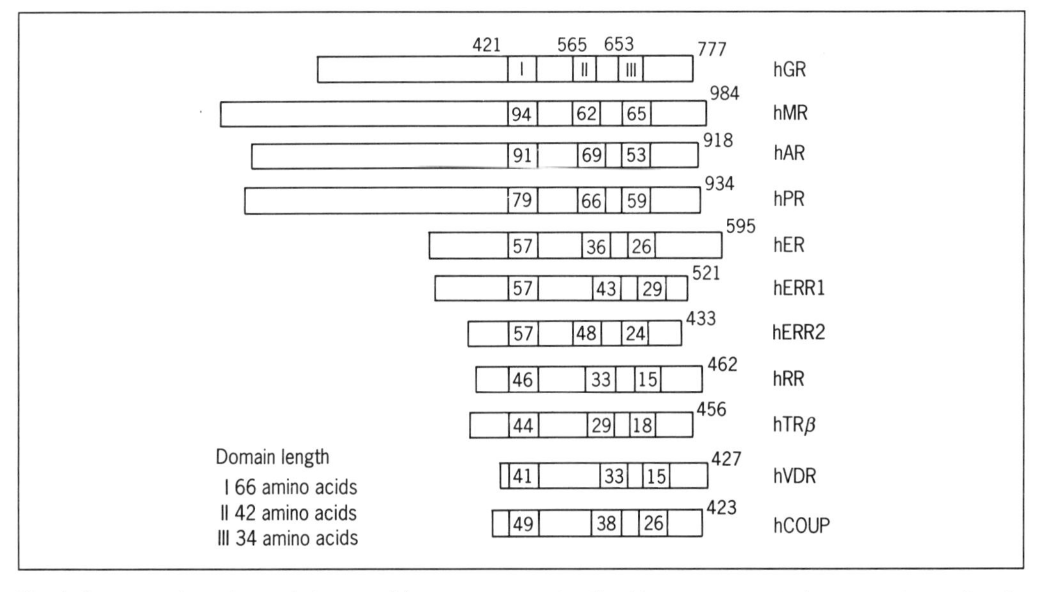

Fig. 1 Sequence homology of the steroid receptor superfamily. Linear representations are shown (top to bottom) for human receptors for glucocorticoid (hGR), mineralcorticoid (hMR), androgen (hAR), progesterone (hPR), eostrogen (hER), oestrogen-related receptors (hERR1, hERR2), retinoic acid (hRR), thyroid hormone (hTRβ), vitamin D (hVDR), and the orphan receptor (COUP–TF, *c*hicken *o*valbumin *u*pstream *p*romoter–*t*ranscription *f*actor). Conserved region I (DNA-binding site) and regions II and III (hormone binding region) are shown for all members, using hGR as a reference point

transcription factors by its subtype (type II). This region of the molecule contains the amino acids that serve to recognize specifically the cognate steroid response elements (SREs) located in genomic DNA adjacent to target genes. This important concept was substantiated by domain swapping experiments which revealed the conversion of a glucocorticoid receptor into an activator of genes containing an oestrogen response element (ERE) when its zinc-finger domain was replaced with the identical region of the oestrogen receptor (16). It has been suggested that the first finger contains primary information for sequence specificity of binding while the second finger stabilizes binding of receptor to its DNA response element. Mutational analyses have also implicated amino acids at the bases and at the tips of both fingers as important for the function of target gene activation at the DNA level.

In a series of subsequent experiments, we learned more about the role of specific amino acids in the DNA-binding domains of receptors (17). The first eight, of nine, cysteines are required for function in this region since mutations abolish all activity. More recently, a series of elegant and more subtle mutational studies has allowed a greater definition of the region of the cysteine-rich domain that appears to be in direct contact with the DNA nucleotides (17–20). Changing the Glu–Gly sequence between the second pair of cysteine residues in the first zinc finger of the oestrogen receptor to the Gly–Ser present in the identical region of the glucocorticoid receptor creates a receptor that no longer activates an ERE reporter but has some stimulatory activity on a glucocorticoid response element (GRE) reporter. Conversely, changing the Gly–Ser of the glucocorticoid receptor to a Glu–Gly as found in the oestrogen receptor creates a hybrid protein which now activates predominantly ERE-containing reporter genes. Evidence implicates the amino acids immediately following the second pair of cysteines in the first zinc finger as playing some additional role in sequence-specific binding (20).

Artificially generated or naturally occurring mutations in the second finger give rise to inactive receptors which bind poorly to DNA (17). The precise role of the second zinc finger is less evident but is thought to stabilize the binding to DNA via interactions with the phosphate backbone of the SRE or, perhaps more likely, by protein–protein interactions which stabilize dimer formation at the SRE. Natural mutations in the zinc finger tips can also inhibit receptor function and have been shown to produce hormone-resistant genetic diseases (21, see Chapter 9). Deletion mutations of steroid receptors were used to identify two major regions necessary for their nuclear localization. A putative signal sequence, homologous to that of the SV40 large T antigen, was localized in the proximal part and the C-terminal domain. When the amino acids were deleted, the receptor became cytoplasmic but could be shifted into the nucleus by the addition of hormone (or anti-hormone) (22). It is clear also that the DNA-binding domain itself plays a role in nuclear localization, especially that induced by hormone. Recent studies on the three-dimensional crystal structure of the DNA-binding region of receptor complexes to DNA have been very informative of the role of specific amino acids and have shown that the α-helix-turn-helix surface is formed when the receptor monomers dimerize, which forms the foundation for interactions with DNA (23).

The *N*-terminal region of the steroid receptors is poorly conserved in both length and amino acid sequence. It is considered to be a transcriptional modulation domain. It contains information that enhances transcriptional stimulatory potency and also contains sequences that allow preferential activation of certain genes (24). These functions are thought to be transmitted not via conserved runs of specific amino acids but rather by aggregate charge and higher order structure. Deletion of this domain reduces gene activation potential in glucocorticoid and progesterone receptors by 50–85% after reintroduction to animal cells. It is clear, however, that sequences in the carboxy-terminal ligand-binding domain also play an important role in trancriptional regulation by most receptors. In fact, in the case of the smaller receptors (e.g. thyroid hormone and vitamin D receptors, orphans, etc.) the *C*-terminal domain appears to play the dominant role.

The existence of functional domains in steroid receptors was first deduced by biochemical analyses of receptors; protease degradation led to elucidation of domains for antibodies, DNA, and ligands (25). These domains correspond to the *N*-terminal, zinc-finger (interior), and *C*-terminal domains of receptor molecules and have been verified by mutational analyses of the receptor cDNAs. They also correspond to the three main functions that were ascribed previously to receptors: (1) modulation of target gene transcription; (2) specific interaction with target gene regulatory sequences in DNA; and (3) specific and high-affinity ligand binding.

3. Regulation of gene expression

The precise mechanism by which steroid hormone receptors regulate gene expression is unknown, but the four main reactions appear to be: (1) ligand-induced activation of the receptor; (2) specific binding to SREs; (3) stable complex formation at these DNA enhancer sites; and (4) recruitment of transcription factors and RNA polymerase to initiate transcription of target genes.

It has been proposed that certain of the unoccupied receptors may exist in association with cellular heat-shock proteins (e.g. hsp90) in a complex that is unable to bind DNA (26–29). This is most notable for glucocorticoid, progesterone, and oestrogen receptors, but appears not to be the case for vitamin D and thyroid hormone receptors. The prevalent scenario suggests that ligand binding causes dissociation of attendant heat-shock proteins so that DNA interactions can now occur (Chapter 4). It appears that dissociation of heat-shock proteins, however, is not a simple activation switch since certain steroid analogues that have little or no effect on transcription will none the less cause dissociation of heat-shock proteins and allow receptors to bind to DNA. Nevertheless, the evidence is strong that hsp90 and other heat-shock proteins must be stripped from receptor before they can bind DNA *in vitro*.

Target genes for steroid receptors possess short (approximately 15 base pairs) *cis*-elements, usually located within their 5′-flanking regions, which confer hormonal regulation upon receptor binding (30–33). Single base changes within these SREs can alter receptor binding and destroy hormone response (33, 34). Surprisingly, as few as two base changes in an SRE can convert a glucocorticoid response gene to an oestrogen-responsive gene by weakening the binding of glucocorticoid receptor while enhancing the affinity of oestrogen receptor for the element.

Although there are certain conserved bases among all SREs, the consensus sequences of the elements vary for different receptors. Surprisingly, the glucocorticoid and progesterone receptors bind to the identical DNA element (GRE–PRE). Since the physiology of glucocorticoid and progestin actions is often quite different, it appears that the specific cellular effects of these two hormones are determined by differential expression of their respective receptors. This conclusion has been substantiated by introduction of progesterone receptor (via an expression vector) into rat hepatoma cells which contain only glucocorticoid receptor, and by observing that the glucocorticoid-responsive genes are now inducible by progesterone (35).

The role of ligand in receptor–DNA interactions at the SRE has been less clear. Ligand has been reported to induce small changes in binding kinetics and affinity of purified receptor for SREs (36, 37). None of these reported effects appears great enough to explain the dependence on hormone for gene activation observed in the intact cell. In a series of recent studies, we have investigated the ability of ligand to alter the conformation on the *C*-terminal end of the receptor (38, 39). We found that a protease-resistant structure is induced by ligand in progesterone, oestrogen, glucocorticoid, retinoic acid, and thyroid hormone receptors, which correlates with the ability of ligand to activate receptors. Importantly, in the case of the human progesterone and oestrogen receptors, an alternative conformation is induced by all biologically active antiprogestins and antioestrogens respectively (38, G. F. Allan *et al.*, unpublished results). This alternative structure is centred at the extreme *C*-terminus (approximately 2–4 kDa) of the hormone-binding domain. It is resistant to protease digestion and unavailable to antibodies when receptors complex with agonists, and is sensitive to protease and available to antibodies when they bind to antagonists. These results suggest that the *C*-terminal hormone-binding domain has distinct conformations when bound with agonists versus antagonists. Mutational analysis has led us to postulate that this *C*-terminus possesses a negative regulatory function for transcription, since when it is available to interact with the transcriptional machinery, the antagonist-bound receptor is not active. Deletion of this region leads to a receptor mutant which can be transactivated by antagonists (39).

Most steroid receptors have the inherent capacity to form dimers in solution (40, 41). Receptors bind to their respective SREs as dimers (42, 43). The SRE is composed of two half-sites, having a dual axis of symmetry (inverted repeat sequences); each half-site binds one monomer of receptor. Only the dimeric form of receptor binds with an affinity (K_d approximately 10^{-9} M) sufficient to influence transcription. Once bound to an SRE, the receptor dimer can couple with another dimer (or other transcription factor) at an adjacent SRE to create a more stable complex with much higher affinity (K_d 10^{-11} M) (44). It is thought that such stable complexes have a sufficient residence time at the gene to influence transcription significantly. These *in vitro* studies of receptor–DNA interactions correlate well with cellular data which reveal synergistic effects on transcription when two or more regulatory elements are placed in the 5′-flanking region of target genes (44–48).

Co-operation between different proteins at the DNA level has been reported. Such heterodimeric interactions at SREs may be of two types. It has been reported that certain nuclear factors, for example retinoid X receptor (RXR), are required for stable binding of thyroid hormone, vitamin D, and retinoic acid receptors to their response elements (49–54). Thus, the transactivation function of these receptors is greatly enhanced by the presence of these nuclear factors. In contrast, ovalbumin upstream promoter transcription factor (COUP-TF) can form an inactive heterodimer with RXR and thus repress the functional activities of thyroid hormone, retinoic acid, and vitamin D receptors (55). Finally, the human thyroid receptor appears to be able to form a heterodimer with the retinoic acid receptor at its DNA response element. This heterodimer exhibits interesting transcriptional properties in that co-expression of both receptors in CV-1 cells results in a positive transcriptional effect on promoters containing a palindromic thyroid hormone response element (TRE) but causes a negative effect on the natural TRE derived from the α-myosin heavy-chain gene (56). These results suggest that, by forming heterodimers, a greater range of control of target gene expression may be achieved. Finally, certain receptors (e.g. glucocorticoid) can interact with activating protein constituents (e.g. c-*jun*) to form a complex which down-regulates receptor target genes (57–59). The exact mechanism of this response is not known at present.

Receptor binding to an SRE sequence may have additional combinatorial effects. For instance, it could lead to a diminished rate of transcription if the SRE overlapped with the binding site for some other positive regulator of transcription. In this way, receptors have been postulated to exclude a strong positive regulator from binding and thereby decrease the transcription rate of certain genes (60, 61). It is possible also that amino acid domains of receptor could bind and stabilize negative regulatory factors at select target genes. Alternatively, a given gene could be silenced by a strong negative regulator bound to a silencer sequence in its 5′-flanking DNA. Formation of a steroid–receptor complex at a nearby SRE could either convert the negative regulator to a positive hybrid complex or exclude the negative regulatory from its DNA-binding site. Albeit by an indirect mechanism, both types of interaction would lead to induction of transcription at the adjacent gene. Since the control of eukaryotic gene transcription appears to be the result of a multitude of combinations of a finite number of transcription factors, it may be safe to say that any imaginable alternative is likely to be found to exist at some gene locus.

A prime question relative to receptor activation of gene transcription is again the role of the ligand. Mutational studies have revealed that most receptors have at least two regions which participate in transcriptional activation. At least one is usually located in the *N*-terminal domain and the other(s) in the C-terminal domain of the receptor. It appears likely that ligand binding induces an allosteric conformation that allows access to these transcription activation regions of receptor to other transcription factors and/or RNA polymerase. These higher-order protein–protein interactions are thought to signal the initiation of transcription. Inactive analogues of oestrogen, such as tamoxifen, or of progesterone, such as RU486, appear to be unable to induce the proper allosteric folding of receptor to promote regulation of transcription (38). The complexities of these reactions make it unlikely that definitive structural information will be forthcoming until receptors are crystallized and studied by X-ray diffraction.

Using crude extracts of amphibian target cell nuclei, the addition of oestrogen has been reported to stimulate vitellogenin gene transcription (62). This stimulation occurs presumably via the oestrogen receptor present in the extracts. Recently, selective stimulation of target gene transcription by purified steroid receptor preparations or derivatives has been accomplished *in vitro*. The initial report showed that fragments of the glucocorticoid receptor (region I, Fig. 1) were able to stimulate cell-free transcription of a reporter gene containing multiple GRE elements (63). This result confirmed the conclusion that the DNA-binding domain and its surrounding peptides constitute an active site for transcriptional regulation.

In our laboratory, we have carried out reconstitution assays using HeLa transcription factor supplements, purified preparations of the native A or B forms of the chicken or human progesterone receptor, and a reporter gene containing an ovalbumin TATA region and PRE elements. Transcriptional stimulation is dependent absolutely on the presence of PRE sequences since deletion or mutation of the PRE prevents the response. Both A and B forms of the progesterone receptor stimulate transcription greater than 25-fold, while internal control genes of adenovirus are unaffected by receptor. Purified bacterial fragments of progesterone receptor (region I) are also able to stimulate transcription of target genes, but only at concentrations 40–80-fold higher than the intact (full-length) receptor (64).

A hormone-dependent cell-free transcriptional system was also developed in our laboratory. Using T47D crude extracts, we were able to show that activation of target genes is regulated by hormone *in vitro* (65). Similar results were obtained with human and chicken progesterone receptor expressed in insect cells (66). Thus, it is possible to activate a hormone response gene in a cell-free transcription system in both a hormone- and receptor-dependent manner. This hormone-dependent cell-free system enabled us to analyse the role of heat-shock proteins in the ligand activation of receptor. We have demonstrated that the human progesterone receptor remains inactive after heat-shock proteins were removed by high salt and ATP treatments. This receptor preparation still required hormone for binding to its response element and for transcriptional activation. Therefore, these results indicate that heat-shock protein removal is not sufficient to activate the receptor and that other hormone-dependent steps are required.

Using a series of preincubation and template competition analyses combined with kinetic analyses, we were able to dissect the mechanism of action of chicken progesterone receptor (cPR) in our *in vitro* transcription system (64). In short, the steroid receptor enhances the formation of a preinitiation complex by RNA polymerase (67). It appears to do this by enhancing the assembly of a template-committed complex of transcription factors at the core promoter (Fig. 2).

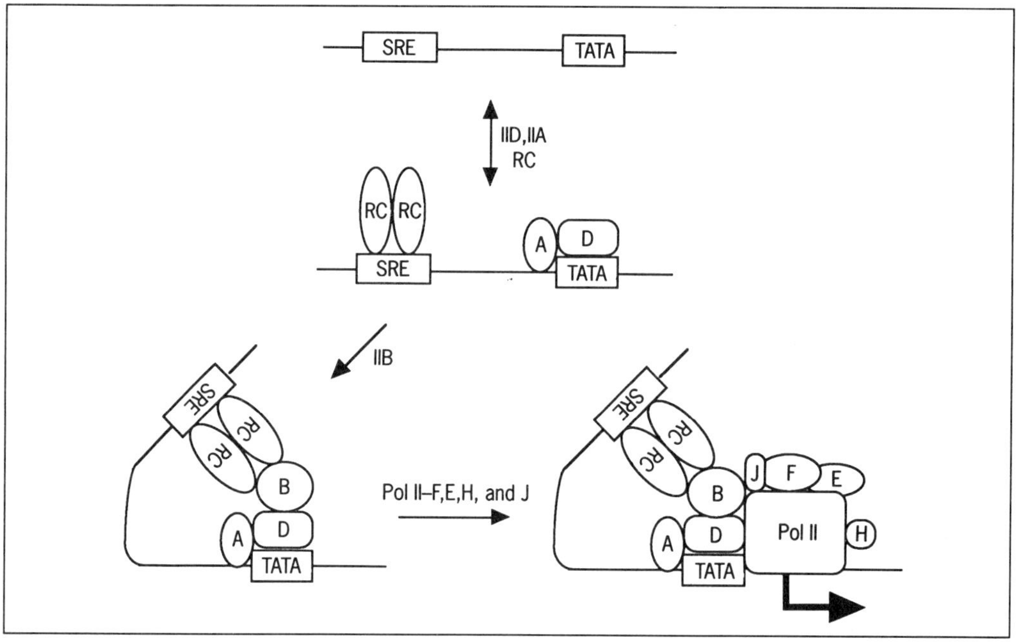

Fig. 2 Role of the receptor in gene activation. The general factor TFIIB is a target for transactivation domains of the steroid hormone receptors. The promoter of a gene regulated by steroid hormone receptors is shown, with *cis*-elements indicated: steroid hormone responsive element (SRE) and TATA box. Activated steroid hormone receptors (RCs) bind their cognate *cis*-element as a dimer. The general transcription factor TFIID binds to the TATA element along with TFIIA. The binding of the factors is reversible until TFIIB joins the protein–DNA complex to stabilize them. DNA looping brings the complexes together. This stable framework allows subsequent binding of RNA polymerase II (Pol II) and other general transcription factors TFIIF, E, H, and J efficiently to complete formation of the preinitiation complex and rapidly transcribe an activated gene

This stable complex is now poised for rapid initiation of transcription by RNA polymerase.

Multiple factors must interact at the proximal promoter (TATA box) of a gene to allow initiation of transcription (68). This sequence of events is thought to be dependent on the initial reversible interaction of the TFIID protein with the TATA sequence itself. This interaction is subsequently stabilized by sequential binding of TFIIA, TFIIB, but the precise roles of these factors in the initiation complex are only partially known (69). Nevertheless, together they form a three-dimensional surface conformation which attracts RNA polymerase and other transcription factors such as TFIIE or F to the gene and transcription begins.

In this cell-free system, cPR acts to enhance formation of a stable template-committed complex of these transcription factors, thereby promoting formation of the rapid-start complex and stimulating transcriptional initiation. Receptor may do this by facilitating recognition of the promoter by such factors or simply by stabilizing the promoter DNA–protein complex once it is formed. In either case, it is likely that the effect is modulated via local protein–protein interactions between the ligand-activated receptor and the promoter complex. Current questions to be asked in this context relate to the precise contacts between the receptor amino acids and the surface of individual promoter factors. Our recent experiments have led us to believe that a significant affinity of receptors for TFIIB may be the key step in promoting transcriptional enhancement (70). This hypothesis is even more interesting in light of evidence that recruitment of TFIIB may be the rate limiting event in transcriptional initiation by RNA polymerase (71).

Figure 3 depicts current understanding of the steps involved in the ligand activation of its cognate receptor. Upon ligand binding, the conformation of the receptor is altered, rendering the inhibitory domain non-functional but allowing the activation domains to interact directly or indirectly with component(s) of the transcriptional machinery. Through this interaction, it is now possible for the stable preinitiation complex to form and transcription to proceed.

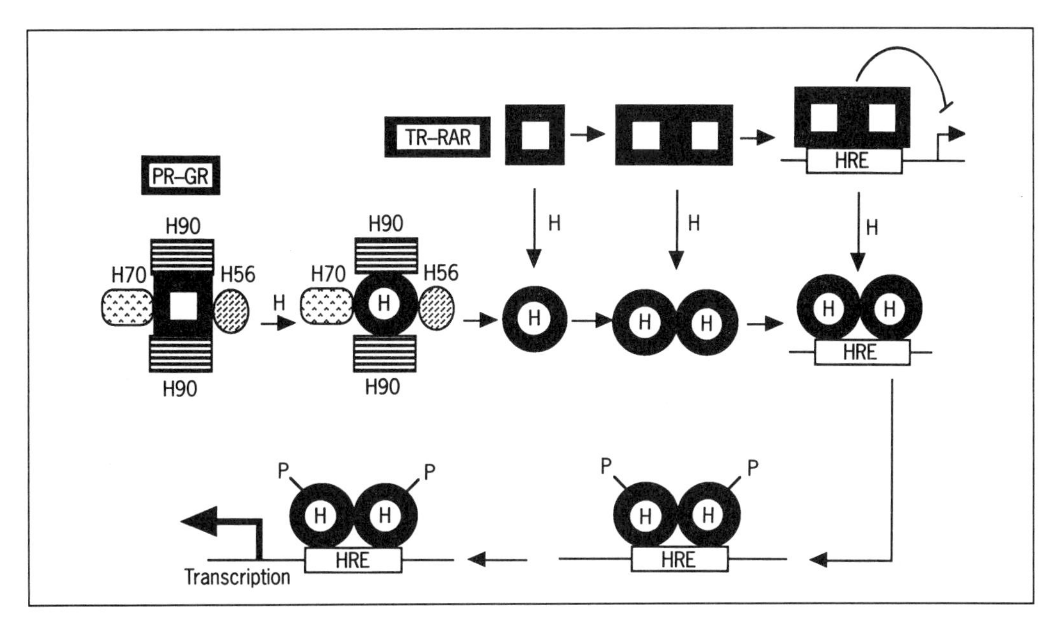

Fig. 3 Model for the role of hormone in steroid receptor activation. For the progesterone–glucocorticoid receptor (PR–GR) subfamily (e.g. PR, GR, oestrogen receptor, and androgen receptor), the receptor exists as an 8–10*S* complex in association with heat-shock proteins 90 (H90), 70 (H70), and 56 (H56). Hormone binding to PR–GR subfamily members results in a conformational change in which the ligand-binding domain becomes unavailable to protease digestion or antibody recognition. This conformational change triggers dissociation of heat-shock proteins. The receptor then undergoes homodimerization and binds to a hormone response element (HRE). In the case of the thyroid hormone–retinoic acid receptor (TR–RAR) subfamily (including vitamin D receptor and most orphan receptors), the process is simplified. In contrast to the classical PR–GR subfamily, this group of receptors is free of heat-shock proteins and ready to homodimerize or heterodimerize with retinoid X receptor in the absence of hormone. The homodimerized or heterodimerized receptor is capable of binding to DNA and inhibiting basal promoter activity. Similar to the PR–GR subfamily, hormone binding changes the receptor to an active conformation. Upon binding to DNA, any of the activated receptors may undergo additional phosphorylation by a nuclear DNA-dependent protein kinase. The resultant conformationally altered and phosphorylated receptor, bound with cognate ligand, will then display the capacity to interact with the transcriptional apparatus and activate the target gene expression

4. 'Orphan' receptors

One of the most fascinating observations to evolve from the cloning of receptor cDNAs has been the unexpectedly large size of this superfamily of transcription factors. The term 'orphan receptors' is used to designate related proteins that may or may not truly be receptors in the endocrinological sense. They nevertheless comprise the vast majority of the steroid–thyroid–vitamin superfamily. They are proteins that contain distinctive homologies in their amino acid sequence (e.g. type II zinc fingers) that clearly mark them as members of this superfamily of gene regulators. Most have been discovered by cross-hybridization with authentic steroid receptor cDNAs. Consequently, when initially identified, they have had no known ligand and usually no known function (72–74).

The supposition that all of these orphan receptors indeed have important cellular functions is based on the following: (1) they are expressed as proteins in cells, sometimes with cell-type and developmental specificity; (2) they are authentic members of a highly evolved and powerful family of eukaryotic transcription factors; (3) certain of the orphan receptors already have been shown to function by regulating known specific genes and developmental processes; (4) selected orphans have been implicated in the mediation of cellular responses to neurotransmitters, retinoic acid, and peroxisome proliferators; and (5) orphans have been reported to act as immediate early mediators of growth factor and/or electrical response in the central nervous system.

When a computer-derived schema of a phylogenetic tree of orphan receptors is constructed (Fig. 4), it appears that the steroid–thyroid–vitamin superfamily appears to be an old family with representations in major species of insects, amphibians, echinoderms, etc. It seems reasonable to guess that the earliest members of this family were the orphan transcription factors, some (or all) of which acquired regulation through evolutionary selective pressures. The most highly evolved members are those regulated by steroid and thyroid hormones and vitamins. As for the large array of other members of this family, some type of molecular activation or regulation is implicit, as these orphans are powerful gene regulators and would preferably be controlled by synthesis, ligands, covalent modification, or some combination thereof.

What would be reasonable yet primitive possibilities for ligands that regulate orphans? Although we would probably not expect highly differentiated and gland-specific hormones, it is difficult to rule out many molecular structures by this approach. Nevertheless, two possibilities come readily to mind. Environmental nutritional agents and/or intracellular metabolic intermediates represent one possibility. The interaction of such ligands with their receptors might be expected to be of lower affinity and specificity than those seen for true hormone receptors, and consequently more difficult to detect or assay. Amino acids and peptides, biogenic amines, etc. are all ligand candidates. A second possibility for activation is covalent modification, most likely via phosphorylation. Phosphorylation is an ancient metabolic process that can have profound effects on the structure and function of

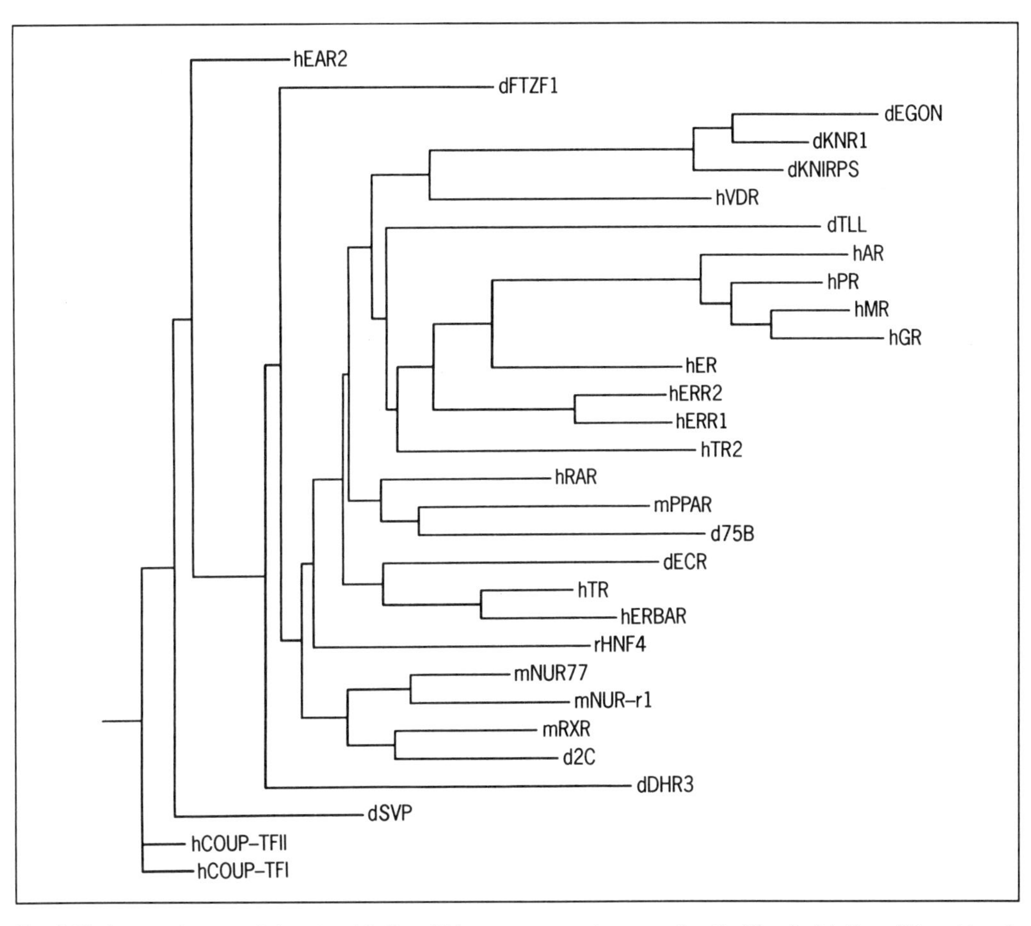

Fig. 4 Phylogenetic tree of the steroid–thyroid hormone receptor superfamily. The first letter of the abbreviated names of all the receptors represents the species (h for human, m for mouse, r for rat, d for *Drosophila*). hAR, androgen receptor; hGR, glucocorticoid receptor; hPR, progesterone receptor; hER, oestrogen receptor; hMR, mineralocorticoid receptor; hVDR, vitamin D receptor; hCOUP-TFI (*ear*3), ovalbumin upstream promoter transcription factor I; hCOUP-TFII (*arp*1), ovalbumin upstream promoter transcription factor II; hRAR, retinoic acid receptor; hTR, thyroid hormone receptor; hERBAR, c-*erb*A; hERR1 and hERR2, oestrogen-related orphan receptors; hTR2, an orphan receptor isolated from testis library, hEAR2, v-*erb*A-related *ear*2 gene; mNUR77, immediate early gene; nNUR-r1, mNUR77-related gene; mPPAR, peroxisome proliferator activated receptor, mRXR, retinoid X receptor; rHNF4, hepatocyte nuclear factor 4; dECR, ecdysone receptor; dDHR3, a *Drosophila* orphan receptor; d75B, an ecdysome-inducible receptor, d2C, a steroid hormone receptor-like gene; dKNIRPS, a hormone-like protein encoded by the gap gene *knirps*; dKNR1, *knirps*-related gene; dFTZF1, a steroid hormone receptor-like protein; dTLL, *Drosophila* tailless gene; dEGON, a *kni* homologous gene encodes an zinc-finger protein; dSVP, seven-up gene. The amino acid sequences of the DNA-binding domain of the receptors were first aligned using PIMA and the nucleic acid sequences were then aligned with reference to the amino acid alignment by the software PIMA-TO-PAUP (76). The phylogenetic tree is constructed by Fitch's distance method of PHYLIP v3.4

biological macromolecules. Phosphorylation reactions are intimately intertwined with metabolic control and cellular growth, and are subject to the same signals for regulation as might be expected for orphan receptors.

5. Activation of receptors

Most recent evidence suggests that both ligands and phosphorylation could play important roles in the activation of classical receptors. It is generally accepted that the classical receptors are phosphoproteins. Consequently, it seems reasonable that phosphorylation regulates function of receptors. Recent evidence from our laboratory (76, 77) and that of Horwitz (78) has revealed that chicken and human progesterone receptors are allosterically activated by ligand and then become a substrate for a series of separate protein kinases, one of which is DNA-dependent kinase. Correlative evidence is consistent with the hypothesis that the final common pathway for gene transactivation involves phosphorylation of the progesterone receptor. In fact, the most recent evidence indicates that phosphorylation mechanisms may exist which can activate this receptor even in the absence of its steroidal ligand. Evidence also exists from mutational studies that thyroid hormone receptor and v-*erb*A require phosphorylation for certain functions. It should be noted, however, that it has not been proven yet that phosphorylation is the key reaction that imparts a terminal transactivation potential to steroid receptor family members. Although the role that phosphorylation plays in regulating the function of these receptor family members remains to be definitively explained, the systems to do so are now available.

Our experiments have shown that under cell transfection conditions certain steroid receptor family members may be activated via pharmacological manipulation of intracellular phosphorylation pathways using 8-bromo-cyclic AMP and okadaic acid (a strong phosphatase inhibitor) (79). A number of laboratories have now confirmed that the state of intracellular phosphorylation can have a marked influence on the biological response to hormone in target cells (80). More recently, we uncovered a rather interesting pathway for activation of selected receptors and/or orphan chimeras which was dependent on dopamine binding to its D_1 receptors on the plasma membrane (81). Although the precise intracellular pathway of activation is presently unclear, phosphorylation may be involved. Our results suggest that members of this family may exist in cells in an inactive form which can be activated to regulate gene transcription by modulators that bind to receptors on the membrane or by internal changes in phosphorylation pathways. This type of cellular cross-talk between membrane and intracellular signalling pathways has not been proposed previously and must now be demonstrated to have physiological relevance in intact animals. A recent ligand-independent response of oestrogen receptor to epidermal growth factor (EGF) in the mammalian uterus could be related to this phenomenon (82).

In the steroid–thyroid–vitamin superfamily, there are orphans whose functions can be controlled by mechanisms other than ligand binding. The PPAR (peroxisome proliferator activated receptor) orphan subfamily is activated by chemicals classified as peroxisome proliferators, which are not ligands (83). These orphans have recently been shown to be activated by fatty acids (e.g. arachidonic and linoleic acid) and perhaps should be renamed *f*atty *a*cid *a*ctivatable *r*eceptors (FAAR) (84). Another group of orphans may be typified by Nur77, a member of the superfamily whose cellular concentration of mRNA and protein level is induced by serum growth factors, EGF or NGF (nerve growth factor) (85). Membrane depolarization also will activate Nur77 in PC12 cells, and this orphan has been postulated to play a role in neuronal function. Recently, we have discovered another orphan receptor (NURR1) whose concentration varies during brain development and is increased following depolarization with potassium. At present Nur77 is the only member of the superfamily whose cellular concentration has been reported to be rapidly and transiently induced by growth factors, but others are likely to follow. Notably, Nur77 may be regulated also by differential phosphorylation (85).

A few general conclusions can be made concerning the pathway for activation of members of the steroid–thyroid–vitamin superfamily of gene regulators. Given the importance, size, and diversity of this family of transcription factors, it may be naïve to think that they will have a single unifying mechanism or pathway for activation. The following generalities appear to apply to this superfamily: (1) it is a very large and probably ancient family which is represented throughout most eukaryotic species; (2) in humans, the family is widely distributed among different chromosomes; (3) the members are transcription factors with varying degrees of tissue specificity; (4) their transcriptional function may be activated by a variety of mechanisms, including synthesis–degradation, phosphorylation, and ligand binding; and (5) activation may be initiated by long-distance signal transduction (e.g. hormones from endocrine glands), by signal transduction at the plasma membrane (e.g. catecholamines, growth factors), and by transduction at the nuclear level (e.g. mRNA synthesis). In a given instance, an orphan receptor may be activated by any one or a combination of the above mechanisms. It is our guess that phosphorylation mechanisms are evolutionarily the most primitive, with high-affinity ligands developing only for the more recently diverging members such as the classical steroid–thyroid–vitamin-regulated receptors. Finally, it is conceivable that certain of these superfamily members could also be inactivated by phosphorylation or by environmental ligands, similar to the manner in which classical steroid receptors can be inactivated by antihormones.

Summary

In summary, it seems logical that orphan members of the steroid receptor superfamily are regulated by diverse mechanisms using both intracellular (intracrine) and extracellular signals (Fig. 5). Because of the diversity and importance of this family of orphan receptors, it will remain a topical focus for modern molecular endocrinology for some time to come. By understanding the regulation pathways and genetic targets of this fascinating new superfamily, we will almost certainly

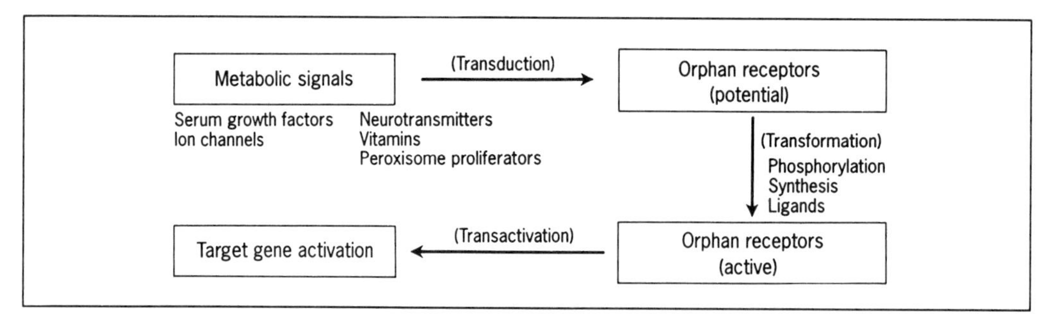

Fig. 5 Schematic diagram of hypothetical pathway for activation of orphan receptors

develop new insights into normal and abnormal cell physiology that will impact on the diagnosis and therapy of human diseases.

References

1. Jensen, E. V., Suzuki, T., Kawashima, T., Stumpf, W. E., Jungblut, P. W., and De Sombre, E. R. (1968). A two-step mechanism for the interaction of estradiol with rat uterus. *Proc. Natl Acad. Sci. USA*, **59,** 632.
2. Gorski, J., Toft, D., Shyamala, G., Smith, D., and Notides, A. (1968) Hormone receptors: studies on the interaction of estrogen with the uterus. *Recent Prog. Horm. Res.*, **24,** 45.
3. O'Malley, B. W. and Means, A. R. (1974) Female steroid hormones and target cell nuclei. *Science*, **183,** 610.
4. O'Malley, B. W., Roop, D. R., Lai, E. C., Nordstrom, J. L., Catterall, J. F., Swaneck, G. E., Colbert, D. A., Tsai, M.-J., Dugaiczyk, A., and Woo, S. L. C. (1979) The ovalbumin gene: organization, structure, transcription and regulation. *Recent Prog. Horm. Res.*, **35,** 1.
5. O'Malley, B. W., McGuire, W. L., Kohler, P. O., and Korenman, S. (1969) Studies on the mechanism of steroid hormone regulation of synthesis of specific proteins. *Recent Prog. Horm. Res.*, **25,** 105.
6. Jensen, E. V., Jacobson, H. I., Flesher, J. W., Saha, N. N., Gupta, G. N., Smith, S., Colucci, V., Shiplacoff, D., Neuman, H. G., DeSombre, E. R., and Jungblut, P. W. (1966) Estrogen receptors in target tissues. In *Steroid Dynamics*. Pincus, G., Nakao, T., and Tait, J. R. (eds). Academic Press, New York, p. 133.
7. Gorski, J. and Gannon, F. (1976) Current models of steroid hormone action: a critique. *Annu. Rev. Biochem.*, **28,** 425.
8. Katzenellenbogen, B. S. (1980) Dynamics of steroid hormone receptor action. *Annu. Rev. Physiol.*, **42,** 17.
9. O'Malley, B. W. (1984) Steroid hormone action in eukaryotic cells. *J. Clin. Invest.*, **74,** 207.
10. Yamamoto, K. (1985) Steroid receptors regulated transcription of specific genes and gene networks. *Annu. Rev. Genet.*, **19,** 209.
11. Ringold, G. (1985) Steroid hormone regulation of gene expression. *Annu. Rev. Pharmacol. Toxicol.*, **25,** 529.

12. Walters, M. R. (1985) Steroid hormone receptors and the nucleus. *Endocr. Rev.*, **6,** 512.
13. Haussler, M. R. (1986) Vitamin D receptors: nature and function. *Annu. Rev. Nutr.*, **6,** 527.
14. Evans, R. M. (1988) The steroid and thyroid hormone receptor superfamily. *Science*, **240,** 889.
15. Beato, M. (1989) Gene regulation by steroid hormones. *Cell*, **56,** 335.
16. Green, S. and Chambon, P. (1987) Oestradiol induction of a glucocorticoid-responsive gene by a chimeric receptor. *Nature*, **325,** 75.
17. Berg, J. M. (1989) Binding specificity of steroid receptors. *Cell*, **57,** 1065.
18. Mader, S., Kumar, V., DeVerneuil, H., and Chambon, P. (1989) Three amino acids of the oestrogen receptor are essential to its ability to distinguish an oestrogen from a glucocorticoid-responsive element. *Nature*, **338,** 271.
19. Umesono, K. and Evans, R. (1989) Determinants of target gene specificity for steroid/thyroid hormone receptors. *Cell*, **57,** 1139.
20. Danielsen, M., Hinck, L., and Ringold, G. (1989) Two amino acids within the knuckle of the first zinc finger specify DNA response element activation by the glucocorticoid receptor. *Cell*, **57,** 1131.
21. Hughes, M. R., Malloy, P. J., Kieback, D. G., Kesterson, R. A., Pike, J. W., Feldman, D., and O'Malley, B. W. (1988) Point mutations in the human vitamin D receptor gene associated with hypocalcemic rickets. *Science*, **242,** 1702.
22. Guichon, M. A., Loosfelt, H., Lescop, P., Saar, S., Atger, M., Perrot-Applanat, M., and Milgrom, E. (1989) Mechanisms of nuclear localization of the 'progesterone' receptor: evidence for interaction between monomers. *Cell*, **57,** 1147.
23. Luisi, B. F., Xu, W. X., Otwinowski, Z., Freedman, L. P., Yamamoto, K. R., and Sigler, P. B. (1991) Crystallographic analysis of the interaction of the glucocorticoid receptor with DNA. *Nature*, **352,** 497.
24. Tora, L., Gronemeyer, H., Turcotte, B., Gaub, M.-P., and Chambon, P. (1988) The *N*-terminal region of the chicken progesterone receptor specifies target gene activation. *Nature*, **333,** 185.
25. Wrange, O., Okret, S., Radojcic, M., Carlstedt-Duke, J., and Gustafsson, J.-A. (1984) Characterization of the purified activated glucocorticoid receptor from rat liver cytosol. *J. Biol. Chem.*, **259,** 4534.
26. Schuh, S. W., Yonemoto, J., Brugge, V. J., Bauer, R. M., Reihl, W. P., Sullivan, W. P., and Toft, D. O. (1985) A 90,000-dalton binding protein common to both steroid promoter and 5′-flanking region of the gene. *Nucl. Acid Res.*, **16,** 5459.
27. Sanchez, E. R., Toft, D. O., Schlesinger, M. J., and Pratt, W. B. (1985) Evidence that the 90-k dalton phosphoprotein associated with the untransformed L-cell glucocorticoid receptor is a murine heat shock protein. *J. Biol. Chem.*, **260,** 12 398.
28. Catelli, M.-G., Binart, N., Jung-Testas, I., Renoir, J. M., Baulieu, E.-E., Feramisco, J. R., and Welch, W. J. (1985) The common 80 kDa protein component of non-transformed 8S steroid receptors is a heat-shock protein. *EMBO J.*, **4,** 3131.
29. Howard, K. J. and Distelhorst, C. W. (1985) Evidence for intracellular association of the glucocorticoid receptor with the 90-kDa heat shock protein. *J. Biol. Chem.*, **263,** 3474.
30. Payvar, F., DeFranco, D., Firestone, G. L., Edgar, B., Wrange, O., Okert, S., Gustafsson, J.-A., and Yamamoto, K. R. (1983) Sequence-specific binding of glucocorticoid receptor to MTV DNA at sites within and upstream of the transcribed region. *Cell*, **35,** 381.
31. Chandler, V. L., Maler, B. A., and Yamamoto, K. R. (1983) DNA sequences bound specifically by glucocorticoid receptor *in vitro* render a heterologous promoter hormone responsive *in vitro*. *Cell*, **33,** 489.

32. Scheidereit, C., Geisse, S., Westphal, H. M., and Beato, M. (1983) The glucocorticoid receptor binds to defined nucleotide sequences near the promoter of mouse mammary tumor virus. *Nature*, **304**, 749.
33. Strahle, U., Klock, G., and Schultz, G. (1987) A DNA sequence of 15 base pairs is sufficient to mediate both glucocorticoid and progesterone induction of gene expression. *Proc. Natl Acad. Sci. USA*, **84**, 7871.
34. Klock, G., Strahle, U., and Schultz, G. (1987) Oestrogen and glucocorticoid responsive elements are closely related but distinct. *Nature*, **329**, 734.
35. Strahle, U., Boshart, M., Klock, G., Stewart, F., and Schultz, G. (1989) Glucocorticoid- and progesterone-specific effects are determined by differential expression of the respective hormone receptors. *Nature*, **339**, 629.
36. Schauer, M., Chalepakis, G., Willman, T., and Beato, M. (1989) Binding of hormone accelerates the kinetics of glucocorticoid and progesterone receptor binding to DNA. *Proc.*
Natl Acad. Sci. USA, **86**, 1123.
37. Rodriguez, R., Carson, M. A., Weigel, N. L., O'Malley, B. W., and Schrader, W. T. (1989) The effect of hormone on the *in vitro* DNA binding activity of the chicken progesterone receptor. *Mol. Endocrinol.*, **3**, 356.
38. Allan, G. F., Leng, X., Tsai, S. Y., Weigel, N. L., Edwards, D. P., Tsai, M.-J. and O'Malley, B. W. (1992) Hormone and anti-hormone induce distinct conformational changes which are central to steroid receptor activation. *J. Biol. Chem.*, **267**, 19513.
39. Vegeto, E., Allan, G. F., Schrader, W. T., Tsai, M.-J., McDonnell, D. P., and O'Malley, B. W. (1992) RU486 antagonism is dependent on the conformation of the carboxy-terminal tail of the human progesterone receptor. *Cell*, **69**, 703.
40. Gordon, M. S. and Notides, A. C. (1986) Computer modeling of estradiol interactions with the estrogen receptor. *J. Steroid Biochem.*, **25**, 177.
41. Wrange, O., Eriksson, P., and Perlmann, T. (1989) The purified activated glucocorticoid is a homodimer. *J. Biol. Chem.*, **264**, 5253.
42. Tsai, S. Y., Carlstedt-Duke, J.-A., Weigel, N. L., Dahlman, K., Gustaffson, J.-A., Tsai, M.-J., and O'Malley, B. W. (1988) Molecular interactions of steroid hormone receptor with its enhancer element: evidence for receptor dimer formation. *Cell*, **55**, 361.
43. Kumar, V. and Chambon, P. (1988) The estrogen receptor binds tightly to its responsive element as a ligand-induced homodimer. *Cell*, **55**, 145.
44. Tsai, S. Y., Tsai, M.-J., and O'Malley, B. W. (1989) Cooperative binding of steroid hormone receptors contributes to transcriptional synergism at target enhancer elements. *Cell*, **57**, 443.
45. Cato, A. C., Heitlinger, E., Ponta, H., Klein-Hitpass, L., Ryffel, G. U., Bailly, A., Rauch, C., and Milgrom, E. (1988) Estrogen and progesterone receptor-binding sites on the chicken vitellogenin II gene: synergism of steroid hormone action. *Mol. Cell. Biol.*, **8**, 5323.
46. Danesch, U., Gloss, B., Schmid, W., Schutz, G., Schule, R., and Renkawitz, R. (1987) Glucocorticoid induction of the rat tryptophan oxygenase gene is mediated by two widely separated glucocorticoid-responsive elements. *EMBO J.*, **6**, 625.
47. Schule, R., Muller, M., Kaltschmidt, C., and Renkawitz, R. (1988) Many transcription factors interact synergistically with the steroid receptors. *Science*, **242**, 1418.
48. Bradshaw, M. S., Tsai, S. Y., Leng, X., Dobson, A. D. W., Conneely, O. M., O'Malley, B. W., and Tsai, M.-J. (1991) Studies on the mechanism of functional cooperativity between progesterone and estrogen receptors. *J. Biol. Chem.*, **266**, 16 684.
49. Marks, M. S., Hallenbeck, P. L., Nagata, T., Segars, J. H., Appella, E., Nikodem, V. M.,

and Ozato, K. (1992) H-2RIIBP (RXRβ) heterodimerization provides a mechanism for combinatorial diversity in the regulation of retinoic acid and thyroid hormone responsive genes. *EMBO J.*, **11,** 1419.
50. Yu, V. C., Delsert, C., Anderson, B., Holloway, J. M., Devary, O. V., Naar, A. M., Kim, S. Y., Boutin, J.-M., Glass, C. K., and Rosenfeld, M. G. (1991) RXRβ: a coregulator that enhances binding of retinoic acid, thyroid hormone, and vitamin D receptors to their cognate response elements. *Cell,* **67,** 1251.
51. Leid, M., Kastner, P., Lysons, R., Nakshatri, H., Saunders, M., Zacharewski, T., Chen, J.-Y., Staub, A., Garnier, J.-M., Mader, S., and Chambon, P. (1992) Purification, cloning, and RXR identity of the HeLa cell factor with which RAR or TR heterodimerizes to bind target sequences efficiently. *Cell,* **68,** 377.
52. Kliewer, S. A., Umesono, K., Mangesldorf, D. J., and Evans, R. M. (1992) Retinoic X receptor interacts with nuclear receptors in retinoic acid, thyroid hormone and vitamin D_3 signalling. *Nature,* **355,** 446.
53. Bugge, T. H., Pohl, J., Lonnoy, O., and Stunnenberg, G. (1992) RXRα, a promiscuous partner of retinoic acid and thyroid hormone receptors. *EMBO J.*, **11,** 1409.
54. Zhang, X.-K., Hoffmann, B., Tran, P. B.-V., Graupner, G., and Pfahl, M. (1992) Retinoic X receptors is an auxiliary protein for thyroid hormone and retinoic acid receptors. *Nature,* **355,** 441.
55. Cooney, A., Tsai, S. Y., O'Malley, B. W., and Tsai, M.-J. (1992) COUP-TF dimers bind to different GGTCA response elements allowing it to repress hormonal induction of VDR, TR and RAR. *Mol. Cell. Biol.*, **12,** 4153.
56. Glass, C., Lipkin, S., Devary, O., and Rosenfeld, M. G. (1989) Positive and negative regulation of gene transcription by a retinoic acid–thyroid hormone receptor heterodimer. *Cell,* **59,** 697.
57. Jonat, G., Rahmsdorf, H. J., Park, K.-K., Cato, A. C. B., Gebel, S., Ponta, H., and Herrlich, P. (1990) Antitumor promotion and antiinflammation: down-modulation of Ap-1 (*Fos/Jun*) activity by glucocorticoid hormone. *Cell,* **62,** 1189.
58. Schule, R., Rangurajan, P., Kliewer, S., Ransone, L. J., Bolado, J., Yang, N., Verma, I. M., and Evans, R. M. (1990) Functional antagonism between oncoprotein c-*Jun* and the glucocorticoid receptor. *Cell,* **62,** 1217.
59. Diamond, M. I., Miner, J. N., Yoshinaga, S. K., and Yamamoto, K. R. (1990) Transcription factor interactions: selections of positive or negative regulation from a single DNA element. *Science* **249,** 266.
60. Akerblom, I., Slater, E., Beato, M., Baxter, J., and Mellon, P. (1988) Negative regulation by glucocorticoids with a cAMP responsive enhancer. *Science,* **241,** 350.
61. Charron, J. and Drouin, J. (1986) Glucocorticoid inhibition of transcription from episomal proopiomelanocortin gene promoter. *Proc. Natl Acad. Sci. USA,* **83,** 8903.
62. Corthesy, B., Hispking, R., Theulaz, I., and Wahli, W. (1988) Estrogen-dependent *in vitro* transcription from the vitellogenin promoter in liver nuclear extracts. *Science,* **239,** 1137.
63. Freedman, L., Yoshinaga, S., Vanderbilt, J., and Yamamoto, K. (1989) *In vitro* transcription enhancement by purified derivatives of the glucocorticoid receptor. *Science,* **245,** 298.
64. Klein-Hitpass, L., Tsai, S. Y., Weigel, N. L., Riley, D., Rodriguez, R., Schrader, W. T., Tsai, M.-J., and O'Malley, B. W. (1990) Native progesterone receptor stimulates cell-free transcription of a target gene. *Cell,* **60,** 247.
65. Bagchi, M. K., Tsai, S. Y., Tsai, M.-J., and O'Malley, B. W. (1991) Progesterone

receptors free of heat shock protein hsp90, and hsp70 require ligand-dependent activation for target gene transcription. *Mol. Cell. Biol.*, **11**, 4998.
66. Elliston, J. F., Beekman, J. M., Tsai, S. Y., O'Malley, B. W., and Tsai, M.-J. (1992) Hormone dependent activation of baculovirus expressed progesterone receptors. *J. Biol. Chem.*, **267**, 5193.
67. Hawley, D. K. and Roeder, R. G. (1985) Separation and partial characterization of three functional steps in transcription initiation by human RNA polymerase II. *J. Biol. Chem.*, **260**, 8163.
68. Van Dyke, M. W., Sawadogo, M., and Roeder, R. G. (1989) Stability of transcription complexes on class II genes. *Mol. Cell. Biol.*, **9**, 342.
69. Buratowski, S., Hahn, S., Guarente, L., and Sharp, P. A. (1989) Five intermediate complexes in transcription initiation by RNA polymerase II. *Cell*, **56**, 549.
70. Ing, N. H., Beekman, J. M., Tsai, S. Y., Tsai, M.-J., and O'Malley, B. W. (1992) Members of the steroid hormone receptor suprfamily interact directly with TFIIB (S300-II) to mediate transcription induction. *J. Biol. Chem.*, **267**, 17617.
71. Lin, Y.-S., Ha, I., Maldonado, E., Reinberg, D., and Green, M. R. (1991) Binding of general transcription factor TFIIB to an acidic activating region. *Nature*, **353**, 569.
72. Giguere, V., Yang, N., Segui, P., and Evans, R. (1988) Identification of a new class of steroid hormone receptors. *Nature*, **331**, 91.
73. O'Malley, B. W. (1989) Did eucaryotic steroid receptors evolve from 'intracrine' gene regulators? *Endocrinology*, **125**, 1119.
74. Wang, L.-H., Tsai, S. Y., Cook, R. G., Beattie, W. G., Tsai, M.-J., and O'Malley, B. W. (1989) COUP transcription factor is a member of the steroid receptor superfamily. *Nature*, **340**, 163.
75. Smith, R. F. and Smith, T. F. (1992) Pattern-induced multi-sequence alignment (PIMA) algorithm employing secondary structure-dependent gap penalties for use in comparative protein modeling. *Protein Eng.*, **5**, 35.
76. Bagchi, M. K., Tsai, S. Y., Tsai, M.-J., and O'Malley, B. W. (1992) Ligand and DNA dependent phosphorylation of human progesterone receptor *in vitro*. *Proc. Natl Acad. Sci. USA*, **89**, 2664.
77. Weigel, N., Carter, T. H., Schrader, W. T., and O'Malley, B. W. (1992) Chicken progesterone receptor is phosphorylated by a DNA-dependent protein kinase during *in vitro* transcription assays. *Mol. Endocrinology*, **6**, 8.
78. Takimoto, G. S., Tasset, D. M., Eppert, E. C., and Horwitz, K. B. (1992) Hormone induced progesterone receptor phosphorylation consists of sequential DNA-independent and DNA-dependent stages. Analysis with zinc-finger mutants and the progesterone antagonist ZK-98299. *Proc. Natl Acad. Sci. USA*, **89**, 3050–4.
79. Denner, L. A., Weigel, N. L., Maxwell, B. L., Schrader, W. T., and O'Malley, B. W. (1990) Regulation of progesterone receptor-mediated transcription by phosphorylation. *Science*, **250**, 1740.
80. Beck, C. A., Weigel, N. L., and Edwards, D. P. (1992) Effect of hormone and cellular modulators of protein phosphorylation on transcriptional activity, DNA binding, and phosphorylation of human progesterone receptors. *Mol. Endocrinol.*, **6**, 607.
81. Power, R. F., Mani, S. K., Codina, J., Conneely, O. M., and O'Malley, B. W. (1991) Dopaminergic and ligand-independent activation of steroid receptors. *Science*, **254**, 1636.
82. Ignar-Trowbridge, D. M., Nelson, K. G., Bidwell, M. C., Curtis, S. W., Washburn, T. F., McLachlan, J. A., and Korach, K. S. (1992) Coupling of dual signaling pathways:

epidermal growth factor action involves the estrogen receptor. *Proc. Natl Acad. Sci. USA,* **89,** 4658.
83. Issemann, I. and Green, S. (1990) Activation of a member of the steroid hormone receptor superfamily by peroxisome proliferators. *Nature,* **347,** 645.
84. Göttlicher, M., Widmark, E., and Gustafsson, J.-A. (1992) Fatty acids activate a chimera of the clofibric acid activated receptor and the glucocorticoid receptors. *Proc. Natl Acad. Sci. USA,* **89,** 4653.
85. Hazel, T. G., Misra, R., Davis, I. J., Greenberg, M. E., and Lau, L. F. (1991) Nur77 is differentially modified in PC12 cells upon membrane depolarization and growth factor treatment. *Mol. Cell. Biol.,* **6,** 3239.

4 Role of heat-shock proteins in steroid receptor function

WILLIAM B. PRATT

1. Introduction

Because of their intrinsic interest as hormone-regulated transcriptional enhancers, the steroid receptors have attracted the curiosity of people with diverse backgrounds in biology. In recent years, the tools of modern molecular biology and protein chemistry have been used to define both steroid receptor primary structure and supramolecular structure. The demonstration that steroid receptors form stable interactions with heat-shock proteins (hsp) has contributed significantly to the understanding of steroid hormone action, and the study of receptor-associated proteins has evolved as a subfield of nuclear receptor research. It should not escape the notice of the general scientific community that this structural knowledge has made the receptor proteins themselves into powerful tools for probing the mechanisms of basic biological processes, such as protein folding, protein transport, and regulation of protein function. This chapter addresses the way in which the study of hsp-mediated receptor folding is leading to the development of some concepts that may have wide application in other biological systems.

The study of receptor–hsp interactions evolved from the observation of Toft and Gorski (1) in 1966 that the oestrogen-binding activity in rat uterine cytosol sedimented at 9 S when analysed by sucrose gradient centrifugation. Subsequently, it was found that all the steroid receptors could be recovered from hormone-free cells largely as 9S complexes, whereas receptors recovered from hormone-treated cells generally sedimented at 4 S. The 9S complex had an estimated molecular mass of 320–350 kDa, whereas that of the 4S receptor monomers ranged from 70 to 110 kDa (2). Because the 9S (untransformed) form of the receptors did not bind to DNA, whereas the smaller (transformed) form bound to DNA with high affinity, the dissociation of the 9S complex was utilized as a potential cell-free model of the initial event resulting from steroid binding to the receptor.

During the 1970s, the composition and significance of the 9S complex remained undefined. It was not known whether 9S receptors represented a tetramer of the steroid binding protein (2) or an association of receptor with other proteins in a heterocomplex. Indeed, many investigators dismissed the 9S complex as a nonspecific molecular aggregate. The demonstration in the late 1970s that molybdate stabilizes steroid receptors in their 9S form (3) permitted affinity purification of the untransformed form of progesterone (PR) and glucocorticoid (GR) receptors. For both of these receptors, the major protein recovered in the molybdate-stabilized complex was a protein of approximately 90 kDa (4–6). It was subsequently shown by site-specific affinity labelling, however, that the 90-kDa protein was different from the progesterone- or glucocorticoid-binding proteins (7–9). An important advance was made when a monoclonal antibody prepared against the molybdate-stabilized PR was shown to react only with the 90-kDa non-steroid-binding protein, but nevertheless to shift the sedimentation velocity of molybdate-stabilized PR, oestrogen (ER), androgen (AR), and GR, suggesting a common protein subunit in the 9S complex (10).

In 1985, this common subunit of the steroid receptor heterocomplexes was identified as the 90-kDa heat-shock protein, hsp90 (11–13). The studies leading to the identification of hsp90 as the common component of the untransformed, non-DNA-binding state of steroid receptors have been reviewed in detail (14). In recent years, it has become clear that hsp90 is not the only receptor-associated protein. Rather, it seems that receptors recovered from hormone-free cells are bound to a large complex containing several proteins, including other heat-shock proteins, and that the 9S form of the receptor is a core unit derived from a larger structure (15, 16). I will first review the proteins that have been identified as receptor-associated proteins and how they are associated in a heat-shock protein complex. I will then discuss the role of the proteins in receptor folding and inactivation of receptor function.

2. Receptor-associated proteins

The proteins that have been reported to co-immunoadsorb with GR, PR, and the avian oncogenic tyrosine kinase pp60^{src} are listed in Table 1. In the early 1980s, pp60^{src} was shown to be associated with hsp90 in a heterocomplex that also contains a 50-kDa protein (p50) of unknown function (see Brugge (17) for a review). Co-immunoadsorption studies with monoclonal antibodies directed against two of the receptor-associated heat shock proteins, hsp56 (18) and hsp90 (19), have shown that these two proteins exist in cytosolic complexes with a third heat-shock protein,

Table 1 Protein heterocomplexes determined by co-immunoadsorption

Immunoadsorbed protein	PR	GR	pp60^{src}	hsp90	hsp56
Associated protein	hsp90	hsp90	hsp90	—	hsp90
	hsp70	hsp70		hsp70	hsp70
				p63	
	hsp56	hsp56		hsp56	—
	p54 (avian)				
	p50		p50	p50	
	p23	p23			p23

hsp70. These three heat-shock proteins are all thought to have protein folding and chaperone functions, and they are associated with each other in cytosols independent of the presence of steroid receptors (18, 19). Although most of the work since 1985 on steroid receptor heterocomplexes has focused on the receptor–hsp90 interaction, considerable interest has recently been generated regarding the functions of other proteins in the multiprotein complex.

2.1 Hsp90

Hsp90 is an ubiquitous, conserved, and essential protein in eukaryotic cells (20, 21). It is present in high concentration in unstressed cells (approximately 1% of cytosolic protein) and is synthesized at an increased rate after heat shock. It has been established that hsp90 is present in untransformed glucocorticoid, mineralocorticoid, oestrogen, progestin, androgen, and dioxin receptor complexes (14, 15).

2.1.1 Site of interaction with receptors

It is clear that hsp90 interacts with the hormone-binding domain (HBD) of the glucocorticoid receptor (22, 23), and that the HBD alone is sufficient to confer the property of hsp90 binding on to chimæric proteins (L. C. Scherrer *et al.*, *Biochemistry*, in press). Initially, it was shown that transfected human GRs deleted for the HBD are not recovered in COS cell cytosol in the 9S, hsp90-bound form (22). In a direct aproach, Denis *et al.* (23) cleaved molybdate-stabilized rat liver GR with trypsin and isolated a 27-kDa fragment from the COOH-terminus of the receptor that was bound to hsp90. Previously, this fragment of the GR was shown to contain the HBD (24).

Although a specific hsp90 binding site has not been defined, several approaches have been used to identify a subregion of the GR HBD that forms a minimal hsp90 binding site (Fig. 1). First, it was shown that rat GR translated in rabbit reticulocyte lysate binds to rabbit hsp90 and is in a 9S form that can be transformed to the 4S, DNA-binding state by steroid (30). Translation of GR containing COOH-terminal deletions of various lengths, followed by immunoadsorption of heterocomplexes with a monoclonal antibody against hsp90, revealed a minimal region (amino acids 604–659, mouse GR) that was required for a high yield of high-affinity hsp90 binding (25). Using the same approach, Howard *et al.* (26) found that the region 556–604 (mouse) was also sufficient to yield some hsp90 binding, consistent with the assignment of an approximately 100-amino-acid region (556–659) for a minimal high-affinity hsp90 binding site using this terminal deletion technique. That the 604–659 segment is at least part of the hsp90 contact site was supported by the results of peptide competition for hsp90 reassociation with the full-length GR in a cell-free system (25). One problem with defining a minimal binding site using truncated receptors is that the contribution to hsp90 binding of portions of the receptor that lie COOH-termnal to the minimal site are not defined. Cadepond *et al.* (31) divided the HBD of the GR into three subregions of roughly equal length

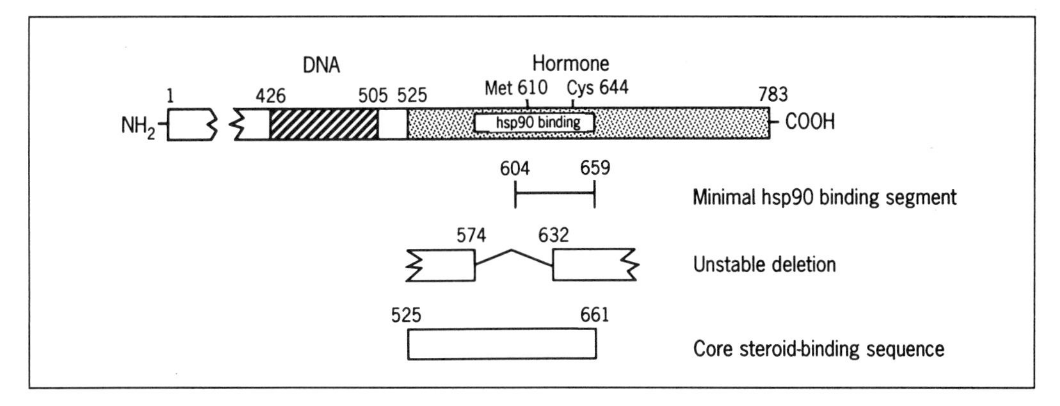

Fig. 1 Mouse glucocorticoid receptor (GR). Localization of a minimal hsp90 binding region within the hormone-binding domain (HBD) of the GR. The numbers represent amino acid assignments for the mouse GR. A minimal hsp90 binding region of approximately 100 amino acids is indicated by the open rectangle in the stippled HBD. The minimal region of Dalman *et al.* (25) for a high yield of stable hsp90 binding as determined from COOH-terminal GR truncations would suggest a COOH-terminal border for the minimal site extending to the region of amino acid 659, and a similar study by Howard *et al.* (26) suggested that the amino-terminal border should not be less than residue 556. The internal deletion that binds hsp90 was described by Housley *et al.* (27). Also shown is the core steroid-binding fragment of Simons *et al.* (28), which was derived by proteolysis of the molybdate-stabilized wild-type GR and remains bound to hsp90 (29)

and showed that any of the three regions by itself was sufficient to yield at least some 9S heterocomplex. Two of these subregions contain major portions of the minimal hsp90 binding site shown in Figure 1, but the third subregion lies completely to the COOH-terminal side of it.

The study by Cadepond *et al.* (31) makes it quite clear that it will be very difficult to define specific contact sites for hsp90 on the GR, and thus it will be difficult to define a specific set of protein–protein interactions. This is a general problem that may be inherent to the fact that folding or chaperone proteins must normally interact with a variety of proteins of widely different structure. It has been speculated that a 20-amino-acid segment (583–602) that is highly conserved throughout the steroid receptor family may be an essential site of interaction with hsp90 (32). Housley *et al.* (27) have shown, however, that a mutant GR deleted for this region (Δ574–632 in Fig. 1) is bound to hsp90, although the mutant receptor is unstable in that it is sensitive to proteolytic cleavage in the intact cell. Whether or not this instability reflects a less stable interaction with hsp90 in the cell is not clear. This 20-amino-acid region is conserved in the thyroid hormone (TR) and retinoic acid (RAR) receptors, neither of which displays stable binding to hsp90 (33, 34).

Simons *et al.* (28) have cleaved the molybdate-stabilized wild-type GR with trypsin to produce the 16-kDa core steroid-binding fragment (525–671) shown in Figure 1. This fragment retains the ability to bind steroid and is bound to hsp90 (29). This core fragment demonstrates that there must be an intimate spatial association between the steroid-binding pocket and a minimal contact region for hsp90. This is important because there is now considerable evidence that hsp90 must be bound to the HBD of the GR for it to be in a high-affinity steroid-binding conformation (25, 35–39).

The site of hsp90 interaction with chicken PR has also been localized to the HBD (40, 41). Carson-Jurica *et al.* (40) found that it was necessary to delete the entire HBD of the PR (A form, PRa) before hsp90 binding was eliminated, and binding of hsp90 was not affected by several short internal deletions within the HBD. In a subsequent detailed analysis, Schowalter *et al.* (41) showed that an internal deletion of amino acids 451–521 of the chicken PRa (a segment similar in location to the minimal hsp90 binding site indicated within the HBD of the mouse GR in Figure 1) yielded a mutant PR that still bound hsp90 and hsp70, although perhaps with less stability. Importantly, when the 451–521 segment was attached to the COOH-terminus of a truncated PR that was deleted for the HBD and did not bind hsp90 or hsp70, the resulting chimæric receptor was isolated in association with both hsp90 and hsp70. Thus, the addition of a region that is presumably non-essential for hsp90 and hsp70 binding by analysis of the PR deletion mutant produces a chimæra that now has hsp-binding activity. These observations are consistent with the observations cited above for the GR system (27, 31), and suggest that there are either multiple contact sites for the heat-shock proteins in the HBD or that hsp90 binding depends on a general property of the unfolded protein that involves a diffuse region of the HBD.

Chambraud *et al.* (42) examined the binding of hsp90 by mutant forms of the human ER expressed in COS cells. Recovery of the receptors as 9S complexes in cytosol was used as the criterion of hsp90 binding. They found that the ER HBD was required for forming the 9S complex but it was not sufficient, and, as with the GR and PR, no specific region of the HBD proved to be uniquely involved. A short sequence between amino acids 251 and 271 at the COOH-terminal end of the DNA-binding domain was also necessary for recovery of the ER in the 9S form, but 251–271 was not sufficient without the presence of the HBD. This behaviour is somewhat different from the GR and PR where the HBD alone is sufficient for stable hsp90 binding. The region 256–303 of the human ER contains a number of basic amino acids that determine the nuclear localization of the receptor (43). Schlatter *et al.* (44) have examined hsp90 binding by co-immunoadsorption of *in vitro* translated HBD with anti-hsp90 antibodies. It was shown that the HBD of the GR was sufficient for stable association with hsp90 but that the HBD of the ER was not. Their data again suggested that a sequence NH_2-terminal to the HBD was required for a stable complex with hsp90, but this region, like the HBD, was not sufficient by itself. We have found that a chimæric protein (Z.HEI4; 43) of β-galactosidase and the HBD (282–595) of the human ER forms a complex with hsp90 (L. C. Scherrer *et al.*, *Biochemistry*, in press). Thus, the region between amino acids 251 and 271 of the human ER is not essential for forming a complex with hsp90. In agreement with others, however, we find the ER HBD–hsp90 complex to be very unstable in comparison to the GR HBD–hsp90 complex. The segment 256–303 in the nuclear localization region of the human ER could stabilize the complex either by binding directly to hsp90 as a second binding site in addition to the HBD or by binding to another protein in the heteroprotein complex that is also associated with hsp90. Hsp56 is an hsp90-associated and receptor-associated protein that may bind to a region of the receptor outside of the HBD.

2.1.2 Steroid-dependent dissociation of hsp90

Mendel *et al.* (45) were the first to provide direct evidence that exposure of intact cells (mouse thymoma cells) to glucocorticoid caused the dissociation of the GR from the 90-kDa protein. A subsequent study by Sanchez *et al.* (46) demonstrated that binding of steroid to the GR in cytosol promotes temperature-dependent dissociation of the receptor from hsp90 and that the physical properties of the untransformed and transformed GR were explained by its association with or dissociation from hsp90, respectively. These results were confirmed by Denis *et al.* (47). Importantly, steroid-dependent and temperature-dependent dissociation from hsp90 was accompanied by conversion of the GR from a non-DNA-binding form to a DNA-binding form. Indeed a careful examination of the time course for both events showed that hsp90 dissociation from the GR is accompanied by the simultaneous acquisition of DNA-binding activity (48).

Kost *et al.* (49) have demonstrated progesterone-mediated dissociation of the avian PR from hsp90, both with *in vivo* progesterone treatment and after addition of steroid to cytosol. In a study of the hormone effect on several PR-associated proteins, Smith *et al.* (50) showed that progesterone treatment of chicks resulted in loss of hsp90, p54, p50, and p23 from the oviduct PR. However, the amount of hsp70 bound to PR was not affected by hormone addition *in vivo* or *in vitro*. Although the relationships between hormone binding, dissociation of hsp90, and acquisition of DNA binding activity have not been examined for other steroid receptors with quite the same detail as for the GR, the general conclusion is that receptors for the adrenocorticoids and the sex hormones undergo similar changes as a result of hormone binding.

Several studies have suggested that the antagonist RU38486 does not promote dissociation of 9S GR to the 4S form in intact cells (51–54) and thus prevents GR translocation to the nucleus (54). Similarly, it has been suggested that binding of RU38486 stabilizes the PR–hsp90 complex (55, 56). It is difficult to evaluate the effects of RU38486 on receptor–hsp90 complex stability. It seems that in some systems the antagonist binds to receptors but does not provoke subsequent nuclear transfer (54), whereas in others it does (57). Also, in some systems, RU38486 has partial agonist activity and in others it does not (Chapter 5). At this time, it certainly cannot be concluded that RU38486 acts as an antagonist solely because it does not promote dissociation of the receptor from hsp90 (Chapter 8).

2.1.3 Classification of receptors by hsp90 binding

The steroid–thyroid receptor family may be divided into three classes according to the avidity of hsp90 binding and the requirement of hsp90 for steroid binding (Table 2). In class I fall the receptors for thyroid hormone, retinoic acid, and

Table 2 Classification of steroid and thyroid receptors by hsp90 binding

Class I	Receptors do not form a stable complex with hsp90
Retinoic acid	Unliganded receptors move directly to tight nuclear binding sites
Thyroid hormone	
Vitamin D	
Class II	Receptors form a stable complex with hsp90
Glucocorticoid	In hormone-free cells, receptor is retained in a cytoplasmic–nuclear 'docking' complex with hsp90
Mineralocorticoid	hsp90 is required for high-affinity steroid-binding conformation
Class III	Receptors form stable complex with hsp90
Progestin	In hormone-free cells, receptor is retained in a nuclear 'docking' complex that is recovered in the cytosolic fraction
Oestrogen	hsp90 is not required for steroid-binding conformation
Androgen	

vitamin D. Receptors in this class are not recovered from cells in association with hsp90. In contrast to the steroid receptors in classes II and III, which remain bound to hsp90 in the multiprotein complex and are located either in the cytoplasm or in loose association with the nucleus in hormone-free cells, the unliganded receptors in class I form tight complexes with nuclei that require high salt conditions for extraction (58–61). We found that TRs (33) and RARs (34) translated in reticulocyte lysate were produced in a form that had full DNA binding activity and that could not be immunoadsorbed with an anti-hsp90 antibody. This led to the conclusion that these receptors did not bind hsp90 at all, but recently we have found that if the immunopellet is washed rapidly with low ionic strength buffer, a very small amount of hsp90 can be recovered with these receptors. Thus, although they do not form a stable complex with hsp90, they are capable of forming a very weak complex.

In contrast to the evolutionarily more primitive receptors in class I, receptors in classes II and III form stable complexes with hsp90. The principal differentiating feature is that the receptors in class II must be bound to hsp90 for the HBD to be in an appropriate high-affinity conformation. If the steroid-bound GR is transformed to the 4S form, the steroid remains bound, but if the hormone-free complex is dissociated, then the hsp90-free receptor has a very low binding affinity for steroid (35–39, 54). Bresnick *et al.* (35) demonstrated a direct relationship between the amount of GR-associated hsp90 and the specific glucocorticoid binding capacity of immunopurified receptors. The requirement of hsp90 for steroid binding makes the GR especially useful for studying receptor reconstitution into a complex with hsp90, because conversion of the receptor from a non-steroid-binding to a steroid-binding form can be used to demonstrate that the reconstituted GR–hsp90 complex is functional (37–39). Although the mineralocorticoid receptor has not yet been demonstrated to require hsp90 for steroid binding, its physical behaviour (62) strongly suggests that this is the case.

The GR and mineralocorticoid receptor are usually (but not always) located in the cytoplasm of hormone-free cells, and after addition of steroid, they are rapidly translocated to the nucleus (57, 62, 63). In contrast, the unliganded sex hormone receptors in class III move directly from their cytoplasmic sites of synthesis into the nucleus (64–66), where they remain 'docked' to hsp90 until bound by hormone (see reference 15 for a review). The cytoplasmic location of the hormone-free GR appears to be a consequence of its association with hsp90. Picard and Yamamoto (57) prepared a number of fusion proteins containing β-galactosidase and portions of the rat GR and determined their cellular localization by immunofluorescence. We have found that, of the fusion proteins possessing the NL1 nuclear localization signal, only those that are cytoplasmic and move to the nucleus under hormonal control are bound by hsp90 (L. C. Scherrer *et al.*, *Biochemistry*, in press). This correlation is consistent with the proposal that when the GR is in the untransformed state, NL1 is either conformationaly inactive or blocked by receptor-associated protein. Thus, Urda *et al.* (67) have shown that the untransformed GR does not react with an antibody directed against NL1, whereas the dissociated receptor does. Using the same antibody, we found that the presence of the GR HBD in a fusion protein prevents reaction with an antibody to NL1 under conditions where the fusion protein is bound to hsp90 but not under conditions where it is dissociated from hsp90 (L. C. Scherrer *et al.*, *Biochemistry*, in press). When the HBD of the GR is replaced by the HBD of the ER, the resulting GR–ER fusion (like the wild-type ER) is nuclear in the absence of hormone (43). The reason why (despite its binding to hsp90) the ER HBD does not inactivate either its own nuclear localization signal in the wild-type ER or the NL1 of the GR in the GR–ER fusion is unknown.

2.2 Hsp70

The hsp70 family of proteins (20) consists of several proteins, some of which are produced constitutively (hsc70), while others are highly inducible by heat shock (hsp70). The protein(s) associated with steroid receptors will be referred to here simply as hsp70. The hsp70 proteins have been implicated as catalysts of protein assembly and are known to bind to newly synthesized but incompletely assembled oligomeric enzymes, but they are not bound to the fully assembled product (see Rothman (68) for review). Hsp70 is also known to have a protein unfoldase activity that is important for translocation of proteins across the membranes of organelles, such as mitochondria. All known members of the family are ATP-binding proteins.

2.2.1 Hsp70 in native receptor heterocomplexes

The first mention of a protein of approximately 70 kDa being associated with a steroid receptor appears to be in a report by Wrange *et al.* (69) who found a 72-kDa protein co-purifying with the transformed GR from rat liver. The 72-kDa protein did not have hormone-binding or DNA-binding activity but appeared to be associated with the receptor when it bound to DNA (70). The GR-associated 72-kDa protein was not further identified, but it is possible that it was hsp70. Similarly,

Estes *et al.* (71) purified hsp90-free PR (type B) from T47D human breast cancer cells and co-purified a 76-kDa non-hormone-binding protein that appeared to remain with the receptor when it bound to DNA.

Kost *et al.* (49) were the first to identify a receptor-associated protein of approximately 70 kDa as the constitutive form of hsp70. They found that chick oviduct PR underwent hormone-dependent or salt-mediated dissociation from hsp90 but a protein that reacted with hsp70-specific antibody remained PR Bound. The PR-associated 70-kDa protein was shown to be an ATP-binding protein by affinity labelling with azido ATP, and when the PR–hsp70 complex was incubated with ATP, the 70-kDa protein dissociated. Subsequently, Onate *et al.* (72) showed that the 76-kDa protein reported to be associated with the human PR was hsp70, and a careful examination of the DNA-binding properties of the transformed human PR suggested that hsp70 is not involved in receptor binding to DNA. In a study of mouse GR that was overexpressed in Chinese hamster ovary cells, Sanchez *et al.* (73) found the GR to be associated with both hsp70 and hsp90. As reported for the PR, salt transformation of this overexpressed GR to the DNA-binding state was accompanied by dissociation of hsp90 but not of hsp70.

2.2.2 Speculations on the role of hsp70 in receptor folding and trafficking

It is important to note that hsp70 has been carefully looked for and not found as a component of native GR heterocomplexes from L cells (73, 74), WEHI mouse lymphoma cells (75), rat thymocytes, and cultured human cells (HeLa) (76). It is not clear why hsp70 is recovered as a component of some native steroid receptor heterocomplexes but not with others. We have sometimes identified trace amounts of hsp70 in immunoadsorbed heterocomplexes from L cell cytosol, much as we have found small amounts of hsp70 in immunoadsorbed $pp60^{src}$–hsp90–p50 complexes (77). The presence of hsp70 could reflect an intermediate state in heterocomplex assembly, and in the case of the L cell GR or $pp60^{src}$, the intermediate may represent only a small fraction of the total complexes recovered from the cell. As will be discussed below, hsp70 is a major component of both steroid receptor (37, 38, 78, 79) and $pp60^{src}$ (80) heterocomplexes reconstituted with reticulocyte lysate, and it is thought that hsp70 unfoldase activity may be required for heterocomplex assembly (38, 79). It is also possible that recovery of receptors with hsp70 is related to their cellular location. The hormone-free GR in L cells and in many other cell types is located in the cytoplasm, but, like the PR (66), the GR overexpressed in Chinese hamster ovary (CHO) cells is entirely nuclear in the absence of hormone (73, 81). Sanchez *et al.* (73) have suggested that hsp70 may play a role in the passage of steroid receptors across the nucler membrane, much as hsp70 unfoldase activity has been implicated in the passage of molecules across the membranes of the endoplasmic reticulum and mitochondria (82–85).

2.2.3 Site of hsp70 interaction with receptors

The site of hsp70 interaction with the PR lies in the HBD. As discussed above for the binding of hsp90, Schowalter *et al.* (41) found it necessary to delete essentially the entire HBD (d369–659) before chicken PR binding to hsp70 was eliminated, and three separate regions of the HBD were found partially to restore hsp70 binding to the d369–659 mutant protein. They proposed that, like hsp90, hsp70 either contacts the receptor at multiple locations or binds to it through some general structural quality that is distributed throughout the HBD. The site of hsp70 interaction with the GR has not been determined. Hsp70 is present in the native Z.HE14–hsp90 complex, implying that the HBD of the ER is sufficient for hsp70 binding (L. C. Scherrer *et al.*, *Biochemistry*, in press). The location of the hsp70 binding site in the HBD of steroid receptors is consistent with the speculation that hsp70 plays a key role in receptor–hsp90 heterocomplex reconstitution.

2.3 Hsp56

Hsp56 is a heat-shock protein that was discovered as a direct result of investigations on receptor-associated proteins (86–88). Hsp56 was recently demonstrated to be an immunophilin of the FK506 binding class (89–91).

2.3.1 Discovery of hsp56

With the goal of preparing antibodies against the 9S PR, Nakao *et al.* (86) purified the molybdate-stabilized PR 10- to 30-fold by chromatography on hydroxylapatite and DEAE-cellulose. The crudely purified PR heterocomplex was then used as antigen to prepare the EC1 monoclonal antibody. By sucrose gradient analysis, the EC1 antibody was shown to interact with the 9S rabbit PR but not with 7S PR and smaller forms (87). In addition to the 59-kDa protein that reacted with EC1, both the PR and a 90-kDa non-progestin-binding protein were co-immunoadsorbed by the antibody, leading to the conclusion that both the 59-kDa and 90-kDa proteins were part of the receptor heterocomplex (87). It was subsequently demonstrated that the EC1 antibody reacted only with the 59-kDa protein, but that it nevertheless shifted the sedimentation of, and co-immunoadsorbed untransformed rabbit receptors for, progestin, oestrogen, androgen, and glucocorticoid (88). The suggestion that the rabbit 59-kDa protein was part of the 9S nuclear PR complex was supported by studies of Renoir *et al.* (92, 93) who identified the 59-kDa protein as a component of multiple steroid receptor complexes prepared from calf uterus (PR, AR, ER, and GR) and human MCF7 cells (PR and GR).

Although the 59-kDa protein has now been identified as part of the untransformed complex for multiple receptors in multiple mammalian species, the EC1 antibody does not recognize an avian protein; thus, p59 has not been identified as a component of the well-studied chicken PR heterocomplex. There is a 54-kDa protein in the chicken PR heterocomplex that consists of several isomorphs and migrates on two-dimensional gels in the same way as the hsp56 component of the human GR heterocomplex (cf. Fig. 3 in reference 50 with Fig. 9 in reference 18). It is possible that chicken p54 and hsp56 are related. From the results of cross-linking experiments, Renoir *et al.* (93) proposed that the mammalian PR-associated 59-kDa protein interacts with hsp90 rather than with the steroid-binding proteins. The results of cross-linking experiments with the GR (75, 76), however, suggest that this proposal is not entirely correct and that the p59 (hsp56) contacts both hsp90 and the steroid-binding protein. The region of this contact with the receptor has not been determined.

Two previously unidentified murine GR-associated proteins with reported $M_r \simeq$ 50 kDa and $\simeq$ 55 kDa have now been identified by their immunological reactivity to be the same protein as rabbit p59. By determining the size of products produced after exposure of the affinity-labelled WEHI cell GR to a cross-linking agent, Rexin *et al.* (94) deduced the existence of a subunit of approximately 50 kDa in the heterocomplex. This protein has recently been identified on Western blots as reacting with an antiserum directed against rabbit p59 (75). Bresnick *et al.* (74) reported that a 55-kDa protein was co-immunoadsorbed with the GR in cytosol prepared from [^{35}S]methionine-labelled L cells and this protein has now been identified as hsp56 by reaction with a specific antiserum directed against hsp56 (K. A. Hutchison *et al.*, *Biochemistry*, in press).

2.3.2 The hsp90–hsp70–hsp56 heterocomplex

As the reader will realize, nomenclature has been a problem because of different M_r reported for this protein in different species. In rabbit uterus cytosol the major species reacting with the EC1 antibody migrates on denaturing gels at 59 kDa, but in other species it usually migrates in the range of 55–57 kDa. Immunoadsorption of human IM-9 cell cytosol with EC1 revealed a major band of 56 kDa (and a minor species at approximately 54 kDa) that resolved into several isomorphs by two-dimensional gel analysis (18). Immunoadsorption of this p56 with EC1 yielded co-immunoadsorption of hsp70 and hsp90 (18). Hsp70 and hsp90 are both abundant proteins and the p56 was found by Coomassie blue staining to be a moderately abundant protein in IM-9 cell cytosol. Because all three proteins were co-immunoadsorbed by EC1 in great stoichiometric excess of the GR, it was proposed that they were present as a cytosolic complex independent of the presence of receptor (18). Because two of the proteins in the complex were known heat-shock proteins, Sanchez (95) submitted IM-9 cells to heat shock and demonstrated that the rate of p56 synthesis increased. Thus, the receptor-associated 56–59-kDa protein is itself a novel heat-shock protein and can be called hsp56. Hsp56 and p59 are clearly the same protein and hopefully the term hsp56, which is more descriptive and more representative of the major M_r in most species, will now be the standard nomenclature. The notion that a cytosolic complex of heat-shock proteins exists was confirmed when Perdew and Whitelaw (19, 96) immunoadsorbed rat hepatocyte cytosol with a monoclonal antibody against hsp90 and co-immunoadsorbed hsp70 and hsp56 (as well as two other proteins that migrated at M_r 50 kDa and 188 kDa).

2.3.3 Hsp56 is an immunophilin of the FK506 binding class

Hsp56 has now been identified as a member of the immunophilin class of proteins. The immunophilins are proteins that bind immunosuppressive agents such as cyclosporin A, FK506, and rapamycin in a high-affinity and specific manner, and all of the members of the family studied to date have rotamase (peptidyl-prolyl *cis-trans*-isomerase) activity *in vitro* (97, 98). Hsp56 was identified as an immunophilin when Yem *et al.* (89) passed cytosol from human Jurkat cells through a matrix of immobilized FK506 and isolated a protein of approximately 60 kDa with a NH_2-terminal sequence identical to that which we had previously published for hsp56 (18). The protein also showed homology to a region near the COOH-terminus of the FK506-binding proteins FKBP-12 and FKBP-13 (89). Simultaneously, Lebeau *et al.* (90) used the EC1 antibody to screen a rabbit liver cDNA library and clone the cDNA for hsp56. They found that a segment between amino acids 41 and 137 of the protein had 55% amino acid homology to rotamase. Subsequently, Tai *et al.* (91) isolated a 59-kDa protein from rat and human cells using immobilized FK506 or rapamycin. The NH_2-terminal sequence of the human protein was again identical to that reported for hsp56. Importantly, Tai *et al.* (91) showed that the FK506 affinity matrix selectively retained hsp90, hsp70, and the GR in addition to hsp56. FK506 inhibits the *in vitro* rotamase activity of immunophilins (97, 98). However, the role of rotamase activity in the cell is undefined and FK506 at high concentration has no effect on the function of the GR *in vitro* or on steroid-mediated transcriptional activation in intact cells (K. A. Hutchison *et al.*, *Biochemistry*, in press).

The presence of FK-binding proteins in many organisms suggests that the enzymes have an important general function in the cell. Thus:

1. It is of intrinsic interest that the three heat-shock proteins—hsp90, hsp70, and hsp56—have been shown by several methods to exist together in a cytosolic complex.
2. Each of these heat-shock proteins is likely to be involved in protein folding or chaparone functions in the cell.
3. The existence of the proteins in a complex suggests that they may act together in a spatially organized and temporally co-ordinated manner.

2.4 Other receptor-associated proteins

Two additional proteins (a 50-kDa and a 23-kDa protein) have been found in association with steroid receptors that were rapidly immunoadsorbed and washed under the gentlest of conditions in order to preserve as many potential weaker binding proteins as possible (50, 74).

2.4.1 p50

To date, the 50-kDa protein has been identified only in the chick PR heterocomplex, where its presence is stabilized by molybdate and it dissociates in a hormone-dependent manner (50). Both of these properties could reflect binding of p50 to hsp90. As mentioned above, Perdew and Whitelaw (19) identified a 50-kDa protein that co-immunoadsorbed from rat liver cell cytosol with monoclonal antibodies directed against hsp90. Whitelaw *et al.* (96) subsequently prepared an antibody to p50 purified from the hsp90 heterocomplex, and demonstrated by both immunoreactivity and peptide cleavage that this was the same as the 50-kDa phosphoprotein that is a long-established component of the pp60^{src}–hsp90 heterocomplex (17). Unfortunately, this anti-50 antibody does not react with chicken proteins; thus, it is not known whether the chick PR-associated p50 is the same or not. Mouse GR heterocomplexes from L cell cytosol and from the overexpressed CHO cell system have been probed with the anti-p50 antibody, and in neither case was the p50 protein detected (96).

2.4.2 p23

The 23-kDa protein in the chick PR heterocomplex is also stabilized by molybdate and undergoes hormone-dependent dissociation *in vivo* (50). The chick p23 is an acidic phosphoprotein with an ATP binding site, and analysis of proteolytic fragments shows no significant homology to known protein sequences (99). Johnson and Toft (99) have prepared a monoclonal antibody to p23 and have demonstrated that it is not a heat-shock protein. A 23-kDa protein that was reported by Bresnick *et al.* (74) to be a component of the L cell GR heterocomplex is recognized by this monoclonal antibody (P. R. Housley, personal communication), suggesting that the mouse GR- and chicken PR-associated 23-kDa proteins are the same. A 23-kDa protein with a similar acidic pI was also found in the hsp56–hsp90–hsp70 complex immunoadsorbed from human IM-9 cytosol with the EC1 antibody (18), but it has not yet been determined whether this protein is the same as the p23 in the receptor heterocomplex.

2.5 Stoichiometry of the heat-shock protein heterocomplex

The stoichiometry of the 9S GR complex of M_r approximately 330 000 has been studied most extensively. Initial hydrodynamic (100, 101) and immunological (102) analyses proved correct in indicating the presence of one molecule of steroid-binding protein in the GR heterocomplex. Hsp90 behaves as a dimer under non-denaturing conditions and can be purified to homogeneity in the dimeric form (103). Denis *et al.* (104) demonstrated that hsp90 released from the GR heterocomplex on transformation was released as a dimer. Mendel and Orti (105) then immunopurified the GR heterocomplex from WEHI cells that were labelled to steady state with [^{35}S]methionine and provided the first direct determination of GR–hsp90 stoichiometry calculated from the respective methionine contents of the purified proteins. They found a stoichiometry of one molecule of receptor to two molecules of hsp90, a ratio confirmed by similar metabolic labelling studies in L cells (74) and HeLa cells (76).

Murine hsp90 consists of two isoforms that are separate gene products and migrate at slightly different M_r. Both of these forms are recovered in the GR heterocomplex (74, 105). One problem that awaits solution is whether the GR binds selectively to one hsp90 gene product or equivalently to both. Lefebvre *et al.* (106)

have shown that one of the epitopes for a monoclonal antibody against hsp90 is inactivated (or blocked) when the hsp90 dimer is bound to the GR. This suggests that the untransformed GR–hsp90 complex is an asymmetrical complex in which the GR is interacting more directly with one molecule of hsp90 in the dimer than the other.

In a series of stoichiometric experiments based on the analysis of products produced after cross-linking of [^{3}H]dexamethasone mesylate-labelled GR, Rexin *et al.* (94) deduced that the untransformed receptor was a tetrameric structure. The proteins cross-linked in the GR complex in WEHI cell cytosol were identified as hsp90 and hsp56, yielding a tetramer containing one GR, two hsp90, and one hsp56 (75). Because hsp56 was readily cross-linked to the GR, it probably contacts the GR directly. Importantly, the receptor can be cross-linked to hsp90 and hsp56 in intact cells (107) with the same stoichiometry as the cytosolic complex — one GR, two hsp90, and one hsp56 (108).

Hsp90 has been reported to bind to actin (109) and microtubules (110), and the untransformed GR has been found to bind to actin through its hsp90 component (111). In a stoichiometric study by Bresnick *et al.* (74), gentle conditions of GR isolation yielded complexes with an average of four molecules of hsp90 per molecule of GR. This ratio was reduced to two hsp90 to one GR when the complex was washed with salt in the presence of molybdate. Given the possibility that the highly abundant hsp90 may bind to cytoskeleton or other structures in the cell, it is not surprising that larger hsp90 to receptor ratios may be detectable under particularly gentle conditions of isolation that do not destroy all of the higher-order structure. Ultimately, one would expect the core tetramer itself to be associated with some structure that retains the receptor in the cytoplasm or nucleus of the hormone-free cell.

The first stoichiometry estimates for any of the receptor–hsp90 complexes were made by Renoir *et al.* (7) who predicted correctly in 1984, on the basis of densitometric scanning and cross-linking data, that the purified molybdate-stabilized PR contained one molecule of steroid-binding protein and two molecules of 90-kDa non-hormone-binding protein. Further studies from the same laboratory were consistent with this stoichiometry and with the model that an hsp90 dimer is released on PR transformation *in vitro* (112, 113). Smith *et al.* (50) have made densitometric estimates of the relative abundance of five proteins in the chicken PR complex, predicting a stoichiometry of two molecules of hsp90 and one molecule each of hsp70, p54, p50, and p23 per molecule of A or B form of the PR.

The only stoichiometric estimates available for the ER heterocomplex are from Redeuilh *et al.* (114), who published a detailed hydrodynamic analysis of the behaviour of the purified, molybdate-stabilized 9S ER bound with monoclonal antibodies directed against hsp90 and the steroid-binding protein in a double-antibody technique. From their calculations, the authors suggested that the 9S ER heterocomplex contains a dimer of the steroid-binding protein bound to a dimer of hsp90. This stoichiometry differs from all the stoichiometric data available for the GR and PR heterocomplexes, and it poses two problems. First, the HBD of the ER contains a strong dimerization site and dimerization is required for high-affinity DNA binding (115). It is important to determine whether dimerization occurs before or after hsp90 dissociation, and the two ER–two hsp90 model of Redeuilh *et al.* (114) suggests that dimerization is a very early event occurring before the receptor is attached to the heat-shock protein complex. As there is clear evidence that the calf uterine ER heterocomplex contains hsp56 as well as hsp90 (93), the presence of an ER dimer in the complex would likely demand a different set of protein–protein interactions than those involved in the GR and PR tetramers containing only one steroid-binding protein and two hsp90 and one hsp56. Given the choices, an economy of logic would suggest a common heterocomplex core unit with the same stoichiometry of one steroid-binding protein, two hsp90, and one hsp56 exists for all of the steroid receptors, dictating that receptor dimerization occurs after hsp90 dissociation.

3. Cell-free receptor heterocomplex assembly

The binding of hsp90 to steroid receptors does not represent a free thermodynamic equilibrium; thus, the heterocomplex cannot be formed simply by mixing solutions of purified hsp90 and purified receptor. Receptor that is dissociated from hsp90 is thought to have undergone a change in conformation (perhaps a massive change) such that reassociation of hsp90 is not possible, as indicated by the inward folding of the HBD in the 4S, transformed GR in Figure 2. Inano *et al.* (116) have reported that purified ER and purified hsp90 incubated in a phosphate buffer containing Chaps, dimethylformamide, thiocyanate, glycerol, and molybdate combine to form a complex of approximately 9S, but stoichiometry of the components was not determined and the oestradiol dissociation rate was not returned to the rapid dissociation characteristic of the native 9S complex. Thus, in the absence of structural confirmation and an assay of 9S function, it is not clear that this represents reconstitution of a native complex. A method has now been developed in which purified receptors can be reconstituted by an enzymatic system in rabbit reticulocyte lysate into heterocomplexes that function like untransformed receptors. This system represents a superb means of probing the mechanisms by which heat-shock proteins function in protein folding and in the assembly of protein complexes.

3.1 Binding of hsp90 to *in vitro* translated steroid receptors

The first *in vitro* assembly of a functional receptor hsp90 complex was carried out by translating the rat GR in rabbit reticulocyte lysate (30). which contains a high concentration (approximately 2 μM) of hsp90 (117). Denis and Gustafsson (30) demonstrated that the translated receptor behaved as a 9S complex, and it did not bind to DNA but was converted to the DNA-binding form following labelling with dexamethasone and heating. Thus, the complex functioned like the native untransformed receptor complex isolated from cells.

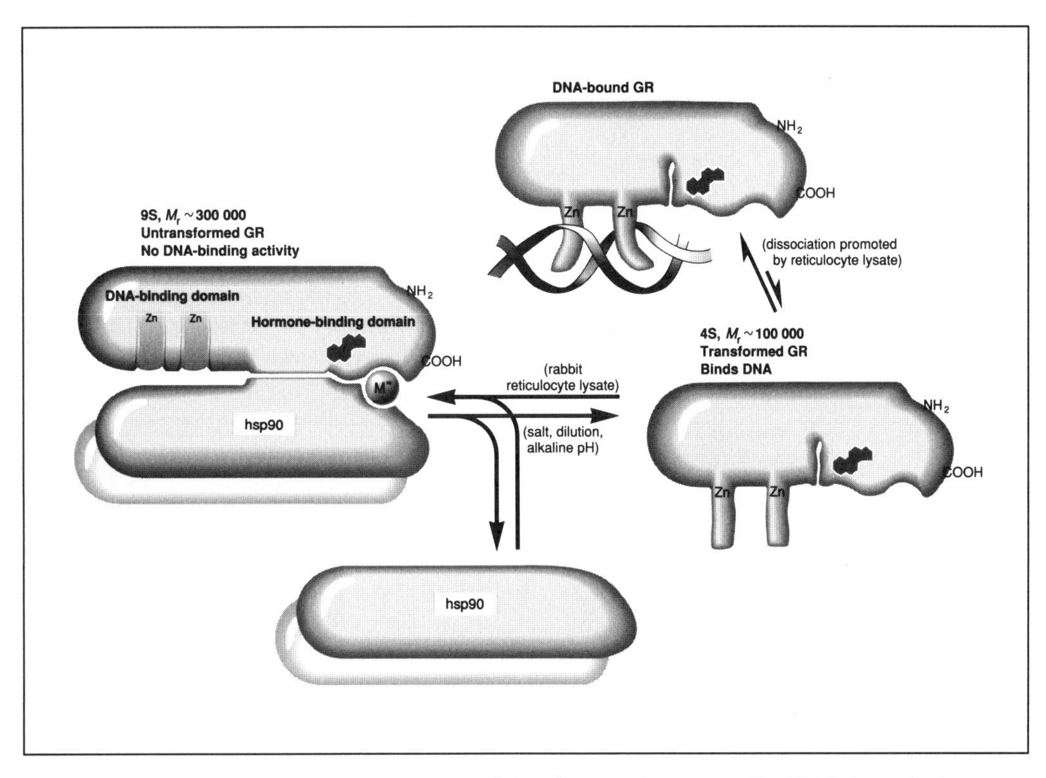

Fig. 2 Model of dissociation and reconstitution of the GR–hsp90 complex. Hsp90 binds to the hormone-binding domain of the receptor to form a 9S complex without DNA-binding activity. This complex is stabilized by molybdate (indicated by the small globe with the $M^{=}$). Addition of salt, dilution of cytosol, or alkaline pH promotes hsp90 dissociation of the hormone-free receptor complex and simultaneous transformation to the DNA-binding state (indicated by the receptor with the exposed 'zinc fingers' on the right). Physiologically, this dissociation is promoted by steroid. Hormone-free receptor that is either immunoadsorbed or DNA bound is reassociated with hsp90 in an ATP-dependent reaction directed by rabbit reticulocyte lysate, to reconstitute the non-DNA binding, untransformed receptor. From Scherrer *et al.* (37), reproduced by permission of the American Society for Biochemistry and Molecular Biology.

Dalman *et al.* (117) showed that rat GR translated in reticulocyte lysate was bound to rabbit hsp90; it bound steroid with high affinity, and it was in a non-DNA-binding form that could be transformed to the DNA-binding state. In contrast, GR translated in wheat-germ extract was not bound to plant protein, did not bind steroid, and bound to DNA without transformation (33, 117). Thus, the hsp90-bound receptor translated in reticulocyte lysate was functionally indistinguishable from the native 9S complex, and the hsp90-free receptor translated in wheat-germ extract behaved like a 4S receptor. Schlatter *et al.* (44) showed that the GR HBD alone was sufficient for hsp90 binding in the lysate translation system. Dalman *et al.* (117) determined that the GR became bound to hsp90 just at the termination of receptor translation in lysate. At that time, it was thought possible that receptor association with hsp90 might be obligatorily coupled to receptor translation.

3.2 Reconstitution of receptor heterocomplexes with reticulocyte lysate

Smith *et al.* (78) established that receptor binding to hsp90 is not obligatorily coupled to the translation process when they showed that incubation of immunoadsorbed hsp90-free chicken PR with reticulocyte lysate resulted in binding of rabbit hsp90 and hsp70 to the avian receptor. No reconstitution with hsp was obtained with a PR mutant lacking the HBD (78). Formation of the PR heterocomplex was temperature and ATP dependent (78, 79). Reconstitution with hsp90 and hsp70 occurred when the PR was hormone-free but not when it was bound with progesterone. The fact that liganded PR was not reassociated with hsp90 was consistent with the overall model of steroid action in which the ligand favours a receptor conformation that results in hsp90 dissociation. In fact, addition of progesterone to the reconstituted PR–hsp90 complex promoted hsp90 dissociation (79).

Scherrer *et al.* (37) incubated immunopurified hormone-free hsp90-free L cell GR with reticulocyte lysate and reconstituted the GR–hsp90 complex. As with the PR, rabbit hsp70 is also part of the hybrid heterocomplex. GR heterocomplex formation is also temperature and ATP dependent (37, 38). In reconstituting the heterocomplex, the mouse GR was converted from a non-steroid-binding state to a conformation that bound steroid with high affinity (37). Reactivation of steroid-binding activity was blocked by peptides (representing portions of the minimal hsp90 binding region of the GR shown in Fig. 1) that blocked hsp90 association with the receptor (25), and a direct relationship was established between the extent of hsp90 binding to the receptor and the number of specific steroid-binding sites that were generated (38). The heterocomplex reconstitution system in lysate also converted the GR from a DNA-binding form back to the non-DNA-binding state typical of the native receptor heterocomplex (37). In fact, hormone-free GR that was salt-transformed and bound to DNA–cellulose was released from DNA by lysate in a temperature-dependent reaction, producing soluble GR–hsp90 heterocomplexes that could bind steroid but not DNA. Thus, the GR in lysate-reconstituted complex had been restored to the same functional state as the native heterocomplex. As with the PR, the steroid-bound receptor was not reassociated with hsp90. Again, the ER HBD alone is sufficient for hsp90 binding (L. C. Scherrer *et al.*, *Biochemistry* , in press).

3.3 General properties of heterocomplex reconstitution

The protein assembly activity of reticulocyte lysate is not limited to the reconstitution of steroid receptor–hsp90 complexes. More than a decade ago, $pp60^{v\text{-}src}$ was found to be associated in a cytosolic complex with hsp90 and p50 (118, 119). As soon as it is translated in the cell, $pp60^{v\text{-}src}$associates with hsp90 and p50, and remains in the heterocomplex while it is transported to the cell membrane (17). We have recently demonstrated that incubating immunopurified pp60$^{v\text{-}src}$ (stripped of hsp90 and p50) with reticulocyte lysate results in ATP-dependent association of the viral tyrosine kinase with rabbit hsp90, hsp70, and p50 (80). As with pp60$^{v\text{-}src}$, the steroid receptor heterocomplexes formed in reticulocyte lysate contain other proteins in addition to hsp90 and hsp70. The reconstituted PR complex, for example, contains p23 (79) and the reconstituted GR complex contains hsp56 (K. A. Hutchison *et. al.*, Biochemistry., in press).

At this time it is not established whether the heterocomplexes are being formed in lysate by a process of ordered addition of individual proteins or whether the receptors and pp60^{src} are being attached to an already existing complex. As documented in previous sections of this chapter, there is substantial evidence that complexes containing hsp90, hsp70, hsp56, and other proteins exist in cytosols independent of the presence of the receptors. Recently, we have shown that a heat-shock protein complex (200–250 kDa) partially purified from rabbit reticulocyte lysate by ammonium sulphate fractionation and Sepharose CL-6B chromatography could reconstitute a functional GR–hsp90 complex, albeit at low efficiency compared with unfractionated lysate (39). The partially purified heat-shock proten complex contained hsp90, hsp70, and hsp56, and it reconstituted the GR heterocomplex in an ATP-dependent manner. These observations might be consistent with a model in which the receptors are attached to a preformed complex in lysate.

3.4 Is protein unfoldase activity of hsp70 required to form the receptor–hsp90 complex?

It is clear that heterocomplex formation is an enzymatic process and several observations suggest that hsp70, which has an established protein unfoldase activity, is required for receptor association with hsp90. First, both our laboratory and that of Toft, have noted that reticulocyte lysate-mediated association of the GR, PR, or pp60^{src} with hsp90 is always accompanied by association with hsp70 (38, 39, 79, 80). Second, dialysis of reticulocyte lysate inactivates its heterocomplex-forming activity, and activity is returned by adding an ATP-generating system and a monovalent cation, with K^+, NH_4^+, and Rb^+ being active and Na^+ and Li^+ being inactive (38). The clathrin unfolding activity of hsp70 has the same monovalent cation selectivity (120), as do some other chaparonin-facilitated folding processes (121). Most importantly, Smith *et al.* (79) have shown that pretreatment of lysate with a monoclonal antibody against hsp70 inhibited binding of hsp90 to the PR.

Taken together, these observations are consistent with a role for hsp70 in permitting the binding of hsp90 to the receptors and pp60^{src}. This is not to say that hsp70 is the only enzymatic component involved in heterocomplex reconstitution. Our work with the partially purified heat-shock protein complex from reticulocyte lysate made it clear that other components are required (39). In their reconstitutions of the PR heterocomplex, Smith *et al.* (79) have consistently observed a 60-kDa chicken protein that they feel is important at an intermediate stage of the assembly process. Further definition of this 60-kDa protein may contribute significantly to our understanding of the heterocomplex assembly mechanism.

3.5 A model of heterocomplex assembly

At this stage in our understanding, we have proposed the most elemental model (Fig. 3; 16), which will undoubtedly expand in complexity (e.g. model of Smith *et al.* (79)) as the protein assembly process becomes better defined. In this model hsp70 and hsp90 are preassociated and act together as a tandem unit, with hsp70 recognizing the folded state of the HBD (transformed GR in Fig. 2) and facilitating an ATP-dependent unfolding process leading to attachment of the receptor to the complex. The process is envisaged as a two-step event in which the second step is the stabilization of the unfolded state of the HBD by the hsp90 component of the heat-shock protein complex (16, 79).

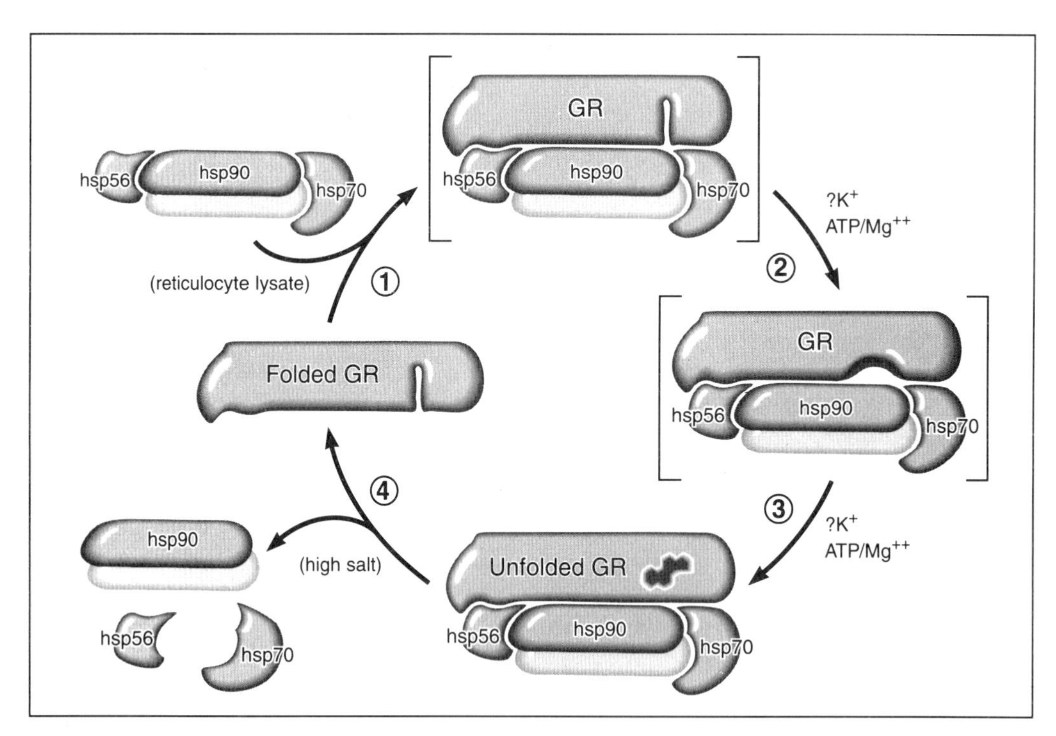

Fig. 3 Model of glucocorticoid receptor (GR) heterocomplex assembly, based on established protein–protein associations and stoichiometry. In step 1, the folded GR binds to a preformed heat-shock protein complex in reticulocyte lysate to form an initial binding complex (in square brackets), followed by hsp70-mediated unfolding of the receptor hormone-binding domain (HBD) (step 2), and stabilization of the unfolded GR by hsp90 (step 3). It is not clear where ATP/Mg^{++} and K^+ are required. The unfolded GR HBD in the heterocomplex has a high-affinity steroid-binding conformation. Dissociation of the hormone-free GR heterocomplex with high salt (step 4) is accompanied by refolding of the HBD and loss of the steroid-binding conformation. The folded GR can be recycled again through the heterocomplex assembly process

4. The heat-shock protein complex and steroid hormone action

In the case of heterocomplex assembly in the cell, association with hsp70 and stabilization of the receptor–hsp90 complex probably occur in a manner that is temporally and spatially linked to receptor translation. Members of the hsp70 family have been shown to interact co-translationally with a variety of proteins (122), an association that is thought to be necessary both for proper protein folding and for protein translocation (68). Hsp90 can be considered as maintaining the receptor HBD in a metastable, partially unfolded state. As the 9S heterocomplex is recovered from both cytoplasmic and nuclear receptors, it is likely that the receptors remain in the heterocomplex as they are transported to and within the nucleus. This heat-shock protein complex may be acting as a 'transportosome' (123), that is, as a structure that serves as a general transport form of steroid receptors, $pp60^{src}$, and probably other proteins in the cell. The steroid receptors may have exploited such a general role of hsp90 as a chaparone component of the protein transport machinery of the cell. The receptor has evolved a high-affinity binding to hsp90 that is under hormonal control, i.e. they remain 'docked' to the transport complex until the binding of hormone triggers their dissociation.

4.1 Model of receptor transformation driven by the trapped folding energy of the HBD

Hsp90 may be conceived as trapping the receptor HBD in a partially unfolded state and thus trapping its inherent folding energy. In the case of the steroid receptors in class II (Table 2) and also the dioxin receptor (124), only the partially unfolded hsp90-bound HBD is readily accessible for hormone binding. Once it has occupied the binding pocket, the steroid favours the folded conformation of the HBD, taking advantage of the stored potential energy of folding to convert the receptor to the active state by releasing it from hsp90 (38).

4.2 Relevance of the heat-shock protein complex to hormone action in the cell

The overriding question regarding steroid receptors and hsp90 is how do we know that binding of the receptor HBD to hsp90 is a significant component of the hormone action in the cell? Originally, there was a question as to whether receptor–hsp90 complexes existed in intact cells at all, but the experiments cited above showing intracellular dissociation by hormone and intracellular cross-linking, combined with a pulse-label study by Howard and Distelhorst (125) providing evidence for intracellular heterocomplex formation, seem to have eliminated this concern. That stable intracellular binding to hsp90 is responsible for recovery of receptors in the cytosolic fraction after hypotonic cell lysis is suggested by the fact that all of the cytosolic receptors are bound to hsp90, whereas the class I receptors, which are tightly bound in the nucleus, do not form 'docking' complexes with hsp90 (33, 34). Recently, Picard *et al.* (126) analysed steroid response in a strain of *Saccharomyces cerevisiae* that was modified to produce only 5% of the hsp90 present in wild-type cells. They showed that receptors were activated with markedly reduced efficacy when hsp90 levels were low. This argues strongly for a role of hsp90 in the signal transduction pathway for steroid receptors.

4.2.1 Hsp90 and receptor inactivation

The presence of the HBD inactivates the transcriptional activating activity of the receptor. Thus, it has been shown that deletion of the HBD of a steroid receptor can yield a constitutive activator of transcription. In the case of the GR (22, 28), the PR (40), and the ER (42), it has been found that these constitutively active receptors are not bound to hsp90, whereas mutant receptors that are steroid-inducible are bound. Thus, the 9S receptor heterocomplex is probably derived from the physiologically inactive state of the receptors that is activated by steroids in the cell, and the 4S dissociated form is derived from receptor that is transcriptionally active.

The inactivating function of the HBD of the steroid receptors is remarkably versatile in that it can confer hormone regulation on to the function of other proteins. Several examples now exist in which the GR (57, 127–130) or ER (128–131) HBD confers hormone regulation on fusion proteins. Thus, the inactivating function of the HBD can be conferred to proteins of different structure, and Picard *et al.* (127) have proposed that the inactivation is due to hsp90. We have recently examined several fusion proteins and find that the presence of the HBD of the GR or ER in a chimæric protein confers intracellular binding to hsp90 (L. C. Scherrer *et al.*, *Biochemistry*, in press).

4.2.2 Models of receptor inactivation by hsp90

Because the receptor HBD carries an inactivation function that operates on activities of structurally different proteins, a model of conferrable hormone regulation based on steric interference of receptor function by hsp90 is not adequate. Two models based on a role for hsp90 in receptor folding do accommodate a range of regulated structures and activities. Picard *et al.* (127) proposed that binding of hsp90 to the HBD within a protein chimæra causes the polypeptide as a whole to assume an 'unfolded' conformation that is reversed on hormone binding and transformation. This model assumes, then, that the function of the chimæra is inactivated because hsp90 determines an unfolded conformation in the region of the chimæra that is regulated as well as in the HBD that is conferring the regulation. The inactive protein is inactive because its active site is not properly folded.

The second model has been called the *docking* model (15, 123) because it proposes that receptors or chimæric proteins are bound to the heat-shock protein heterocomplex, to which they remain 'docked' until released from hsp90 by steroid-mediated reversal of their unfolded conformation. In the docking model, hsp90 has to determine an unfolded conformation of only the HBD, and the unfolded state does not have to be propagated to the rest of the fusion protein. In this case, the protein that is brought under hormonal control may have a properly folded active site but is biologically inactive until it is released from the heat-shock protein structure and allowed to proceed to its ultimate site of action.

References

1. Toft, D. and Gorski, J. (1966) A receptor molecule for estrogens; isolation from the rat uterus and preliminary characterization. *Proc. Natl Acad. Sci. USA,* **55,** 1574.
2. Sherman, M. R. (1984) Structure of mammalian steroid receptors. *Annu. Rev. Physiol.,* **46,** 83.
3. Dahmer, M. K., Housley, P. R., and Pratt, W. B. (1984) Effects of molybdate and endogenous inhibitors on steroid-receptor inactivation, transformation, and translocation. *Annu. Rev. Physiol.,* **46,** 67.
4. Renoir, J. M., Yang, C. R., Formstecher, P., Lustenberger, P., Wolfson, A., Redeuilh, G., Mester, J., Richard-Foy, H., and Baulieu, E. E. (1982) Progesterone receptors from chick oviduct: purification of the molybdate-stabilized form and preliminary characterization. *Eur. J. Biochem.,* **127,** 71.
5. Puri, R. K., Grandics, P., Dougherty, J. J., and Toft, D. O. (1982) Purification of 'nontransformed' avian progesterone receptor and preliminary characterization. *J. Biol. Chem.,* **257,** 10831.
6. Housley, P. R. and Pratt, W. B. (1983) Direct demonstration of glucocorticoid receptor phosphorylation by intact L cells. *J. Biol. Chem.,* **258,** 4630.
7. Renoir, J. M., Buchou, T., Mester, J., Radanyi, C., and Baulieu, E. E. (1984) Oligomeric structure of molybdate-stabilized nontransformed 8S progesterone receptor from chicken oviduct cytosol. *Biochemistry,* **23,** 6016.
8. Dougherty, J. J., Puri, R. K., and Toft, D. O. (1984) Polypeptide components of two 8S forms of chicken oviduct progesterone receptor. *J. Biol. Chem.,* **259,** 8004.
9. Housley, P. R., Sanchez, E. R., Westphal, H. M., Beato, M., and Pratt, W. B. (1985) The molybdate-stabilized L cell glucocorticoid receptor isolated by affinity chromatography or with a monoclonal antibody is associated with a 90–92-kDa non-steroid-binding phosphoprotein. *J. Biol. Chem.,* **260,** 13810.
10. Joab, I., Radanyi, C., Renoir, J. M., Buchou, T., Catelli, M. G., Binart, N., Mester, J., and Baulieu, E. E. (1984) Common non-hormone binding component in non-transformed chick oviduct receptors of four steroid hormones. *Nature,* **308,** 850.
11. Sanchez, E. R., Toft, D. O., Schlesinger, M. J., and Pratt, W. B. (1985) Evidence that the 90-kDa phosphoprotein associated with the untransformed L-cell glucocorticoid receptor is a murine heat shock protein. *J. Biol. Chem.,* **260,** 12398.
12. Schuh, S., Yonemoto, W., Brugge, J., Bauer, V. J., Riehl, R. M., Sullivan, W. P., and Toft, D. O. (1985) A 90 000-dalton binding protein common to both steroid receptors and the Rous sarcoma virus transforming protein, $pp60^{v\text{-}src}$. *J. Biol. Chem.,* **260,** 14292.
13. Catelli, M. G., Binart, N., Jung-Testas, I., Renoir, J. M., Baulieu, E. E., Feramisco, J. R., and Welch, W. J. (1985) The common 90-kD protein component of non-transformed '8S' steroid receptors is a heat-shock protein. *EMBO J.,* **4,** 3131.

14. Pratt, W. B. (1987) Transformation of glucocorticoid and progesterone receptors to the DNA-binding state. *J. Cell. Biochem.*, **35,** 51.
15. Pratt, W. B. (1990) Interaction of hsp90 with steroid receptors; organizing some diverse observations and presenting the newest concepts. *Mol. Cell. Endocrinol.*, **74,** C69.
16. Pratt, W. B., Scherrer, L. C., Hutchison, K. A., and Dalman, F. C. (1992) A model of glucocorticoid receptor unfolding and stabilization by a heat shock protein complex. *J. Steroid Biochem. Mol. Biol.*, **41,** 223.
17. Brugge, J. S. (1986) Interaction of the Rous sarcoma virus protein pp60^{src} with cellular proteins pp50 and pp90. *Curr. Top. Microbiol. Immunol.*, **123,** 1.
18. Sanchez, E. R., Faber, L. E., Henzel, W. J., and Pratt, W. B. (1990) The 56–59 kilodalton protein identified in untransformed steroid receptor complexes is a unique protein that exists in cytosol in a complex with both the 70- and 90-kilodalton heat shock proteins. *Biochemistry*, **29,** 1990.
19. Perdew, G. H. and Whitelaw, M. L. (1991) Evidence that the 90-kDa heat shock protein (hsp90) exists in cytosol in heteromeric complexes containing hsp70 and three other proteins with M_r 63 000, 56 000 and 50 000. *J. Biol. Chem.*, **266,** 6708.
20. Linquist, S. and Craig, E. A. (1988) The heat shock proteins. *Annu. Rev. Genet.*, **22,** 631.
21. Schlesinger, M. J. (1990) Heat shock proteins. *J. Biol. Chem.*, **265,** 12111.
22. Pratt, W. B., Jolly, D. J., Pratt, D. V., Hollenberg, S. M., Giguere, V., Cadepond, F. M., Schweizer-Groyer, G., Catelli, M. G., Evans, R. M., and Baulieu, E. E. (1988) A region in the steroid binding domain determines formation of the non-DNA-binding, 9S glucocorticoid receptor complex. *J. Biol. Chem.*, **263,** 267.
23. Denis, M., Gustafsson, J. A., and Wikstrom, A. C. (1988) Interaction of the M_r = 90 000 heat shock protein with the steroid binding domain of the glucocorticoid receptor. *J. Biol. Chem.*, **263,** 18520.
24. Wrange, O. and Gustafsson, J. A. (1978) Separation of other hormone- and DNA-binding sites of the hepatic glucocorticoid receptor by means of proteolysis. *J. Biol. Chem.*, **253,** 856.
25. Dalman, F. C., Scherrer, L. C., Taylor, L. P., Akil, H., and Pratt, W. B. (1991) Localization of the 90-kDa heat shock protein-binding site within the hormone-binding domain of the glucocorticoid receptor by peptide competition. *J. Biol. Chem.*, **266,** 3482.
26. Howard, K. J., Holley, S. J., Yamamoto, K. R., and Distelhorst, C. W. (1990) Mapping the hsp90 binding region of the glucocorticoid receptor. *J. Biol. Chem.*, **265,** 11928.
27. Housley, P. R., Sanchez, E. R., Danielsen, M., Ringold, G. M., and Pratt, W. B. (1990) Evidence that the conserved region in the steroid binding domain of the glucocorticoid receptor is required for both optimal binding of hsp90 and protection from proteolytic cleavage. A two-site model for hsp90 binding to the steroid binding domain. *J. Biol. Chem.*, **265,** 12 778.
28. Simons, S. S., Sistare, F. D., and Chakraborti, P. K. (1989) Steroid binding activity is retained in a 16-kDa fragment of the steroid binding domain of rat glucocorticoid receptors. *J. Biol. Chem.*, **264,** 14 493.
29. Chakraborti, P. K. and Simons, S. S. (1991) Association of heat shock protein 90 with the 16 kDa steroid binding core fragments of rat glucocorticoid receptors. *Biochem. Biophys. Res. Commun.*, **176,** 1338.
30. Denis, M. and Gustafsson, J. A. (1989) Translation of glucocorticoid receptor mRNA *in vitro* yields a nonactivated protein. *J. Biol. Chem.*, **264,** 6005.
31. Cadepond, F., Schweizer-Groyer, G., Segar-Maurel, I., Jibard, N., Hollenberg, S. M.,

Giguere, V., Evans, R. M., and Baulieu, E. E. (1991) Heat shock protein 90 as a critical factor in maintaining glucocorticosteroid receptor in a nonfunctional state. *J. Biol. Chem.*, **266,** 5834.

32. Danielsen, M., Northrop, J. P., and Ringold, G. M. (1986) The mouse glucocorticoid receptor: mapping of functional domains by cloning, sequencing and expression of wild-type and mutant receptor proteins. *EMBO J.*, **5,** 2513.
33. Dalman, F. C., Koenig, R. J., Perdew, G. H., Massa, E., and Pratt, W. B. (1990) In contrast to the glucocorticoid receptor, the thyroid hormone receptor is translated in the DNA-binding state and is not associated with hsp90. *J. Biol. Chem.*, **265,** 3615.
34. Dalman, F. C., Sturzenbecker, L. J., Levin, A. A., Lucas, D. A., Perdew, G. H., Petkovitch, M., Chambon, P. Grippo, J. F., and Pratt, W. B. (1991) The retinoic acid receptor belongs to a subclass of nuclear receptors that do not form 'docking' complexes with hsp90. *Biochemistry*, **30,** 5605.
35. Bresnick, E. H., Dalman, F. C., Sanchez, E. R., and Pratt, W. B. (1989) Evidence that the 90-kDa heat shock protein is necessary for the steroid binding conformation of the L cell glucocorticoid receptor. *J. Biol. Chem.*, **264,** 4992.
36. Nemoto, T., Ohara-Nemoto, Y., Denis, M., and Gustafsson, J. A. (1990) The transformed glucocorticoid receptor has a lower steroid binding affinity than the nontransformed receptor. *Biochemistry*, **29,** 1880.
37. Scherrer, L. C., Dalman, F. C., Massa, E., Meshinchi, S., and Pratt, W. B. (1990) Structural and functional reconstitution of the glucocorticoid receptor–hsp90 complex. *J. Biol. Chem.*, **265,** 21 397.
38. Hutchison, K. A., Czar, M. J., Scherrer, L. C., and Pratt, W. B. (1992) Monovalent cation selectivity for ATP-dependent association of the glucocorticoid receptor with hsp70 and hsp90. *J. Biol. Chem.*, **267,** 3/90.
39. Scherrer, L. C., Hutchison, K. A., Sanchez, E. R., Randall, S. K., and Pratt, W. B. (1992) A heat shock protein complex isolated from rabbit reticulocyte lysate can reconstitute a functional glucocorticoid receptor–hsp90 complex. *Biochemistry*, **31,** 7325.
40. Carson-Jurica, M. A., Lee, A. T., Dobson, A. W., Conneely, O. M., Schrader, W. T., and O'Malley, B. W. (1989) Interaction of the chicken progesterone receptor with heat shock protein (hsp) 90 *J. Steroid Biochem.*, **34,** 1.
41. Schowalter, D. B., Sullivan, W. P., Maihle, N. J., Dobson, A. D. W., Conneely, O. M., O'Malley, B. W., and Toft, D. O. (1991) Characterization of progesterone receptor binding to the 90- and 70-kDa heat shock proteins. *J. Biol. Chem.*, **266,** 21 165.
42. Chambraud, B., Berry, M., Redeuilh, G., Chambon, P., and Baulieu, E. E. (1990) Several regions of human estrogen receptor are involved in the formation of receptor-heat shock protein 90 complex. *J. Biol. Chem.*, **265,** 20 686.
43. Picard, D., Kumar, V., Chambon, P., and Yamamoto, K. R. (1990) Signal transduction by steroid hormones: nuclear localization is differentially regulated in estrogen and glucocorticoid receptors. *Cell Regul.*, **1,** 291.
44. Schlatter, L. K., Howard, K. J., Parker, M. G., and Distelhorst, C. W. (1992) Comparison of the 90-kilodalton heat shock protein interaction with *in vitro* translated glucocorticoid and estrogen receptors. *Mol. Endocrinol.*, **6,** 132.
45. Mendel, D. B., Bodwell, J. E., Gametchu, B., Harrison, R. W., and Munck, A. (1986) Molybdate-stabilized nonactivated glucocorticoid-receptor complexes contain a 90-kDa nonsteroid-binding phosphoprotein that is lost on activation. *J. Biol. Chem.*, **261,** 3758.

46. Sanchez, E. R., Meshinchi, S., Tienrungroj, W., Schlesinger, M. J., Toft, D. O., and Pratt, W. B. (1987) Relationship of the 90-kDa murine heat shock protein to the untransformed and transformed states of the L cell glucocorticoid receptor. *J. Biol. Chem.*, **262,** 6986.
47. Denis, M., Poellinger, L., Wilkstrom, A. C., and Gustafsson, J. A. (1988) Requirement of hormone for thermal conversion of the glucocorticoid receptor to a DNA-binding state. *Nature*, **333,** 686.
48. Meshinchi, S., Sanchez, E. R., Martell, K. J., and Pratt, W. B. (1990) Elimination and reconstitution of the requirement for hormone in promoting temperature-dependent transformation of cytosolic glucocorticoid receptors to the DNA-binding state. *J. Biol. Chem.*, **265,** 4863.
49. Kost, S. L., Smith, D. F., Sullivan, W. P., Welch, W. J., and Toft, D. O. (1989) Binding of heat shock proteins to the avian progesterone receptor. *Mol. Cell. Biol.*, **9,** 3829.
50. Smith, D. F., Faber, L. E., and Toft, D. O. (1990) Purification of unactivated progesterone receptor and identification of novel receptor-associated proteins. *J. Biol. Chem.*, **265,** 3996.
51. Groyer, A., Schweizer-Groyer, G., Cadepond, F., Mariller, M., and Baulieu, E. E. (1987) Antiglucocorticoid effects suggest why steroid hormone is required for receptors to bind DNA *in vivo* but not *in vitro*. *Nature*, **328,** 624.
52. Lefebvre, P., Formstecher, P., Richard, C., and Dautrevaux, M. (1988) RU486 stabilizes a high molecular weight form of the glucocorticoid receptor containing the 90K non-steroid binding protein in intact thymus cells. *Biochem. Biophys. Res. Commun.*, **150,** 1221.
53. Lefebvre, P., Danze, P. M., Sablonniere, B., Ricahrd, C., Formstecher, P., and Dautrevaux, M. (1988) Association of the glucocorticoid receptor binding subunit with the 90K nonsteroid-binding component is stabilized by both steroidal and nonsteroidal antiglucocorticoids in intact cells. *Biochemistry*, **27,** 9186.
54. Segnitz, B. and Gehring, U. (1990) Mechanism of action of a steroidal antiglucocorticoid in lymphoid cells. *J. Biol. Chem.*, **265,** 2789.
55. Moudgil, V. K. and Hurd, C. (1987) Transformation of calf uterine progesterone receptor: analysis of the process when receptor is bound to progesterone and RU38486. *Biochemistry*, **26,** 4993.
56. Renoir, J. M., Radanyi, C., and Baulieu, E. E. (1989) The antiprogesterone RU486 stabilizes the heterooligomeric, non-DNA-binding, 8S-form of the rabbit uterus cytosol progesterone receptor. *Steroids*, **53,** 1.
57. Picard, D. and Yamamoto, K. R. (1987) Two signals mediate hormone-dependent nuclear localization of the glucocorticoid receptor. *EMBO J.*, **6,** 3333.
58. Samuels, H. H., Tsai, J. S., Casanova, J., and Stanley, F. (1974) *In vitro* characterization of solubilized nuclear receptors from rat liver and cultured GH_1 cells. *J. Clin. Invest.*, **54,** 853.
59. Spindler, S. R., MacLeod, K. M., Ring, J., and Baxter, J. D. (1975) Thyroid hormone receptors: binding characteristics and lack of hormonal dependency for nuclear localization. *J. Biol. Chem.*, **250,** 4113.
60. Casanova, J., Horowitz, Z. D., Copp, R. P., McIntyre, W. R., Pascual, A., and Samuels, H. H. (1984) Photoaffinity labeling of thyroid hormone nuclear receptors: influence of *n*-butyrate and analysis of the half-lives of the 57 000 and 47 000 molecular weight receptor forms. *J. Biol. Chem.*, **259,** 12084.
61. Nervi, C., Grippo, J. F., Sherman, M. I., George, M. D., and Jetten, A. M. (1989)

Identification and characterization of nuclear retinoic acid-binding activity in human myeloblastic leukemia HL-60 cells. *Proc. Natl Acad. Sci. USA*, **86**, 5854.

62. Binart, N., Lombes, M., Rafestin-Oblin, M. E., and Baulieu, E. E. (1991) Characterization of human mineralocorticosteroid receptor expressed in the baculovirus system. *Proc. Natl Acad. Sci. USA*, **88**, 10 681.
63. Qi, M., Hamilton, B. J., and DeFranco, D. (1989) v-*mos* oncoproteins affect the nuclear retention and reutilization of glucocorticoid receptors. *Mol. Endocrinol.*, **3**, 1279.
64. King, W. J. and Greene, G. L. (1984) Mononuclear antibodies localize oestrogen receptor in the nuclei of target cells. *Nature*, **307**, 745.
65. Welshons, W. V., Lieberman, M. E., and Gorski, J. (1984) Nuclear localization of unoccupied oestrogen receptors. *Nature*, **307**, 747.
66. Perrot-Applanat, M., Logeat, F., Groyer-Picard, M. T., and Milgrom, E. (1985) Immunocytochemical study of mammalian progesterone receptor using monoclonal antibodies. *Endocrinology*, **116**, 1473.
67. Urda, L. A., Yen, P. M., Simons, S. S., and Harmon, J. M. (1989) Region-specific antiglucorticoid receptor antibodies selectively recognize the activated form of the ligand-occupied receptor and inhibit the binding of activated complexes to deoxyribonucleic acid. *Mol. Endocrinol.*, **3**, 251.
68. Rothman, J. E. (1989) Polypeptide chain binding proteins: catalysts of protein folding and related processes in cells. *Cell*, **59**, 591.
69. Wrange, O., Okret, S., Radojcic, M., Carlstedt-Duke, J., and Gustafsson, J. A. (1984) Characterization of the purified activated glucocorticoid receptor from rat liver cytosol. *J. Biol. Chem.*, **259**, 4534.
70. Wrange, O., Carlstedt-Duke, J., and Gustafsson, J. A. (1986) Stoichiometric analysis of the glucocorticoid receptor with DNA. *J. Biol., Chem.*, **261**, 11 770.
71. Estes, P. A., Suba, E. J., Lawler-Heavner, J., Elashry-Stowers, D., Wei, L. L., Toft, D. O., Sullivan, W. P., Horwitz, K. B., and Edwards, D. P. (1987) Immunologic analysis of human breast cancer progesterone receptors. 1. Immunoaffinity purification of transformed receptors and production of monoclonal antibodies. *Biochemistry*, **26**, 6250.
72. Onate, S. A., Estes, P. A., Welch, W. J., Nordeen, S. K., and Edwards, D. P. (1991) Evidence that heat shock protein-70 associated with progesterone receptors is not involved in receptor-DNA binding. *Mol. Endocrinol.*, **5**, 1993.
73. Sanchez, E. R., Hirst, M., Scherrer, L. C., Tang, H. Y., Welsh, M. J., Harmon, J. M., Simons, S. S., Ringold, G. M., and Pratt, W. B. (1990) Hormone-free mouse glucocorticoid receptors overexpressed in Chinese hamster ovary cells are localized to the nucleus and are associated with both hsp70 and hsp90. *J. Biol. Chem.*, **265**, 20 123.
74. Bresnick, E. H., Dalman, F. C., and Pratt, W. B. (1990) Direct stoichiometric evidence that the untransformed M_r300 000, 9S, glucocorticoid receptor is a core unit derived from a larger heteromeric complex: *Biochemistry*, **29**, 520.
75. Rexin, M., Busch, W., and Gehring, U. (1991) Protein components of the nonactivated glucocorticoid receptor. *J. Biol. Chem.*, **266**, 24 601.
76. Alexis, M. N., Mavridou, I., and Mitsiou, D. J. (1992) Subunit composition of the untransformed glucocorticoid receptor in the cytosol and in the cell, *Eur. J. Biochem.*, **204**, 75.
77. Hutchison, K. A., Stancato, L. F., Jove, R., and Pratt, W. B. (1992) The protein–protein complex between pp60$^{v\text{-}src}$ and hsp90 is stabilized by molybdate, vanadate, tungstate, and an endogenous cytosolic metal. *J. Biol. Chem.*, **267**, 13952.

78. Smith, D. F., Schowalter, D. B., Kost, S. L., and Toft, D. O. (1990) Reconstitution of progesterone receptor with heat shock proteins. *Mol. Endocrinol.*, **4,** 1704.
79. Smith, D. F., Stensgard, B. A., Welch, W. J., and Toft, D. O. (1992) Assembly of progesterone receptor with heat shock proteins and receptor activation are ATP-mediated events. *J. Biol. Chem.*, **267,** 1350.
80. Hutchison, K. A., Brott, B. K., De Leon, J. H., Perdew, G. H., Jove, R., and Pratt, W. B. (1992) Reconstitution of the multiprotein complex of pp60^{src}, hsp90 and p50 in a cell-free system. *J. Biol. Chem.*, **267,** 2902.
81. Martins, V. R., Pratt, W. B., Terracio, L., Hirst, M. A., Ringold, G. M., and Housley, P. R. (1991) Demonstration by confocal microscopy that unliganded overexpressed gucocorticoid receptors are distributed in a nonrandom manner throughout all planes of the nucleus. *Mol. Endocrinol.*, **5,** 217.
82. Chirico, W. J., Waters, M. G., and Blobel, G. (1988) 70 K heat shock related proteins stimulate protein translocation into microsomes. *Nature*, **332,** 805.
83. Deshaies, R. J., Koch, B. D., Werner-Washburne, M., Craig, E. A., and Schekman, R. (1988) A subfamily of stress proteins facilitates translocation of secretory and mitochondrial precursor polypeptides. *Nature*, **332,** 800.
84. Pelham, H. (1988) Coming in from the cold. *Nature*, **332,** 776.
85. Shefield, W. P., Shore, G. C., and Randall, S. K. Mitochondrial precursor protein: effects of 70-kilodalton heat shock protein on polypeptide folding, aggregation, and import competence. *J. Biol. Chem.*, **265,** 11 069.
86. Nakao, K., Myers, J. E., and Faber, L. E. (1985) Development of a monoclonal antibody to the rabbit 8.5S uterine progestin receptor. *Can. J. Biochem. Cell Biol.*, **63,** 33.
87. Tai, P. K. and Faber, L. E. (1985) Isolation of dissimilar components of the 8.5S nonactivated uterine progestin receptor. *Can. J. Biochem. Cell Biol.*, **63,** 41.
88. Tai, P. K., Maeda, Y., Nakao, K., Wakim, N. G., Duhring, J. L., and Faber, L. E. (1986) A 59-kilodalton protein associated with progestin, estrogen, androgen, and glucocorticoid receptors. *Biochemistry*, **25,** 5269.
89. Yem, A. W., Tomasselli, A. G., Heinrikson, R. L., Zurcher-Neely, H., Ruff, V. A., Johnson, R. A., and Deibel, M. R. (1992) The hsp56 component of steroid receptor complexes binds to immobilized FK506 and shows homology to FKBP-12 and FKBP-13. *J. Biol. Chem.*, **267,** 2868.
90. Lebeau, M. C., Massol, N., Herrick, J., Faber, L. E., Renoir, J. M., Radanyi, C., and Baulieu, E. E. (1992) P59, an hsp90-binding protein: cloning and sequencing of its cDNA and preparation of a peptide-directed polyclonal antibody. *J. Biol. Chem.*, **267,** 4281.
91. Tai, P. K., Albers, M. W., Chang, H., Faber, L. E., and Schreiber, S. L. (1992) Association of a 59-kilodalton immunophilin with the glucocorticoid receptor complex. *Science*, **256,** 1315.
92. Renoir, J. M., Radanyi, C., Jung-Testas, I., Faber, L. E., and Baulieu, E. E. (1990) The nonactivated progesterone receptor is a nuclear heterooligomer. *J. Biol. Chem.*, **265,** 14 402.
93. Renoir, J. M., Radanyi, C., Faber, L. E., and Baulieu, E. E. (1990) The non-DNA-binding heterooligomeric form of mammalian steroid hormone receptors contains a hsp90-bound 59-kilodalton protein. *J. Biol. Chem.*, **265,** 10 740.
94. Rexin, M., Busch, W., Segnitz, B., and Gehring, U. (1988) Tetrameric structure of the nonactivated glucocorticoid receptor in cell extracts and intact cells. *FEBS Lett.*, **241,** 234.

95. Sanchez, E. R. (1990) Hsp56: a novel heat shock protein associated with untransformed steroid receptor complexes. *J. Biol. Chem.*, **265**, 22 067.
96. Whitelaw, M. L., Hutchison, K., and Perdew, G. H. (1991) A 50-kDa cytosolic protein complexes with the 90-kDa heat shock protein (hsp90) is the same protein complexed with pp60$^{v\text{-}src}$ hsp90 in cells transformed by the Rous sarcoma virus. *J. Biol. Chem.*, **266,** 16 436.
97. Schreiber, S. L. (1991) Chemistry and biology of the immunophilins and their immunosuppressive ligands. *Science*, **251,** 283.
98. Walsh, C. T., Zydowsky, L. D., and McKeon, F. D. (1992) Cyclosporin A, the cyclophilin class of petidylpropyl isomerases, and blockade of T cell signal transduction. *J. Biol. Chem.*, **267,** 13 115.
99. Johnson, J. L. and Toft, D. O. (1992) Characterization of the 23 kDa protein associated with the progesterone receptor. *Abstracts of the 74th Annual Meeting of the Endocrinology Society*, p. 231.
100. Gehring, U. and Arndt, H. (1985) Heteromeric nature of glucocorticoid receptors. *FEBS Lett.*, **179,** 1985.
101. Gehring, U., Mugele, K., Arndt, H., and Bush, W. (1987) Subunit dissociation and activation of wild-type and mutant glucocorticoid receptors. *Mol. Cell. Endocrinol.*, **53,** 33.
102. Okret, S., Wikstrom, A. C., and Gustafsson, J. A. (1985) Molybdate-stabilized glucocorticoid receptor: evidence for a receptor heteromer. *Biochemistry*, **24,** 6581.
103. Minami, Y., Kawasaki, H., Miyata, Y., Suzuki, K., and Yahara, I. (1991) Analysis of native forms and isoform compositions of the mouse 90-kDa heat shock protein, hsp90. *J. Biol. Chem.*, **266,** 10 099.
104. Denis, M., Wikstrom, A. C., and Gustafsson, J. A. (1987) The molybdate-stabilized nonactivated glucocorticoid receptor contains a dimer of M_r 90 000 non-hormone-binding protein. *J. Biol. Chem.*, **262,** 11 803.
105. Mendel, D. B. and Orti, E. (1988) Isoform composition and stoichiometry of the ~90-kDa heat shock protein associated with glucocorticoid receptors. *J. Biol. Chem.*, **263,** 6695.
106. Lefebvre, P., Sablonniere, B., Tbarka, N., Formstecher, P., and Dautravaux, M. (1989) Study of the heteromeric structure of the untransformed glucocorticoid receptor using chemical cross-linking and monoclonal antibodies against the 90K heat-shock protein. *Biochem. Biophys. Res. Commun.*, **159,** 677.
107. Rexin, M., Bush, W., and Gehring, U. (1988) Chemical cross-linking of heteromeric glucocorticoid receptors. *Biochemistry*, **27,** 5593.
108. Rexin, M., Busch, W., Segnitz, B., and Gehring, U. (1992) Structure of the glucocorticoid receptor in intact cells in the absence of hormone. *J. Biol. Chem.*, **267,** 9619.
109. Koyasu, S., Nishida, E., Kadowaki, T., Matsuzaki, F., Iida, K., Harada, F., Kasuga, M., Sakai, H., and Yahara, I. (1986) Two mammalian heat shock proteins, hsp90 and hsp100, are actin-binding proteins. *Proc. Natl Acad. Sci. USA*, **83,** 8054.
110. Redmond, T., Sanchez, E. R., Bresnick, E. H., Schlesinger, M. J., Toft, D. O., Pratt, W. B., and Welsh, M. J. (1989) Immunofluorescence colocalization of the 90-kDa heat shock protein and microtubules in interphase and mitotic mammalian cells. *Eur. J. Cell Biol.*, **50,** 66.
111. Miyata, Y. and Yahara, I. (1991) Cytoplasmic 8S glucocorticoid receptor binds to actin filaments through the 90-kDa heat shock protein moiety. *J. Biol. Chem.*, **266,** 8779.
112. Aranyi, P., Radanyi, C., Renoir, M., Devin, J., and Baulieu, E. E. (1988) Covalent

stabilization of the nontransformed chick oviduct progesterone receptor by chemical cross-linking. *Biochemistry*, **27**, 1330.
113. Radanyi, C., Renoir, J. M., Sabbah, M., and Baulieu, E. E. (1989) Chick heat shock protein of $M_r = 90\,000$, free or released from progesterone receptor, is in a dimeric form. *J. Biol. Chem.*, **264**, 2568.
114. Redeuilh, G., Moncharmont, B., Secco, C., and Baulieu, E. E. (1987) Subunit composition of the molybdate-stabilized '8–9S' nontransformed estradiol receptor purified from calf uterus. *J. Biol. Chem.*, **262**, 6969.
115. Fawell, S. E., Lees, J. A., White, R., and Parker, M. G. (1990) Characterization and colocalization of steroid binding and dimerization activities in the mouse estrogen receptor. *Cell*, **60**, 953.
116. Inano, K., Haino, M., Iwasaki, M., Ono, N., Horigome, T., and Sugano, H. (1990) Reconstitution of the 9S estrogen receptor with heat shock protein 90. *FEBS Lett.*, **267**, 157.
117. Dalman, F. C., Bresnick, E. H., Patel, P. D., Perdew, G. H., Watson, S. J., and Pratt, W. B. (1989) Direct evidence that the glucocorticoid receptor binds hsp90 at or near the termination of receptor translation *in vitro*. *J. Biol. Chem.*, **264**, 19 815.
118. Oppermann, H., Levinson, W., and Bishop, J. M. (1981) A cellular protein that associates with the transforming protein of Rous sarcoma virus is also a heat shock protein. *Proc. Natl Acad. Sci. USA*, **78**, 1067.
119. Brugge, J. S., Erikson, E., and Erikson, R. L. (1981) The specific interaction of the Rous sarcoma virus transforming protein, $pp60^{src}$, with two cellular proteins. *Cell*, **25**, 363.
120. Schlossman, D. M., Schmid, S. L., Braell, W. A., and Rothman, J. E. (1984) An enzyme that removes clathrin coats: purification of an uncoating ATPase. *J. Cell Biol.*, **99**, 723.
121. Viitanen, P. V., Lubben, T. H., Reed, J., Goloubinoff, P., O'Keefe, D. P., and Lorimer, G. H. (1990) Chaparonin-facilitated refolding of ribulose bisphosphate carboxylase and ATP hydrolysis by chaparonin 60 (gro EL) are K^+ dependent. *Biochemistry*, **29**, 5665.
122. Beckman, R. P., Mizzen, L. A, and Welch, W. J. (1990) Interaction of hsp70 with newly synthesized proteins: implications for protein folding and assembly. *Science*, **248**, 850.
123. Pratt, W. B. (1992) Control of steroid receptor function and cytoplasmic-nuclear transport by heat shock proteins. *BioEssays*, **14**, 841.
124. Pongratz, I., Mason, G. G. F., and Poellinger, L. (1992) Dual roles of the 90-kDa heat shock protein in modulating functional activities of the dioxin receptor. *J. Biol. Chem.*, **267**, 13 728.
125. Howard, K. J. and Distelhorst, C. W. (1988) Evidence for intracellular association of the glucocorticoid receptor with the 90-kDa heat shock protein. *J. Biol. Chem.*, **263**, 3474.
126. Picard, D., Khursheed, B., Garabedian, M. J., Fortin, M. G., Lindquist, S., and Yamamoto, K. R. (1990) Reduced levels of hsp90 compromise steroid receptor action *in vivo*. *Nature*, **348**, 166.
127. Picard, D., Salser, S. J., and Yamamoto, K. R. (1988) A movable and regulable inactivation function within the steroid binding domain of the glucocorticoid receptor. *Cell*, **54**, 1073.
128. Eilers, M., Picard, D., Yamamoto, K. R., and Bishop, J. M. (1989) Chimæras of Myc oncoprotein and steroid receptors cause hormone-dependent transformation of cells. *Nature*, **340**, 66.
129. Umek, R. M., Friedman, A. D., and McKnight, S. L. (1991) CCAAT-enhancer binding protein: a component of a differentiation switch. *Science*, **251**, 288.

130. Superti-Furga, G., Bergers, G., Picard, D., and Busslinger, M. (1991) Hormone-dependent transcriptional regulation and celular transformation by Fos-steroid receptor fusion proteins. *Proc. Natl Acad. Sci. USA*, **88**, 5114.
131. Burk, O. and Klempnauer, K. H. (1991) Estrogen-dependent alterations in differentiation state of myeloid cells caused by a v-*myb*/estrogen receptor fusion protein. *EMBO J.*, **10**, 3713.

5 Nuclear hormone receptors as transcriptional activators

HINRICH GRONEMEYER

1. Introduction

Multicellular organisms have established sophisticated techniques of information transfer between cells and body compartments. The information that is transmitted can be highly complex and can result in the alteration of gene programmes involved in cellular differentiation, proliferation, or reproduction. Despite the complexity of these events, the inducing signals can be relatively simple molecules, such as small peptides or cholesterol derivatives. Apparently, the need for a high speed of information transfer means that the signals must be small, highly diffusable molecules. However, only limited structural information can be stored in the chemical structures of, for example, steroids, thyroids, or retinoids, and therefore the interpretation of the signal is of major importance and requires a highly specialized system that specifies the appropriate gene programmes. This interpretation can be based on signal-specific nuclear receptors that are expressed in target cells — those cells that are genetically determined to respond to the signal. Understanding the molecular mechanisms by which the information stored in the signal is converted into gene programming is one of the exciting areas of molecular biology.

In eukaryotes, two major signal transduction pathways can be distinguished (Fig. 1). One relies on the presence of receptors anchored in the cytoplasmic membrane. Ligand binding induces a cascade of events that are still not fully understood, and which result in a modulation of the activity of certain transcription factors and thus gene programmes. The second pathway represents a 'short-cut' of the first one, in tht the receptors for small hydrophobic molecules, such as steroid and thyroid hormones, vitamin D, and retinoids, are already located in the nucleus and act as ligand-inducible transcription factors (Fig. 1). While the concept of signal transduction originates from work carried out on endocrine systems, where the signal is produced in specified secretory organs and transported in the bloodstream, the identification of a family of nuclear receptors, including those for steroid and thyroid hormones, vitamin D, and retinoids (1–4), has led to an extension of this concept, obviating the need for endocrine production of the ligand, and has pointed to a major role of this signalling pathway in development, differentiation, and homoeostasis. Members of the nuclear receptor family (Fig. 2) are now known for which

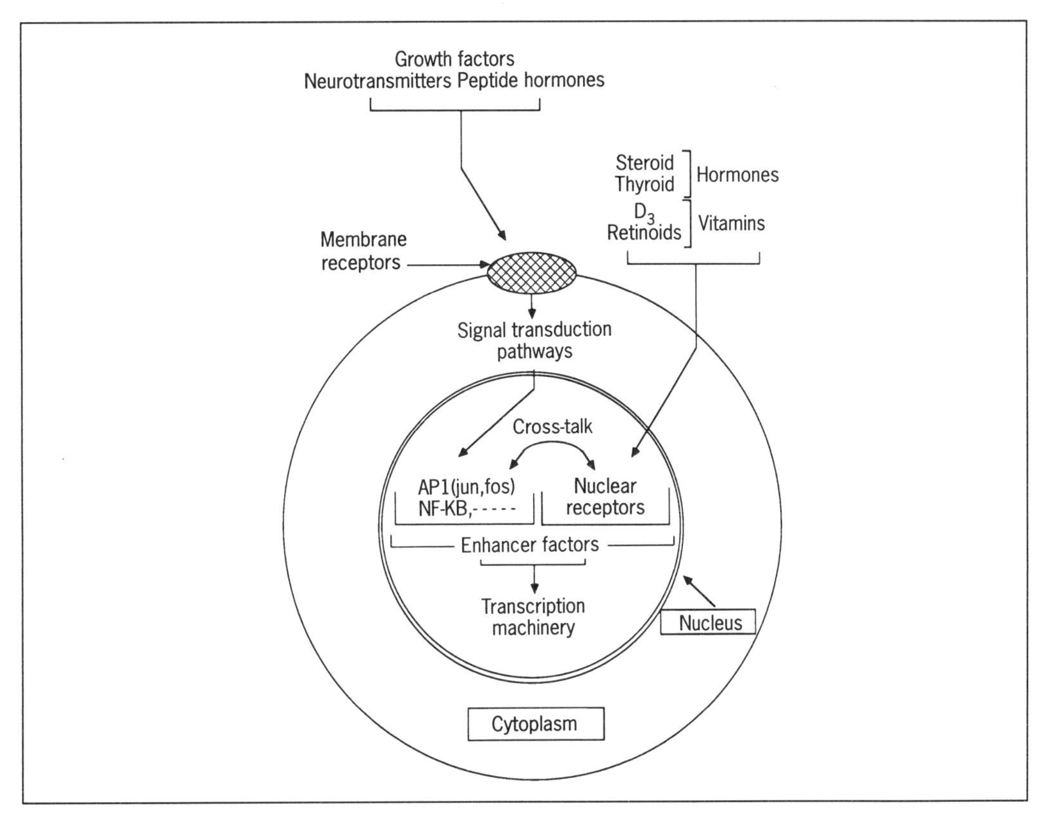

Fig. 1 Enhancer factors as signal transducers: two major signal transduction pathways in higher organisms utilize enhancer factors as signal transducers. Growth factors, neurotransmitters, and peptide hormones act via intermediary membrane-bound receptors and, subsequent to a poorly understood cascade of events, specific gene programmes are initiated by the action of specific (sets of) enhancer factors. The second pathway is a short-cut, in that receptors for certain small diffusible ligands (hormones and vitamins) are located in the nucleus and act as inducible enhancer factors themselves. Note that the cross-talk between nuclear receptors and other transcription factors is not discussed here but is addressed in Chapter 6

only artificial (5) or no ligands (the so-called 'orphan receptors') have been identified (6–14). Several *Drosophila* orphan receptors have been cloned (15–22), indicating the evolutionary conservation of signal transduction via nuclear receptors.

In this chapter I will discuss the current knowledge about several regulatory mechanisms that ensure that signal transduction results in an accurate regulation of the respective gene networks. The mechanisms involved in the binding of a nuclear receptor to its cognate hormone response element and cell- and promoter-specific transcription activation will be emphasized. Different gene programmes may be initiated in response to one ligand as a consequence of the different target gene specificities of nuclear receptor isoforms. This illustrates that signal transduction by nuclear receptors involves a multitude of interactive elements, as would be expected from the central role these signals have in homoeostasis, embryonic development, and differentiation.

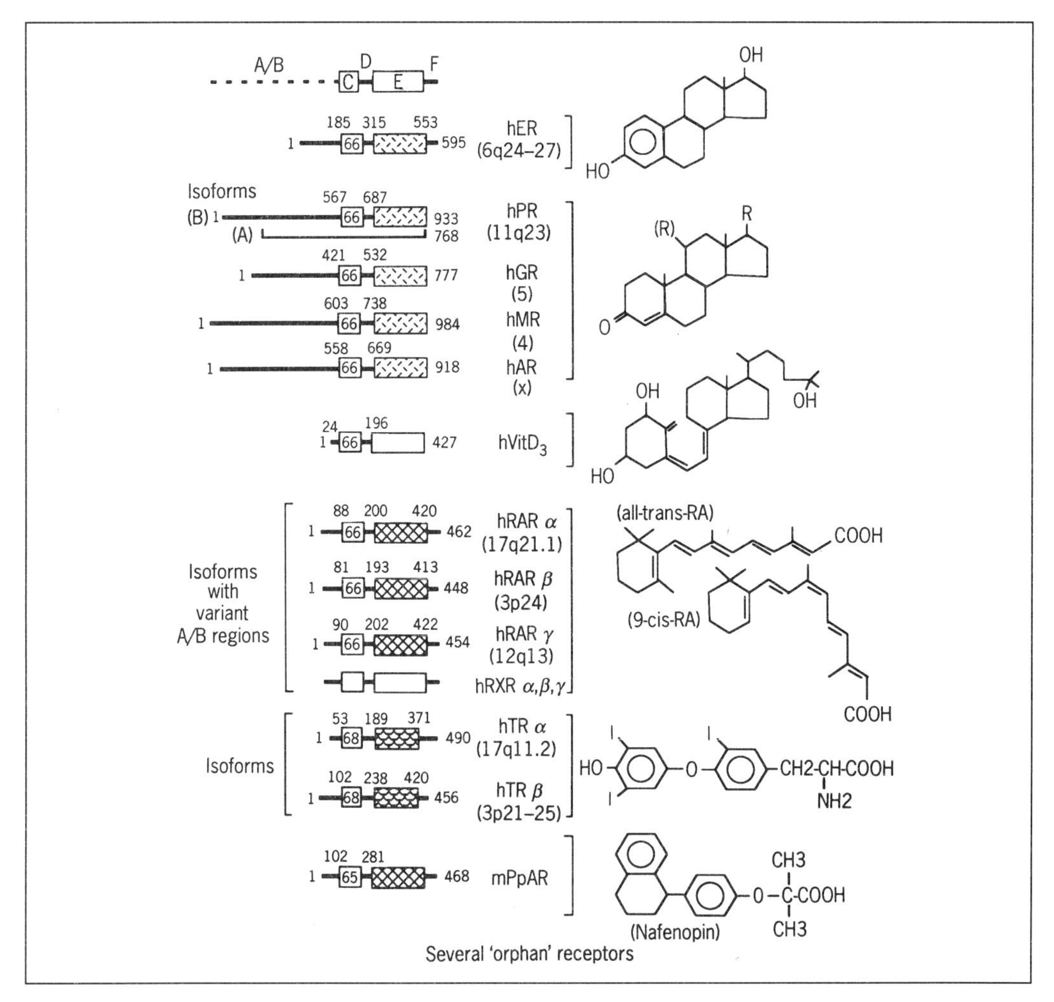

Fig. 2 The superfamily of nuclear receptor genes, and their cognate ligands, if known. The general subdivision into five or six differentially conserved regions (A–F) is shown at the top, with the two boxes illustrating the zinc finger-containing DNA-binding domain (region C) and the ligand-binding domain (region E). The chromosomal location of the human homologue is given in parenthesis. Note the presence of isoforms with different N-terminal regions for the PR, RAR, RXR, and TR, and the presence of two retinoic acid isomers which differentially interact with RARs and RXRs (see text). Note also that nafenopin is not the natural ligand for the PpAR, which, if existing, is unknown at present

2. Nuclear receptors are composed of structural elements that contain redundant and overlapping functions

The differential evolutionary conservation of six regions (denoted A–F) within the primary structure of nuclear receptors suggests that they may have a modular structure (Figs 2 and 3; note that some receptors, such as the progesterone and

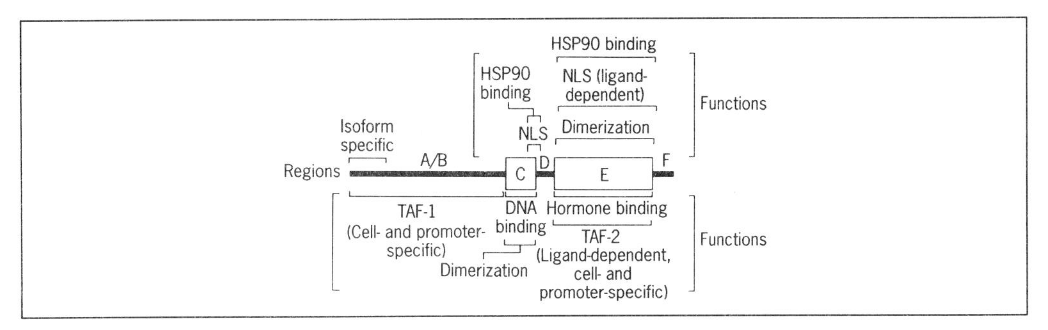

Fig. 3 Schematic illustration of the structural and functional organization of oestrogen and progesterone receptors. The evolutionary conserved regions C and E are indicated as boxes, and a black bar illustrates regions A/B, D, and F. Domain functions are depicted above and below the scheme (see text). NLS, nuclear localization signal; TAF, transcription activation function

retinoid X receptor, do not possess region F). Regions C and E are conserved and therefore likely to correspond to functions common to all members of the family (23). Indeed, analyses of *in vitro* mutated oestrogen (ER) and progesterone (PR) receptors revealed that these two regions contain the DNA- (DBD) and hormone-binding domains (HBD), respectively (see below). Apart from specifying ligand binding, region E of steroid receptors carries additional functions, most of which are ligand inducible: a transcription activation function (TAF-2), a dimerization domain, a signal for nuclear accumulation, and sequences that interact with heat-shock protein 90 (hsp90) *in vitro*. Thus, region E is a multifunction domain, and it will be interesting to understand the contribution of particular (common) sequences to these functions.

A number of functional domains may be present twice in a receptor, or even more often, as is the case for the multiple proto-nuclear localization signals (pNLSs; see below). However, this is not a mere duplication of domains, as the conditions under which two such domains function in a receptor may be different. For example, the two TAFs of the ER exert a different cell and promoter specificity. As depicted schematically in Figure 3, multiple functions also trigger dimerization, nuclear localization, and hsp90 binding.

2.1 Co-operation of proto-signals is responsible for nuclear localization

Both gain- and loss-of-function approaches have demonstrated the presence of nuclear localization signals (NLSs) in steroid hormone receptors. Similar sequences are also present in other nuclear receptors, although their function as NLSs has not yet been experimentally demonstrated. Nuclear targeting of β-galactosidase fusion proteins has provided evidence for the existence in the human ER and rat glucocorticoid receptor (GR) of a constitutive NLS resembling the SV40 T-antigen NLS prototype (24, 25). In addition, the existence of a hormone-dependent NLS has been reported for the rat GR (24).

Recently, these results have been extended in a detailed loss-of-function analysis (26). In this study, the steroid hormone receptor NLSs were maintained in their natural amino acid sequence context, because: (i) it was known that the efficiency of NLSs is sensitive to variations in the protein context (27); (ii) the β-galactosidase 'marker' used for heterologous nuclear targeting studies is, in contrast to nuclear receptors, a tetrameric protein and the consequences of multimerization on NLS efficiency are unknown; and (iii) the possible co-operation between several nuclear targeting sequences and the effect of other domains of the protein on their activity can be assessed only in their natural environment.

The analysis demonstrated that multiple proto-signals (pNLSs) for nuclear targeting, none of which suffices on its own, co-operate in the ER and PR. In the ER, an oestrogen-inducible pNLS was found in the hormone-binding domain, in addition to three lysine–arginine-rich motifs resembling prototype constitutive NLSs. The inducible and constitutive ER proto-NLSs co-operate in the presence of oestrogen and hydroxy-tamoxifen, but not in the presence of ICI164 384. In the PR, a signal corresponding to ER pNLS3 and two other pNLSs, located within and directly *C*-terminal of the second zinc finger of the DNA-binding domain, co-operate with each other and with a weak hormone-inducible pNLS in the PR HBD.

The existence of several pNLSs is not peculiar to nuclear receptors: other nuclear proteins have been reported to contain more than one nuclear targeting signal (28–32). It appears, therefore, that co-operation between pNLSs is a common theme among nuclear proteins. The presence of multiple co-operating pNLSs supports the recently proposed concept of co-operation of basic clusters to bind to a single receptor of the nuclear pore complex (32). Two recently characterized nuclear pore complex proteins (33, 34) have central domains consisting of a series of degenerate repeats. These proteins are candidate receptors and it is possible that the pNLSs of ER and PR interact with repeated structures of their cognate receptor(s) in a cooperative manner.

No 'masking' of pNLSs by the HBD was observed for ER and PR, while the ligand-free glucocorticoid receptor (GR) HBD inhibited the activity of both homo- and heterologous NLSs. This masking may explain why the intracellular localization of ligand-free GR, in contrast to that of ER and PR, has remained a matter of controversy (24, 35–38).

2.2 DNA binding, dimerization, and response element selection

2.2.1 Two closely related binding motifs constitute the response elements of nuclear receptors

Analysis of the promoter regions of oestrogen- and glucocorticoid-inducible genes led to the identification of their cognate hormone-responsive enhancer elements (HREs). The consensus sequence of the oestrogen response element (ERE; 5′-GGTCAnnnTGACC) is palindromic and that of the glucocorticoid response element

(GRE; 5′-GGTACAnnTGTYCT, in which Y = T or C, with T found in 65% of GREs (1)) is an 'imperfect' palindrome; both elements contain a three-nucleotide non-conserved spacer between two halves of the palindrome (1, 39–42). The 3′ half-site of the consensus GRE is more conserved than the 5′ counterpart among the various GREs, and a palindromic GRE maintaining the 3′ motif (5′-AGAACAnnnTGTTCT) activates GRE-dependent transcription as does a consensus GRE (42, 43). In the 3′ motifs, an ERE (5′-TGACC) and a GRE (5′-TGTYCT) differ at positions 3 and 4, and in the GRE there is an additional T at position 6. A systematic study of the effects of point mutations in the GRE revealed that T is the only possible nucleotide in position 3 and that position 4 is entirely unconstrained, while position 6 must be occupied by a pyrimidine (43) (note that a T is present in the same position in some EREs; because no systematic study has been performed, it is unclear whether there is any nucleotide preference for position 6 of a palindromic ERE). Thus, the major difference between the two response elements is at position 3, which must be a T in GREs and is usually an A in EREs. Note, however, that the nature of the base present at position 3 of an ERE is less restricted than that of a GRE. For example, the PS2 gene ERE (which is less efficient in transcriptional activation than is the consensus ERE) has 5′-TGGCC as a 3′ motif (44, 45). In conclusion, GREs and EREs may differ by only one base in each half of the palindromic recognition sequence, and this was in fact demonstrated with the corresponding reporter genes in transient transfection assays (40, 42).

A given HRE can in some cases be recognized by several receptors. For example, it has been shown that the consensus GRE can also mediate induction of transcription by progestins, androgens, and mineralocorticoids (42, 46, 47). This promiscuity is not restricted to palindromic GREs, since the complex GRE in the long terminal repeat (LTR) of the mouse mammary tumour virus (MMTV) can also be activated by all these steroids receptors (47–49). Although systematic studies are lacking, there appears to be no difference between the four receptors with respect to their specificity of GRE recognition. Indeed, 'false' activation is possible, since progestins can stimulate the transcription of the endogenous tyrosine aminotransferase gene (a glucocorticoid-induced gene) in stably transfected liver cells expressing the PR (50). Thus, regulatory mechanisms must exist which ensure receptor-specific activation of target genes in cells where more than one of these receptors is expressed (see below).

A different type of promiscuity has also been observed for the ER subfamily of nuclear receptors, since the ERE half-site motif 5′-GGTCA (or 5′-TGACC), when present as a direct or inverted repeat, is also recognized by thyroid hormone (TR), retinoic acid (RAR) and retinoid X (RXR), vitamin D (VDR), and some orphan receptors (see below). Note, however, that some natural retinoic acid response elements may also contain the 5′-GTTCA motif (51–53), which is not recognized by the ER (54).

2.2.2 Steroid receptors bind as dimers to palindromic response elements

The appearance of an intermediary retarded band in gel shift assays with palindromic EREs and *in vivo* co-expressed wild-type and receptor mutants truncated for the A/B region indicates that two receptor molecules bind to a single ERE (55). In support of the possible existence of a dimerization function in the HBD, it was found by gel shift assays that a HBD-truncated ER molecule does not co-bind with a wild-type molecule to a single ERE, whereas two such HBD-truncated human ER (hER) mutants do. In addition, HBD-truncated ERs bind considerably more weakly to a palindromic ERE than to the wild-type receptor (55). These experiments suggest that the HBD of the wild-type ER may stabilize ER homodimers which dissociate less rapidly than those (see below) of mutants lacking the HBD, and that dimerization may stabilize ER–ERE complexes. Indeed, the presence in the ER HBD of a dimerization domain was demonstrated by identifying amino acids within this domain which are critically involved in both high-affinity DNA binding *in vitro* and dimerization in solution, as demonstrated by gel retardation and co-immunoprecipitation assays, respectively (56). Although most of the amino acids are required not only for dimerization and high-affinity DNA binding but also for hormone binding, hormone binding is not an absolute requirement for specific DNA binding *in vitro* (56, 57). Moreover, a 22-amino-acid peptide of the mouse ER HBD confers high-affinity DNA binding on to a HBD-truncated mouse ER, and therefore may correspond to the dimerization domain (58).

That ER mutants lacking the HBD do not bind to ERE half-sites (55) but generate intermediary bands in gel shift assays with palindromic EREs suggests that a second dimerization domain, located in region A/B or C, may stabilize the corresponding ERE–protein complexes. Since mutants lacking both the N-terminal A/B and the HBD regions produce similar results, the A/B region of ER does not appear to contribute to the stability of HBD-truncated ER–DNA complexes, in contrast to the case with GR (59) and PR (see below). Thus, the stabilization must originate from region C itself. While the existence of dimers of HBD-truncated ERs has not yet been demonstrated, the solution structures of the ER and GR DBDs and the crystal structure of the GR DBD–DNA complex (60) are in agreement with the existence of a dimerization interface consisting of the *N*-terminal portion of the *C*-terminal 'finger' (61, 62). In addition, evidence supporting the existence of protein–protein interactions between GR DBDs bound to a GRE has been presented (63) and it has been shown that the substitution of the five *N*-terminal amino acids of the *C*-terminal GR finger by the corresponding ones of the β-thyroid hormone receptor abrogates co-operative binding to a GRE (64).

It has been reported that oestrogen is required for dimerization and binding to an ERE *in vitro* of hER expressed from the complementary DNA (cDNA) which was cloned initially (HEO (65, 66)). This ER cDNA contains a mutation (Gly400 → Val) which decreases the stability of the corresponding receptor protein in the absence of oestrogen, thus resulting in hormone requirement for DNA binding *in vitro* (67, 67). In fact, oestrogen does not appear to be required for DNA binding *in vitro* or for dimerization of the wild-type ER (HEGO; see references 67) irrespective of the salt (and temperature) conditions at which the ER-containing cell extracts were prepared (68; Metzger *et al.*, unpublished), suggesting that the ligand-free '8S complex' (presumed to be a hetero-oligomer with hsp90 (69)) contains ER dimers

(68, 70). *In vitro* DNA binding of wild-type mouse ER also occurs independently of hormone binding (56). In conclusion, the ER appears to contain two dimerization interfaces, a strong one located in the HBD and a second one located in the DBD, perhaps encompassing the five *N*-terminal amino acids of the *C*-terminal zinc finger.

The human and chicken PRs are expressed as two isoforms, A and B (71), which are translated from two messenger RNA (mRNA) classes. At least in the case of the human PR, these are transcribed from two promoters (72–75). For the chicken PR, it has been also proposed that the two isoforms arise from a single class of mRNAs by initiation of translation at two in-frame ATGs (76, 77). In contrast to the wild-type ER, no retarded PR–PRE complex can be seen in the absence of hormone in gel shift assays when using cell extracts containing recombinant PR. However, in the presence of progestins or the antagonist RU486, the PR isoforms form homo- and heterodimers (47, 78–80); no retarded complexes are seen with half-palindromic elements (H. Gronemeyer *et al.*, unpublished results). Note that, similar to the PR, the formation of RXR homodimers and their binding to certain response elements is enhanced by exposure to the cognate ligand 9-*cis* retinoic acid (81). When purified from target issue, neither GR (82) nor chicken and rabbit PR (83, 84) requires hormone for specific DNA binding. However, in these experiments the chicken PR (cPR) was purified by DNA-affinity chromatography after exposure to high salt at a previous step of the purification, and cPR prepared in low salt did not bind to DNA (83, 85). The salt exposure during purification may alter the structure of the PR, thus mimicking hormone exposure in the above studies. Interestingly, no 'mixed-ligand' isoform heterodimers (i.e. dimers containing, for example, form A bound to the progestin R5020 and form B bound to RU486) could be detected, suggesting that hormone and anti-hormone form incompatible dimerization interfaces or, alternatively, inhibit co-binding of differently liganded PR isoforms to the palindromic PRE (47). Thus, as in the case of the ER, dimerization may stabilize PR–PRE complexes, but in contrast to the ER, the hormone or the anti-hormone may be required for dimerization. Immunoprecipitation of hPR form B with form B-specific antibodies using extracts of the human breast cancer cell line T47D (which expresses the two PR isoforms, A and B) results in the co-precipitation of hPR form A, provided that high-salt nuclear extracts of hormone-treated cells or cytoplasmic extracts treated *in vitro* with high salt and hormone are used (79). Co-isolation of form A could not be demonstrated with cytosolic extracts of non-hormone-treated T47D cells containing the 8S 'unactivated' PR (79), indicating that the hormone is required for dimer formation.

While the role of the ligand cannot be easily deduced from the above studies, there is strong evidence that hormone is in fact required for PR dimerization *in vivo*. While studying the nuclear localization signals of the rabbit PR, Guiochon-Mantel *et al.* (86) identified a constitutive NLS and observed that a mutant receptor with a deleted NLS and a non-functional DBD but with an intact HBD was cytoplasmic, both in the absence and the presence of hormone. However, this mutant was co-translocated into the nucleus in a hormone-dependent fashion when co-expressed with a second mutant which lacked both the epitope used for immunocytochemical detection and a functional DBD but not the NLS, and was located in the nucleus in the presence of hormone (86). Ylikomi *et al.* (26) used this nuclear co-translocation assay and investigated the effect of ligand on ER and PR dimerization *in vivo*. Their data showed that co-translocation of translocation-deficient ER or PR mutants is strongly hormone dependent for the PR, while significant co-translocation of the ER mutants occurred in the absence of oestrogen. However, maximal co-translocation required exposure to oestrogen and was inhibited only partially by hydroxy-tamoxifen and ICI164 384. These data indicated that the *in vivo* stability of ER and PR dimers is hormonally controlled, but that in absence of the cognate ligand ER dimers are more stable than PR dimers.

Co-operative binding of the isolated GR DBD expressed in *Escherichia coli* has been reported to an 'imperfect palindromic' GRE or PRE (5′-TGTACAggaTGTTCT) derived from the tyrosine aminotransferase gene (87). Methylation interference studies have indicated that only the half-site TGTTCT is bound at low concentrations of GRE or PRE. At higher concentrations another DBD molecule binds to the second 'imperfect' TGTACA site in a co-operative manner. That this is due to the formation of GR DBD dimers on the DNA was supported by experiments showing that co-operativity is dependent on the distance and relative orientation of the two half-sites, but not on the integrity of the DNA backbone (63). Moreover, co-operativity, but not DNA binding, was lost upon mutating the five *N*-terminal amino acids (AGRND in both GR and PR) of the *C*-terminal zinc finger (64). There are, however, regions located outside region C that are essential for the binding of PR to a PRE. Gel shift studies with transiently expressed wild-type and mutated PRs have indeed revealed that both the HBD and sequences within region A/B are essential for the stability of PR–PRE complexes *in vitro* ((88, 89) and unpublished results from the author's laboratory).

In conclusion, in the presence of progestins and certain anti-hormones, the PR has the capacity to dimerize in solution in the absence of a cognate response element both *in vitro* and *in vivo*. Ligand-induced dimerization is important for high-affinity binding of PR to a PRE *in vitro* and for the RXR to certain response elements. In contrast, the ER forms dimers in the absence of ligand, which may explain the efficient binding of non-liganded ER to EREs *in vitro*. ER dimerization is further stimulated by oestrogen. Two dimerization domains are apparently present in steroid receptors, a weak one in the *C*-terminal zinc finger of the DBD and a strong one in the HBD. However, efficient DNA binding of the PR requires additional sequences located in the A/B region, which most likely provides an additional dimerization interface.

2.2.3 Retinoic acid and thyroid hormone receptors bind to differentially spaced, direct, and inverted repeats of the same motif that constitutes the oestrogen response element

The ER subgroup includes the receptors for oestrogen, thyroids (α and β TRs), retinoids (α, β, and γ RARs and RXRs) and vitamin D (VDR). The DBDs of all these receptors utilize (at least two of) the same discriminating amino acids as does the ER (see below) and bind to ERE-related core motifs that are assembled either as differentially spaced, inverted, or direct repeats. The ER binds as a dimer only to palindromes with a three-base-pair (3-bp) spacing (1), while TR and RAR bind, in addition, to an identical motif with 0-bp spacing. Moreover, TR activates transcription from such an element, but not significantly from the parental ERE (90). Recent studies have addressed the role of spacing and orientation of the core motif for receptor binding and transcription activation (91–93). The results can be summarized as follows: (i) 0-bp spaced palindromes are activated by TR and RAR, but not at all by ER. 3-bp spaced palindromes are activated by ER, but not by RAR or TR; (ii) 3-bp spaced inverted palindromes are activated by TR, weakly by RAR and not by ER; (iii) direct repeats with 1-bp, 3-bp, 4-bp, and 5-bp spacings are preferential response elements for RXR, VDR, TR, and RAR, respectively; (iv) surprisingly, Näär *et al.* (92) found that a 2-bp spaced direct repeat of the core motif confers activation in the absence, and repression in the presence, of thyroid hormone by its cognate receptor (note that the hexanucleotide core motif (5′-TGACCT or its antisense 5′-AGGTCA) is used to describe the spacing, according to Umesono *et al.* (91)).

There are a number of well-defined natural response elements that fit into these categories, such as the classical RARE of the RARβ gene (a 5-bp spaced direct repeat of 5′-GTTCA (51–53)). However, there are other natural response elements that are not consistent with the above hypothesis of a 'response element code'. The RARE of the mouse cellular retinoic acid binding protein I (CRABPI) gene (94), for example, has a 2-bp spaced core motif (5′-AGGTCAaaAGGTCA) and, according to Näär *et al.* (92), should be a negative thyroid RE in the absence, and a positive one in the presence, of the hormone. Similarly, also the RAREs of the laminin B1 gene (95), CRABPII gene (96), and putative elements in the apolipoprotein AI (97), the phosphoenolpyruvate carboxykinase (98), and the oxytocin genes (99) deviate from the postulated response element code. In addition, it has been shown that several nuclear receptors can form both homodimers and heterodimers with RXRs, which may be more stable than the corresponding homodimers (100–102), and there is evidence that the response element repertoires of homo- and heterodimers may be distinct (unpublished results from the author's laboratory).

2.2.4 The DNA-binding domains of nuclear receptors contain a small number of residues that are critically involved in the differential recognition of response elements

Chimeric receptors with altered DNA-binding specificity have been created by swapping the 65- to 68-amino-acid core of region C among the various members of the nuclear receptor family (5, 103, 104), indicating that this sequence contains all the determinants necessary for specific response element recognition. For example, a hER chimera, in which this core region has been replaced by the corresponding one of hGR, activates a glucocorticoid response gene in the presence of oestrogen

(103). Further analysis, in which the two zinc fingers of the ER and GR were individually swapped, revealed that the amino acids critical for the specificity of GRE versus ERE recognition are located within the *N*-terminal zinc finger (105). ER mutants, in which only a few amino acids of *N*-terminal finger have been mutated to the corresponding ones of the GR, indicate that three amino acids (Glu-203, Gly-204, Ala-207) located at the *C*-terminal knuckle of the *N*-terminal finger are critically involved in discriminating between the ERE and the GRE (106). However, ER mutants that exhibit a change in HRE specificity (e.g. Glu-203, Gly-204, Ala-207 of hER to Gly, Ser, Val of hGR, respectively) interact less efficiently with a GRE than do the wild-type hGR, since their affinity for a GRE is lower and they are less efficient at activating transcription from glucocorticoid responsive reporter genes. Thus, additional amino acids of the GR region C must be involved in stabilizing GR–GRE complexes.

While the substitution of the three critical amio acids of ER (Glu-Gly-Ala) by the corresponding ones of GR (Gly-Ser-Val) completely changes the specificity of RE recognition, an exchange of any two, but not one, of these residues is sufficient to gain GRE recognition (106). The mutations Glu-Gly- to Gly-Ser and Glu-Ala to Gly-Val produces ER mutants that recognize the GRE (and not the ERE), but the mutation Gly-Ala to Ser-Val yields a mutant with reduced specificity that recognizes both the GRE and ERE (although it is a poor tanscriptional activator of both oestrogen and glucocorticoid target genes).

Similar results to those described above have been obtained when the specific HRE recognition of the GR is changed by mutagenesis to that of the ER (107, 108). Moreover, further mutagenesis of the GR that had acquired ERE-binding activity suggested that five amino acids in the stem of the *C*-terminal zinc finger are responsible for the differential recognition (see above) by the ER and TR of 0-bp and 3-bp spaced palindromes (107, 109). Together with the observation that for the GR DBD the co-operativity, but not the DNA binding, is lost upon mutating the five *N*-terminal amio acids (Ala-Gly-Arg-Asn-Asp) of the *C*-terminal zinc finger (64), these data suggest that these amino acids may correspond to the weak dimerization domain present in the steroid receptor DBD. Such an interpretation is further supported by the three-dimensional structure of the GR DBD–GRE complex (60). It remains, however, to be established whether these amino acids are also involved in differential binding of the ER subgroup of nuclear receptors to directly repeated ERE half-site motifs.

Structural analysis of complexes between overexpressed receptor DBDs and cognate response elements has provided additional evidence that the three amino acids identified in the *N*-terminal zinc fingers of ER and GR are central to the discrimination between the bases that differ in EREs and GREs. Nuclear magnetic resonance spectroscopy of the GR and ER DBDs (62, 110) and crystallographic analysis of the GR DBD (60) have confirmed that these domains bind as dimers to the palindromic response element, placing its subunits into adjacent major grooves. Each of the subunits forms two different types of zinc-nucleated fingers, with the *C*-terminal zinc finger constituting the dimer interface and the discriminating amino acids being located in a helical segment of the *N*-terminal finger. In the above crystallographic study (60), the 5-methyl group of T at position 3 of the GRE contacts the Val side-chain, which is likely to result in a stabilization of the complex. This stabilization should not occur after substituting Val with Ala, which is found at the corresponding position in the ER. Similarly, a substitution of T at position 3 in the GRE with A (which is found at the corresponding position in an ERE) should result in a decreased stability of the complex. The critical Gly in the GR does not contact any base of the GRE. A Glu at this position (as found in the ER DBD) could possibly lead to hydrogen bonding with a C found at position 4 of an ERE, which may contribute to the stabilization of ER–ERE complexes. The discriminating role of the third critical amino acid (Ser in the GR, Gly in the ER) is unclear in the absence of similar structural data for the ER. Some of the conserved amino acids within the two fingers of GR are apparently important for establishing contacts with the phosphate backbone at both specific and non-specific sites (60). The non-specific nature of the *C*-terminal zinc finger is also illustrated by experiments showing that synthetic peptides encompassing this region of GR bind to both a GRE and a ERE, whereas a peptide comprising the *N*-terminal finger binds only to the GRE (111).

3. Transcriptional activation

3.1 Nuclear receptors contain two transcription activation functions that act in a cell- and promoter-specific fashion

Two types of experiment have demonstrated the existence within the PR of two domains responsible for the transcriptional activation of target genes. On the one hand, deletion mutants lacking either the *N*-terminal region A/B (or fragments thereof) or the HBD activate transcription much less than does the wild-type receptor, but, importantly, not all of their activity has been lost (47, 72). These experiments suggested the possible existence in the PR of more than one autonomous TAF. Indeed, when region A/B or the HBD of PR is linked to the (transcriptionally inactive) DBD of the yeast transcription factor Gal4, the resulting chimerae efficiently activate transcription from Gal4-responsive genes (47). Similarly, the existence of an autonomous TAF1 and TAF2 in the ER (and GR) regions A/B and the HBD, respectively, has been demonstrated (112, 113). The existence of TAF1 is not obvious from experiments performed with ER deletion mutants (114), since TAF1 is only weakly active in some cell types (see below). The nature of the two PR and ER TAFs is unclear. In contrast to a report showing that an acidic 30-amino-acid peptide (designated 'tau-2') of the GR HBD fused to the Gal4 DBD is able to activate transcription from Gal4 reporter genes (115), no such sequence could be identified in the ER TAF2 when individual or combinations of exons constituting the HBD were assayed for their ability to activate transcription (116). Also, a Gal4 chimera containing hPR sequences corresponding to the GR tau-2 did not measurably activate transcription (unpublished results from the author's laboratory). Short sequences containing (the core of) TAF1 of ER and PR have been delineated (117) (unpublished results from the author's laboratory), but they do not exhibit common characteristics, nor do they seem to resemble prototypic acidic activation domains (118).

Initial work concerning the location and number of TAFs within steroid receptors led to apparently contradictory results. While it was reported that a deletion of the HBD within the rat GR creates a constitutively acting 'super-receptor', which activates transcription at least as well as the full-length GR (119), such a deletion in hER or cPR abolishes 95% of the transcriptional activity (72, 114). Furthermore, the deletion of hER region A/B has no effect on the induction of the *vit-tk*-CAT reporter gene (114), while a cPR lacking the corresponding region has lost most of its activation capacity (72). Finally, one group found that the cPR truncated for the HBD has lost most of its transcriptional activity (72), while another group (120) reported that the same deletion has no effect. Consequently, it was concluded that the HBD of the GR (119) and of the PR (120) has a repressor type of function in the absence of ligand.

A systematic study of the different parameters involved in the various experimental protocols used by the different groups finally explained these discrepancies (121). It was demonstrated that two TAFs could be indeed observed in both the PR and GR, provided that the transfections were carried out in HeLa cells. In CV1 cells, TAF2 is weakly active relative to TAF1 when high amounts of receptor expression vectors have been transfected, such that both wild-type and HBD-less GR or PR activated transcription equally well. A similar cell specificity was demonstrated for ER TAF1 and TAF2: chimeric receptors containing the Gal4 DBD and the ER region A/B activate transcription from Gal4-responsive promoters in chicken embryo fibroblasts (CEF) but not in HeLa cells (113, 122). In contrast, ER TAF2 activates transcription in HeLa and CEF cells, but not in yeast, where TAF1 appears to be the only functional TAF of hER when using PGK promoter-based reporter genes (122–124). Collectively, these data indicate that each of the TAFs of a steroid receptor may have a distinct pattern of cell specificity. It is possible that the TAFs of various receptors exhibit different cell specificities, which may be related to their function *in vivo*.

The nature of the promoter context of a given target gene may also influence ER TAF activities. Transcription from reporter genes (e.g. *vit-tk*-CAT, ERE-*tk*-CAT) which contain complex promoters with several upstream elements (such as the HSV *tk*) is strongly enhanced by TAF2 in the presence of oestradiol. In contrast, TAF2 stimulates very weakly a minimal promoter composed of the adenovirus major late TATA box region placed downstream of a palindromic ERE. While transcriptional activation by TAF2 depends strongly on the promoter context, the activity of TAF1 is, in general, less affected by variations of this context (113, 122).

3.2 Evidence for transcription intermediary factors which mediate the enhancer function of nuclear receptors

What could be the basis for cell-specific activation of transcription by the steroid receptor TAFs? One possibility is that TAFs may interact, directly or indirectly, with other cell-specific transcription factors that are necessary to mediate TAF activity to the basic transcriptional machinery. We have termed these hypothetical coupling proteins, which do not have to bind on their own to the target gene promoter and can interact with multiple cognate (classes of) TAFs, trancriptional intermediary factors (TIFs, also called co-activators or adaptors).

Initial evidence for the existence of TIFs came from the so-called 'squelching' or 'transcriptional interference' experiments performed with the yeast transcription factor Gal4 and steroid hormone receptors (121, 125–127). Gill and Ptashne (125) observed that high levels of Gal4 inhibit (squelch) transcription from other genes that do not contain cognate Gal4 binding sites. The Gal4 TAF was found to be responsible for this inhibition, which was proposed to result from the interaction of the transactivator with a titratable non-DNA-binding (i.e. expressed in limiting amounts and interacting with the activator off the template) TIF, thus resulting in a decrease of transcription from promoters whose enhancer factor TAFs require the same TIF as does Gal4 (127). Moreover, if expressed at high concentrations under conditions where the template is limiting, an activator may actually auto-inhibit transcription from its reporter gene, since it will sequester its cognate TIF in solution.

Consistent with the TIF hypothesis, expression of increasing amounts of hER or hER TAF2 (mutant HE19) in HeLa or CV1 cells results in bell-shaped dose–response curves of transcriptional activation (121). The mediating factor(s) that are presumably titrated off the template in this experiment may be commonly 'used' by different steroid receptors, since both hER TAF1 and TAF2 act as efficient squelchers of transcriptional activation by PR and GR, with PR and GR reporter genes that lack EREs (126). Conversely, both PR and GR expression inhibit activation of transcription by the hER from a reporter gene lacking PR or GR binding sites (126). In all of these experiments squelching occurred only when the particular TAF was in its 'active' state, i.e. able to activate transcription when a cognate reporter gene was provided. While TAF1 was constitutively active, transcriptional activation and squelching by the full-length receptor or mutants/chimera containing the HBD required the presence of the cognate hormone. In the absence of a steroid agonist or in the presence of some steroid antagonists (devoid of any agonistic activity), squelching by the full-length receptor or by TAF2 present in HBD-containing mutants/chimerae was severely reduced or abolished (126). It is important to stress that cell-specific TIFs introduce additional steps into the nuclear receptor signal transduction pathway, and, therefore, additional combinatorial possibilities for regulating the transcription of target genes.

Recently, *in vitro* transcription assays have confirmed the existence of factors whose presence is required for stimulation of transcription by purified recombinant TAF1 of the hPR (129). As for other transcriptional activators, these intermediary factors are apparently associated with the TATA-binding protein in the transcription factor (TF) IID fraction (130–133). Cloning, expression, and characterization of these factors will be a breakthrough in the understanding not only of nuclear receptor action but also of mechanisms of transcription activation in general.

3.3 Isoform-specific activation of transcription

The differential activation of target genes by the two isoforms of the PR is the paradigm of isoform-specific activation of transcription (for a review, see reference 134). The human PR isoforms originate from two promoters, such that translation initiation occurs at two in-frame ATGs and results in the addition of 164 amino acids *N*-terminally of form A to give form B (74). While no isoforms are known for the other steroid hormone receptors, multiple isoforms are generated by both differential splicing and differential promoter usage for each of the three RAR subtypes, α, β, and γ, (reviewed in reference 135) as well as for the TR (reviewed in reference 141). In fact, isoform-specific activation of target genes has been recently demonstrated for the various RAR and RXR isoforms (136).

To understand the mechanism by which isoform-specific transcriptional activation by human PR isoforms A and B is generated, we first mapped the minimal sequence within region A/B which gave rise to activated transcription and subsequently investigated the role of the additional *N*-terminal 164 amino acids that are lacking in isoform A (117). No autonomous TAF could be detected in this region, since the only TAF in region A/B was found between amino acids 456 and 546, close to the DBD. Thus, the *N*-terminal region (designated AB3) that confers isoform specificity on transcription activation contains a modulatory function which, on its own, does not promote activated transcription. Because a reduction of the target gene promoter complexity did not alter isoform and promoter specificity, it was clear that additional promoter-bound factors were not responsible for the effect. The finding of a differential effect of region AB3 in HeLa and CV1 cells suggested the implication of a cell-borne factor and, consequently, we investigated by 'squelching' assays (125, 126) whether the AB3 region could bind an intermediary factor that is required for transcription activation by isoform B. Indeed, co-expression of Gal-hPR(AB3), severely reduced hPR form B/R5020-induced transcription (117). Interestingly, also hPR form A/R5020-induced transcription was squelched by Gal-hPR(AB3), suggesting that the AB3 cognate factor binds to one or more intermediary factors that are required for TAF1 and/or TAF2 activity.

4. Perspectives

Obviously, as far as transcription activation is concerned, the isolation, cloning, and characterization of TIFs and co-activators, which provide the link between the activator and components of the transcriptional machinery, will be a major achievement. TBP-associated co-activators have been identified and cloned in Tjian's laboratory, but it remains to be demonstrated that they are indeed involved in nuclear receptor action. SWI proteins (which are apparently not among the cloned TBP-associated factors) may play a role in GR function (137, 138), and co-activators have been identified that are apparently not associated with TBP (139). Also, it remains to be seen whether negatively acting TIFs (140) are involved in nuclear receptor action. Apart from these aspects, the role of chromatin as a modulator of nuclear receptor-induced gene transcription is a field of research where little is known.

Similarly, little is known about the target gene repertoires of the various nuclear receptors and the sequential, spatial, and temporal regulation of their transcription. It will be important to devise efficient methods of isolating target genes. A particularly exciting area is the question of specificity and promiscuity of response element recognition by the members of the ER subfamily of nuclear receptors and the role of homo- and heterodimerization. It is particularly intriguing that RAR, RXR, and TR can bind to, and activate transcription from, elements composed of either directly repeated or invertedly repeated identical motifs. It is unclear which and how many dimerization domains are involved in these processes.

Finally, elucidation of the three-dimensional structure of nuclear receptors alone or in complexes with ligands, various response elements, and heteromeric partners will be a major breakthrough for our molecular understanding of receptor functions.

Acknowledgements

The author is grateful to Lirim Shemshedini for critically reading this manuscript, and thanks members of the steroid receptor group at the LGME for communicating unpublished results. The work performed at the LGME that has been cited in this review was supported by funds from the Association pour la Recherche sur la Cancer, the Institut National de la Santé et de la Recherche Médicale (INSERM), and the Centre National de la Recherche Scientifique (CNRS).

References

1. Green, S. and Chambon, P. (1988) Nuclear receptors enhance our understanding of transcription regulation. *Trends Genet.*, **4**, 309.
2. Evans, R. M. (1988) The steroid and thyroid hormone receptor superfamily. *Science*, **240**, 889.
3. Beato, M. (1989) Gene regulation by steroid hormones. *Cell*, **56**, 335.
4. Gronemeyer, H. (1991) Transcription activation by estrogen and progesterone receptors. *Annu. Rev. Genet.*, **25**, 89.
5. Issemann, I. and Green, S. (1990) Activation of a member of the steroid hormone receptor superfamily by peroxisome proliferators. *Nature*, **347**, 645.
6. Giguere, V., Yang, N., Segui, P., and Evans, R. M. (1988) Identification of a new class of steroid hormone receptors. *Nature*, **331**, 91.
7. Hazel, T. G., Nathans, D., and Lau, L. F. (1988) A gene inducible by serum growth factors encodes a member of the steroid and thyroid hormone receptor superfamily. *Proc. Natl Acad. Sci. USA*, **85**, 8444.
8. Kazushige, H., Gleason, S. L., Levi, B. Z., Hirschfeld, S., Appella, E., *et al.* (1989) H-2RIIBP, a member of the nuclear hormone receptor superfamily that binds to both

the regulatory element of major histocompatibility class I genes and the estrogen response element. *Proc. Natl Acad. Sci. USA,* **86,** 8289.

9. Milbrandt, J. (1988) Nerve growth factor induces a gene homologous to the glucocorticoid receptor gene. *Neuron,* **1,** 183.
10. Watson, M. A. and Milbrandt, J. (1989) The NGFI-B gene, a transcriptionally inducible member of the steroid receptor gene superfamily: genomic structure and expression in rat brain after seizure induction. *Mol. Cell. Biol.,* **9,** 4213.
11. Nakai, A., Kartha, S., Sakurai, A., Toback, F. G., and DeGroot, L. J. (1990) A human early response gene homologous to murine nur77 and rat NGFI-B, and related to the nuclear receptor superfamily. *Mol. Endocrinol.,* **4,** 1438.
12. Wang, L. H., Tsai, S. Y., Cook, R. G., Beattie, W. G., Tsai, M. J., *et al.* (1989) COUP transcription factor is a member of the steroid receptor superfamily. *Nature,* **340,** 163.
13. Sladek, F. M., Zhong, W., Lai, E., and Darnell, J. E. (1990) Liver-enriched transcription factor HNF-4 is a novel member of the steroid hormone receptor family. *Genes Dev.,* **4,** 2353.
14. Ryseck, R. P., Macdonald-Bravo, H., Mattei, M. G., Ruppert, S., and Bravo, R. (1989) Structure, mapping and expression of a growth factor inducible gene encoding a putative nuclear hormonal binding receptor. *EMBO J.,* **8,** 3327.
15. Shea, M. J., King, D. L., Conboy, M. J., Mariani, B. D., and Kafatos, F. C. (1990) Proteins that bind to *Drosophila* chorion *cis*-regulatory elements: a new C2H2 zinc finger protein and a C2C2 steroid receptor-like component. *Genes Dev.,* **4,** 1128.
16. Segraves, W. A. and Hogness, D. S. (1990) The E75 ecdysone-inducible gene responsible for the 75B early puff in *Drosophila* encodes two new members of the steroid receptor superfamily. *Genes Dev.,* **4,** 204.
17. Rothe, M., Nauber, U., and Jäckle, H. (1989) Three hormone receptor-like *Drosophila* genes encode an identical DNA-binding finger. *EMBO J.,* **8,** 3087.
18. Pignoni, F., Baldarelli, R. M., Steingrimsson, E., Diaz, R. J., Patapoutian, A., *et al.* (1990) The *Drosophila* gene tailless is expressed at the embryonic termini and is a member of the steroid receptor superfamily. *Cell,* **62,** 151.
19. Oro, A. E., Ong, E. S., Margolis, J. S., Posakony, J. W., McKeown, M., *et al.* (1988) The *Drosophila* gene *knirps*-related is a member of the steroid receptor gene superfamily. *Nature,* **336,** 493.
20. Nauber, U., Pankratz, M. J., Kienlin, A., Seifert, E., Klemm, U., *et al.* (1988) Abdominal segmentation of the *Drosophila* embryo requires a hormone receptor-like protein encoded by the gap gene *knirps*. *Nature,* **336,** 489.
21. Henrich, V. C., Sliter, T. J., Lubahn, D. B., Macintyre, A., and Gilbert, L. I. (1990) A steroid/thyroid hormone receptor superfamily member in *Drosophila melanogaster* that shares extensive sequence similarity with a mammalian homologue. *Nucleic Acids Res.,* **18,** 4143.
22. Feigl, G., Gram, M., and Pongs, O. (1989) A member of the steroid hormone receptor gene family is expressed in the 20-OH-ecdysone inducible puff 75B in *Drosophila melanogaster*. *Nucleic Acids Res.,* **17,** 7167.
23. Krust, A., Green, S., Argos, P., Kumar, V., Walter, P., *et al.* (1986) The chicken oestrogen receptor sequence: homology with v-*erbA* and the human oestrogen and glucocorticoid receptors. *EMBO J.,* **5,** 891.
24. Picard, D. and Yamamoto, K. R. (1987) Two signals mediate hormone-dependent nuclear localization of the glucocorticoid receptor. *EMBO J.,* **6,** 3333.
25. Picard, D., Kumar, V., Chambon, P., and Yamamoto, K. R. (1990) Signal transduction

by steroid hormones: nuclear localization is differentially regulated in estrogen and glucocorticoid receptors. *Cell Regul.*, **1,** 291.
26. Ylikomi, T., Bocquel, M. T., Berry, M., Gronemeyer, H., and Chambon, P. (1992) Cooperation of proto-signals for nuclear accumulation of estrogen and progesterone receptors. *EMBO J.*, **11,** 3681.
27. Roberts, B. L., Richardson, W. D., and Smith, A. E. (1987) The effect of protein context on nuclear location signal function. *Cell,* **50,** 465.
28. Dang, C. V. and Lee, W. M. (1988) Identification of the human c-*myc* protein nuclear translocation signal. *Mol. Cell. Biol.*, **8,** 4048–4054.
29. Morin, N., Delsert, C., and Klessig, D. F. (1989) Nuclear localization of the adenovirus DNA-binding protein: requirement for two signals and complementation during viral infection. *Mol. Cell. Biol.*, **9,** 4372.
30. Eckhard, S. G., Milich, D. R., and McLachlan, A. (1991) Hepatitis B core antigen has two nuclear localization sequences in the arginine-rich carboxyl terminus. *J. Virol.*, **65,** 575.
31. Tratner, I. and Verma, I. M. (1991) Identification of a nuclear targeting sequence in the Fos protein. *Oncogene,* **6,** 2049.
32. Robbins J., Dilworth, S. M., Laskey, R. A., and Dingwall, C. (1991) Two interdependent basic domains in nucleoplasmin nuclear targeting sequence: identification of a class of bipartite nuclear targeting sequence. *Cell,* **64,** 615.
33. Davis, L. I. and Fink, G. R. (1990) The NUP1 gene encodes an essential component of the yeast nuclear pore complex. *Cell,* **61,** 965.
34. Nehrbass, U., Kern, H., Mutvei, A., Horstmann, H., Marshallsay, B., *et al.* (1990) NSP1: a yeast nuclear envelope protein localized at the nuclear pores exerts its essential function by its carboxyterminal domain. *Cell,* **61,** 979.
35. Gasc, J. M., Delahaye, F., and Baulieur, E. E. (1989) Compared intracellular localization of the glucocorticosteroid and progesterone receptors: an immunocytochemical study. *Exp. Cell Res.*, **181,** 492.
36. Wikstrom, A. C., Bakke, O., Okret, S., Bronnegard, M., and Gustafsson J. A. (1987) Intracellular localization of the glucocorticoid receptor: evidence for cytoplasmic and nuclear localization. *Endocrinology,* **120,** 1232.
37. Antakly, T., Thompson, E. B., and O'Donnell, D. (1989) Demonstration of the intracellular localization and up-regulation of glucocorticoid receptor by *in situ* hybridization and immunocytochemistry. *Cancer Res.*, **49,** 2230S.
38. Brink, M., Humbel, B. M., De Kloet, R. E., and Van Driel, R. (1992) The unliganded glucocorticoid receptor is localized in the nucleus, not in the cytoplasm. *Endocrinology,* **130,** 3575.
39. Klein-Hitpass, L., Ryffel, G. U., Heitlinger, E., and Cato, A. C. B. (1988) A 13 bp palindrome is a functional estrogen responsive element and interacts specifically with estrogen receptor. *Nucl. Acids Res.*, **16,** 647.
40. Martinez, A., Givel, F., and Wahli, W. (1987) The estrogen-responsive element as an inducible enhancer: DNA sequence requirements and conversion to a glucocorticoid responsive element. *EMBO J.*, **6,** 3719.
41. Klein-Hitpass, L., Schorpp, M., Wagner, U., and Ryffel, G. U. (1986) An estrogen-responsive element derived from the 5′ flanking region of the *Xenopus* vitellogenin A2 gene functions in transfected human cells. *Cell,* **46,** 1053.
42. Klock, G., Strähle, U., and Schütz, B. (1987) Oestrogen and glucocorticoid responsive elements are closely related but distinct. *Nature,* **329,** 734.

43. Nordeen, S. K., Suh, B. J., Kühnel, B., and Hutchinson III, C. A. (1990) Structural determinants of a glucocorticoid receptor recognition element. *Mol. Endocrinol.*, **4,** 1866.
44. Berry, M., Nunez, A. M., and Chambon, P. (1989) The estrogen-responsive element of the human pS2 gene is an imperfectly palindromic sequence. *Proc. Natl Acad. Sci. USA,* **86,** 1218.
45. Nunez, A. M., Berry, M., Imler, J. L., and Chambon, P. (1989) The 5′ flanking region of the pS2 gene contains a complex enhancer region responsive to oestrogens, epidermal growth factor, a tumor promoter (TPA), the c-Ha-ras oncoprotein and the c-jun protein. *EMBO J.*, **8,** 823.
46. Ham, J., Thompson, A., Needham, M., Webb, P., and Parker, M. (1988) Characterization of response elements for androgens, glucocorticoids and progestins in mouse mammary tumour virus. *Nucl. Acids Res.*, **16,** 5263.
47. Meyer, M. E., Pornon, A., Ji, J., Bocquel, M. T., Chambon, P., *et al.* (1990) Agonistic and antagonistic activities or RU486 on the functions of the human progesterone receptor. *EMBO J.*, **12,** 3923.
48. Arriza, J. L., Weinberger, C., Cerelli, G., Glaser, T. M., Handelin, B. L., *et al.* (1987) Cloning of human mineralocorticoid receptor complementary DNA: structural and functional kinship with the glucocorticoid receptor. *Science,* **237,** 268.
49. Simental, J. A., Sar, M., Lane, M. V., French, F. S., and Wilson, E. M. (1991) Transcriptional activation and nuclear targeting signals of the human androgen receptor. *J. Biol. Chem.*, **266,** 510.
50. Strähle, U., Boshart, M., Klock, G., Stewart, F., and Schütz, G. (1989) Glucocorticoid and progesterone-specific effects are determined by differential expression of the respective hormone receptors. *Nature,* **339,** 629.
51. de The, H., Vivanco-Ruiz, M. d. M., Tiollais, P., Stunnenberg, H., and Dejean, A. (1990) Identification of a retinoic acid response element in the retinoic acid receptor β gene. *Nature,* **343,** 177.
52. Sucov, H. M., Murakami, K. K., and Evans, R. M. (1990) Characterization of an autoregulated response element in the mouse retinoic acid receptor type β gene. *Proc. Natl Acad. Sci. USA,* **87,** 5392.
53. Hoffmann, B., Lehmann, J. M., Zhang, X. K., Hermann, T., Husmann, M., *et al.* (1990) A retinoic acid receptor-specific element controls the retinoic acid receptor–β promoter. *Mol. Endocrinol.*, **4,** 1727.
54. Mader, S., Leroy, P., Chen, J. Y., and Chambon, P. (1993) Multiple parameters control the selectivity of nuclear receptors for their response element. *J. Biol. Chem.*, **268,** 591.
55. Kumar, V. and Chambon, P. (1988) The estrogen receptor binds tightly to its responsive element as a ligand-induced homodimer. *Cell,* **55,** 145.
56. Fawell, S. E., Lees, J. A., White, R., and Parker, M. G. (1990) Characterization and colocalization of steroid binding and dimerization activities in the mouse estrogen receptor. *Cell,* **60,** 953.
57. Danielian, P. S., White, R., Lees, J. A., and Parker, M. G. (1992) Identification of a conserved region required for hormone dependent transcriptional activation by steroid hormone receptors. *EMBO J.*, **11,** 1025.
58. Lees, J. A., Fawell, S. E., White, R., and Parker, M. G. (1990) A 22-amino-acid peptide restores DNA-binding activity to dimerization defective mutants of the estrogen receptor. *Mol. Cell. Biol.*, **10,** 5529.
59. Eriksson, P. and Wrange, O. (1990) Protein–protein contacts in the glucocorticoid

receptor homodimer influence its DNA binding properties. *J. Biol. Chem.*, **265,** 3535.
60. Luisi, B. F., Xu, W. X., Otwinowski, Z., Freedman, L. P., Yamamoto, K. R., *et al.* (1991) Crystallographic analysis of the interaction of the glucocorticoid receptor with DNA. *Nature,* **352,** 497.
61. Härd, T., Kellenbach, E., Boelens, R., Maler, B. A., Dahlman, K., *et al.* (1990) Solution structure of the glucocorticoid receptor DNA-binding domain. *Science,* **249,** 157.
62. Schwabe, J. W. R., Neuhaus, D. D., and Rhodes, D. (1990) Solution structure of the DNA-binding domain of the oestrogen receptor. *Nature,* **348,** 458.
63. Dahlman-Wright, K., Siltala-Roos, H., Carlstedt-Duke, J., and Gustafsson, J. A. (1990) Protein–protein interactions facilitate DNA binding by the glucocorticoid receptor DNA binding domain. *J. Biol. Chem.,* **265,** 14 030.
64. Dahlman-Wright, K., Wright, A., Gustafsson, J. A., and Carlstedt-Duke, J. (1991) Interaction of the glucocorticoid receptor DNA-binding domain with DNA as a dimer is mediated by a short segment of five amino acids. *J. Biol. Chem.,* **266,** 3107.
65. Walter, P., Green, S., Greene, G., Krust, A., Bornert, J. M., *et al.* (1985) Cloning of the human estrogen receptor cDNA. *Proc. Natl Acad. Sci. USA,* **82,** 7889.
66. Green, S., Walter, P., Kumar, V., Krust, A., Bornert, J. M., *et al.* (1986) Human oestrogen receptor cDNA: sequence, expression and homology to v-*erb*-A. *Nature,* **320,** 134.
67. Tora, L., Mullick, A., Metzger, D., Ponglikitmongkol, M., Park, I., *et al.* (1989) The cloned human oestrogen receptor contains a mutation which alters its hormone binding properties. *EMBO J.,* **8,** 1981.
68. Sabbah, M., Gouilleux, F., Sola, B., Redeuilh, G., and Baulieu, E. E. (1991). Structural differences between the hormone and antihormone estrogen receptor complexes bound to the hormone response element. *Proc. Natl. Acad. Sci,* **88,** 390.
69. Chambraud, B., Berry, M., Redeuilh, G., Chambon, P., and Baulieu, E. E. (1990) Several regions of human estrogen receptor are involved in the formation of receptor-heat shock protein 90 complexes. *J. Biol. Chem.,* **265,** 20 686.
70. Sabbah, M., Redeuilh, G., and Baulieu, E. E. (1989) Subunit composition of the estrogen receptor. *J. Biol. Chem.,* **264,** 2397.
71. Schrader, W. T. and O'Malley, B. W. (1972) Progesterone-binding components of chick oviduct. *J. Biol. Chem.,* **217,** 51.
72. Gronemeyer, H., Turcotte, B., Quirin-Stricker, C., Bocquel, M. T., Meyer, M. E., *et al.* (1987) The chicken progesterone receptor: sequence, expression and functional analysis. *EMBO J.,* **6,** 3985.
73. Jeltsch, J. M., Turcotte, B., Garnier, J. M., Lerouge, T., and Krozowski, H., *et al.* (1990) Characterization of multiple mRNAs originating from the chicken progesterone receptor gene. *J. Biol. Chem.,* **265,** 3967.
74. Kastner, P., Krust, A., Turcotte, B., Stropp, U., Tora, L., *et al.* (1990) Two distinct estrogen-regulated promoters generate transcripts encoding the two functionally different human progesterone receptor forms A and B. *EMBO J.,* **5,** 1603.
75. Kastner, P., Bocquel, M. T., Turcotte, B., Garnier, J. M., Horwitz, K. B., *et al.* (1990) Transient expression of human and chicken progesterone receptors does not support alternative translational initiation from a single mRNA as the mechanism generating two receptor isoforms. *J. Biol. Chem.,* **265,** 12 163.
76. Conneely, O. M., Maxwell, B. L., Toft, D. O., Schrader, W. T., and O'Malley, B. W. (1987) The A and B forms of the chicken progesterone receptor arise by alternative initiation of translation of a unique mRNA. *Biochem. Biophys. Res. Commun.,* **149,** 493.
77. Conneely, O. M., Kettelberger, D. M., Tsai, M. J., Schrader, W. T., and O'Malley,

B. W. (1989) The chicken progesterone receptor A and B isoforms are products of an alternate translation initiation event. *J. Biol. Chem.*, **264,** 14 062.

78. El-Ashry, D., Onate, S. A., Nordeen, S. K., and Edwards, D. P. (1989) Human progesterone receptor complexed with the antagonist RU486 binds to a hormone response element in a structurally altered form. *Mol. Endocrinol.*, **3,** 1545.
79. DeMarzo, A. M., Beck, C. A., Onate, S. A., and Edwards, D. P. (1991) Dimerization of mammalian progesterone receptors occurs in the absence of DNA and is related to the release of the 90-kDa heat shock protein. *Proc. Natl Acad. Sci. USA*, **88,** 72.
80. Bagchi, M. K., Elliston, J. F., Tsai, S. Y., Edwards, D. P., Tsai, M. J., *et al.* (1988) Steroid hormone-dependent interaction of human progesterone receptor with its target enhancer element. *Mol. Endocrinol.*, **2,** 1221.
81. Zhang, X. K., Lehmann, J., Hoffmann, B., Dawson, M. I., Cameron, J., *et al.* (1992) Homodimer formation of retinoid X receptor induced by 9-*cis* retinoic acid. *Nature*, **358,** 587.
82. Willmann, T. and Beato, M. (1986) Steroid-free glucocorticoid receptor binds specifically to mouse mammary tumour virus DNA. *Nature*, **324,** 688.
83. Rodriguez, R., Carson, M. A., Weigel, N. L., O'Malley, B. W., and Schrader, W. T. (1989) Hormone-induced changes in the *in vitro* DNA-binding activity of the chicken progesterone receptor. *Mol. Endocrinol.*, **3,** 356.
84. Bailly, A., Le Page, C., Rauch, M., and Milgrom, E. (1986) Sequence-specific DNA binding of the progesterone receptor to the uteroglobin gene: effects of hormone, antihormone and receptor phosphorylation. *EMBO J.*, **5,** 3235.
85. Rodriguez, R., Weigel, N. L., O'Malley, B. W., and Schrader, W. T. (1990) Dimerization of the chicken progesterone receptor *in vitro* can occur in the absence of hormone and DNA. *Mol. Endocrinol.*, **4,** 1782.
86. Guiochon-Mantel, A., Loosfelt, H., Lescop, P., Star, S., Atger, M., *et al.* (1989) Mechanisms of nuclear localization of the progesterone receptor: evidence for interaction between monomers. *Cell*, **57,** 1147.
87. Tsai, S. Y., Carlstedt-Duke, J., Weigel, N. L., Dahlman, K., Gustafsson, J. A., *et al.* (1989) Molecular interactions of steroid hormone receptor with its enhancer element: evidence for receptor dimer formation. *Cell*, **55,** 361.
88. Turcotte, B., Meyer, M. E., Bocquel, M. T., Bélanger, L., and Chambon, P. (1990) Repression of the alpha-fetoprotein gene promoter by progesterone and chimeric receptors in the presence of hormones and anti-hormones. *Mol. Cell. Biol.*, **10,** 5002.
89. Eul, J., Meyer, M. E., Tora, L., Bocquel, M. T., Quirin-Stricker, C., *et al.* (1989) Expression of active hormone and DNA-binding domains of the chicken progesterone receptor in *E. coli*. *EMBO J.*, **8,** 83.
90. Glass, C. K., Holloway, J. M., Devary, O. V., and Rosenfeld, M. G. (1988) The thyroid hormone receptor binds with opposite transcriptional effects to a common sequence motif in thryoid hormone and estrogen response elements. *Cell*, **54,** 313.
91. Umesono, K., Murakami, K. K., Thompson, C. C., and Evans, R. M. (1991) Direct repeats as selective response elements for the thyroid hormone, retinoic acid, and vitamin D_3 receptors. *Cell*, **65,** 1255.
92. Näär, A. M., Boutin, J. M., Lipkin, S. M., Yu, V. C., Holloway, J. M., *et al.* (1991) The orientation and spacing of core DNA-binding motifs dictate selective transcriptional responses to three nuclear receptors. *Cell*, **65,** 1276.
93. Mangelsdorf, D. J., Umesono, K., Kliewer, S. A., Borgmeyer, U., Ong, E. S., *et al.* (1991) A direct repeat in the cellular retinol-binding protein type II gene confers differential regulation by RXR and RAR. *Cell*, **6,** 555.

94. Smith, W. C., Nakshatri, H., Leroy, P. Rees, J., and Chambon, P. (1991) A retinoic acid response element is present in the mouse cellular retinol binding protein I (mCRBPI) promoter. *EMBO J.*, **10,** 2223.
95. Vasios, G. W., Gold, J. D., Petkovitch, M., Chambon, P., and Gudas, L. J. (1989) A retinoic acid responsive element is present in the 5′ flanking region of the laminin B1 gene. *Proc. Natl Acad. Sci. USA,* **86,** 9099.
96. Durand, B., Saunders, M., Leroy, P., Leid, M., and Chambon, P. (1992) All-trans and 9-cis retinoic acid induction of mouse CRABPII gene transcription is mediated by RAR/RXR heterodimers. *Cell,* **71,** 73.
97. Rottmann, J. N., Widom, R. L., Nadal-Ginard, B., Mahdavi, V., and Karathanasis, S. K. (1991) A retinoic acid-responsive element in the apolipoprotein AI gene distinguishes between two different retinoic acid response pathways. *Mol. Cell. Biol.*, **11,** 3814.
98. Lucas, P. C., O'Brien, R., Mitchell, J. A., Davis, C. M., Imai, E., *et al.* (1991) A retinoic acid response element is part of a pleiotropic domain in the phosphoenolpyruvate carboxykinase gene. *Proc. Natl Acad. Sci. USA,* **88,** 2184.
99. Richard, S. and Zingg, H. H. (1991) Identification of a retinoic acid response element in the human oxytocin promoter. *J. Biol. Chem.*, **266,** 21 428.
100. Leid, M., Kastner, P., Lyons, R., Nakshatri, H., Saunders, M., *et al.* (1992) Purification, cloning and RXR identity of the HeLa cell factor with which RAR or TR heterodimerizes to efficiently bind target sequences. *Cell,* **68,** 377.
101. Yu, V. C., Delsert, C., Andersen, B., Holloway, J. M., Devary, O. V., *et al.* (1992) RXR beta: a coregulator that enhances binding of retinoic acid, thyroid hormone, and vitamin D receptors to their cognate response elements. *Cell,* **67,** 1251.
102. Zhang, X. K., Hoffmann, B., Tran, P. B. V., Graupner, G., and Pfahl, M. (1992) Retinoid X receptoris an auxiliary protein for thyroid hormone and retinoic acid receptors. *Nature,* **355,** 441.
103. Green, S. and Chambon, P. (1987) Oestradiol induction of a glucocorticoid-responsive gene by a chimeric receptor. *Nature,* **325,** 75.
104. Petkovich, M., Brand, N. J., Krust, A., and Chambon, P. (1987) A human retinoic acid receptor which belongs to the family of nuclear receptors. *Nature,* **330,** 444.
105. Green, S., Kumar, V., Theulaz, I., Wahli, W., and Chambon, P. (1988) The *N*-terminal DNA-binding 'zinc-finger' of the oestrogen and glucocorticoid receptors determines target gene specificity. *EMBO J.*, **7,** 3037.
106. Mader, S., Kumar, V., de Verneuil, H., and Chambon, P. (1989) Three amio acids of the oestrogen receptor are essential to its ability to distinguish an oestrogen from a glucocorticoid-responsive element. *Nature,* **338,** 271.
107. Umesono, K. and Evans, R. M. (1989) Determinants of target gene specificity for steroid/thyroid hormone receptors. *Cell,* **57,** 1139.
108. Danielsen, M., Hinck, L., and Ringold, G. M. (1989) Two amio acids within the knuckle of the first zinc finger specific DNA response element activation by the glucocorticoid receptor. *Cell,* **57,** 1131.
109. Hirst, M. A., Hinck, L., Danielsen, M., and Ringold, G. M. (1992) Discrimination of DNA response elements for thyroid hormone and estrogen is dependent on dimerization of receptor DNA binding domains. *Proc. Natl Acad. Sci. USA,* **89,** 5527.
110. Härd, T., Kellenbach, E., Boelens, R., Kapstein, R., Dahlman, K., *et al.* (1990) ^{1}HNMR studies of the glucocorticoid receptor DNA-binding domain sequential assignments and identification of secondary structure elements. *Biochemistry,* **29,** 9015.
111. Archer, T. K., Hager, G. L., and Omichinski, J. G. (1990) Sequence-specific DNA

binding by glucocorticoid receptor 'zinc finger peptides'. *Proc. Natl Acad. Sci. USA*, **87,** 7560.

112. Webster, N. J. G., Green, S., Jin, J. R., and Chambon, P. (1988) The hormone-binding domains of the estrogen and glucocorticoid receptors contain an inducible transcription activation function. *Cell*, **54,** 199.
113. Tora, L., White, J., Brou, C., Tasset, D., Webster, N., *et al.* (1989) The human estrogen receptor has two independent nonacidic transcriptional activation functions. *Cell*, **59,** 477.
114. Kumar, V., Green, S., Stack, G., Berry, M., Jin, J. R., *et al.* (1987) Functional domains of the human estrogen receptor. *Cell*, **51,** 941.
115. Hollenberg, S. M. and Evans, R. M. (1988) Multiple and cooperative transactivation domains of the human glucocorticoid receptor. *Cell*, **55,** 899.
116. Webster, N. J. G., Green, S., Tasset, D., Ponglikitmongkol, M., and Chambon, P. (1989) The transcriptional activation function located in the hormone-binding domain of the human oestrogen receptor is not encoded in a single exon. *EMBO J.*, **8,** 1441.
117. Meyer, M. E., Quirin-Stricker, C., Lerouge, T., Bocquel, M. T., and Gronemeyer, H. (1992) A limiting factor mediates the differential activation of promoters by the human progesterone receptor isoforms. *J. Biol. Chem.*, **267,** 10882.
118. Ptashne, M. and Gann, A. F. (1990) Activators and targets. *Nature*, **346,** 329.
119. Godowski, P. J., Rusconi, S., Miesfeld, R., and Yamamoto, K. R. (1987) Glucocorticoid receptor mutants that are constitutive activators of transcriptional enhancement. *Nature*, **325,** 365.
120. Carson, M. A., Tsai, M. J., Conneely, O. M., Maxwell, B. L., Clark, J. H., *et al.* (1987) Structure–function properties of the chicken progesterone receptor A synthesized from complementary deoxyribonucleic acid. *Mol. Endocrinol.*, **1,** 791.
121. Bocquel, M. T., Kumar, V., Stricker, C., Chambon, P., and Gronemeyer, H. (1989) The contribution of the *N*- and *C*-terminal regions of steroid receptors to activation of transcription is both receptor and cell-specific. *Nucl. Acids Res.*, **17,** 2581.
122. Berry, M., Metzger, D., and Chambon, P. (1990) Role of the two activating domains of the oestrogen receptor in the cell-type and promoter-context dependent agonistic activity of the anti-oestrogen 4-hydroxytamoxifen. *EMBO J.*, **9,** 2811.
123. Metzger, D., White, J. H., and Chambon, P. (1988) The human oestrogen receptor functions in yeast. *Nature*, **334,** 31.
124. White, J. H., Metzger, D., and Chambon, P. (1988) Expression and function of the human estrogen receptor in yeast. *Cold Spring Harb. Symp. Quant. Biol.*, **LIII,** 819.
125. Gill, G. and Ptashne, M. (1988) Negative effect of the transcriptional activator GAL4. *Nature*, **334,** 721.
126. Meyer, M. E., Gronemeyer, H., Turcotte, B., Bocquel, M. T., Tasset, D., *et al.* (1989) Steroid hormone receptors compete for factors that mediate their enhancer function. *Cell*, **57,** 433.
127. Tasset, D., Tora, L., Fromental, C., Scheer, E., and Chambon, P. (1990) Distinct classes of transcriptional activating domains function by different mechanisms. *Cell*, **62,** 1177.
128. Ptashne, M. (1988) How eukaryotic transcriptional activators work. *Nature*, **335,** 683.
129. Shemshedini, L., Ji, J., Brou, C., Chambon, P., and Gronemeyer, H. (1992) *In vitro* activity of the transcription activation functions of the progesterone receptor. *J. Biol. Chem.*, **267,** 1834.
130. Pugh, B. F. and Tjian, R. (1990) Mechanism of transcriptional activation by Sp1: evidence for coativators. *Cell*, **61,** 1187.

131. White, J. H., Brou, C., Wu, J., Burton, N., Egly, J. M., *et al.* (1991) Evidence for a factor required for transcriptional stimulation by the chimeric acidic activator GAL-VP16 in HeLa cell extracts. *Proc. Natl Acad. Sci. USA,* **88,** 7674.
132. Dynlacht, B. D., Hoey, T., and Tjian, R. (1991) Isolation of coactivators associated with the TATA-binding protein that mediate transcriptional activation. *Cell,* **66,** 563.
133. Tanese, N., Pugh, B. F., and Tjian, R. (1991) Coactivators for a proline-rich activator purified from the multisubunit human TFIID complex. *Genes Dev.,* **5,** 2212.
134. Gronemeyer, H., Meyer, M. E., Bocquel, M. T., Kastner, P., Turcotte, B., *et al.* (1991) Progestin receptors: isoforms and antihormone action. *J. Steroid Biochem. Mol. Biol.,* **40,** 271.
135. Leid, M., Kastner, P., and Chambon, P. (1992) Multiplicity generates diversity in the retinoic acid signalling pathways. *Trends Biochem. Sci.,* **17,** 427.
136. Nagpal, S., Saunders, M., Kastner, P., Durand, B., Nakshatri, H., *et al.* (1992) Promoter-context and response element dependent specificity of the transcriptional activation and modulating functions of retinoic acid receptors (RARs and RXRs). *Cell,* in press.
137. Peterson, C. L. and Herskowitz, I. (1992) Characterization of the yeast SWI1, SWI2, and SWI3 genes, which encode a global activator of transcription. Cell 68: 573–583.
138. Tamkun, J. W., Scott, M. P., Kissinger, M., Pattatucci, A. M., Kaufman, T. C., *et al.* (1992) Brahma: a regulator of *Drosophila* homeotic genes structurally related to the yeast transcriptional activator SNF2/SWI2. *Cell,* **68,** 561.
139. Zhu, H. and Prywes, R. (1992) Identification of a coactivator that increases activation of transcription by serum response factor and GAL4–VP16 *in vitro. Proc. Natl Acad. Sci. USA,* **89,** 5291.
140. Inostroza, J. A., Mermelstein, F. H., Ha, I., Lane, W. S., and Reinberg, D. (1992) Dr1, a TATA-binding protein-associated phosphoprotein and inhibitor of class II gene transcription. *Cell,* **70,** 477.
141. Lazar, M. A. (1993) Thyroid hormone receptors: Multiple forms, multiple possibilities. *Endocrine Reviews,* **14,** 184.

6 Repression of transcription by nuclear receptors

JACQUES DROUIN

1. Introduction

We now have a fairly detailed understanding of the mechanisms by which nuclear receptors recognize their target genes and activate transcription (Chapters 3 and 5). Much less can be said about the ways in which these receptors repress transcription of other target genes. At the onset, it is interesting to note that early studies of the action of steroids on their target tissues indicated, for example, that glucocorticoids decrease the level of about the same number of liver proteins as they stimulate. In all likelihood, many of these changes will be ascribed to transcriptional regulation.

Studies of glucocorticoid-inducible genes such as mouse mammary tumour virus or hepatic tyrosine aminotransferase provided initial insight into the general mechanism of receptor-dependent activation of transcription (1–3). The availability of purified glucocorticoid receptor (GR) and of cloned complementary DNAs (cDNAs) for GR led t the demonstration that the genetically defined glucocorticoid response element (GRE) is a binding site for GR and that this binding site is sufficient to confer hormone responsiveness. The identification of well-conserved palindromic GREs in the promoters or enhancers of many glucocorticoid-inducible genes led to the simple concept that these hormone response elements (HREs) are like any other gene regulatory element, with the particularity that the activity of the transcription factor is ligand (hormone) dependent.

In contrast, a simple unique model could not be formulated to account for receptor-dependent repression of target genes. Indeed, it was not possible even to define a consensus negative GRE (nGRE) or negative thyroid hormone response element (nTRE); furthermore, it appeared in many cases that receptor binding to DNA was not in itself sufficient to confer hormone-dependent repression. Instead, repression appeared to involve the interaction of receptor with various proteins, either free or bound to DNA. General models are even more difficult to develop for the action of other receptors, for example the thyroid hormone receptor (TR), which appear to have much less well-conserved response elements, both positive and negative (4).

To contrast the variety of mechanisms that appear to be involved in transcriptional repression with the apparent simplicity of those involved in transcriptional activation by nuclear receptors, this review has been structured according to models of repression which provide a conceptual framework for discussion. There seems to be no single mechanism of repression that applies to all receptors; instead, receptors appear to use different strategies for repression of different target genes.

2. Mechanisms of transcriptional repression

A brief review of the various mechanisms that have been proposed for transcription repression is presented below (Figs 1 and 2) within the broader context of

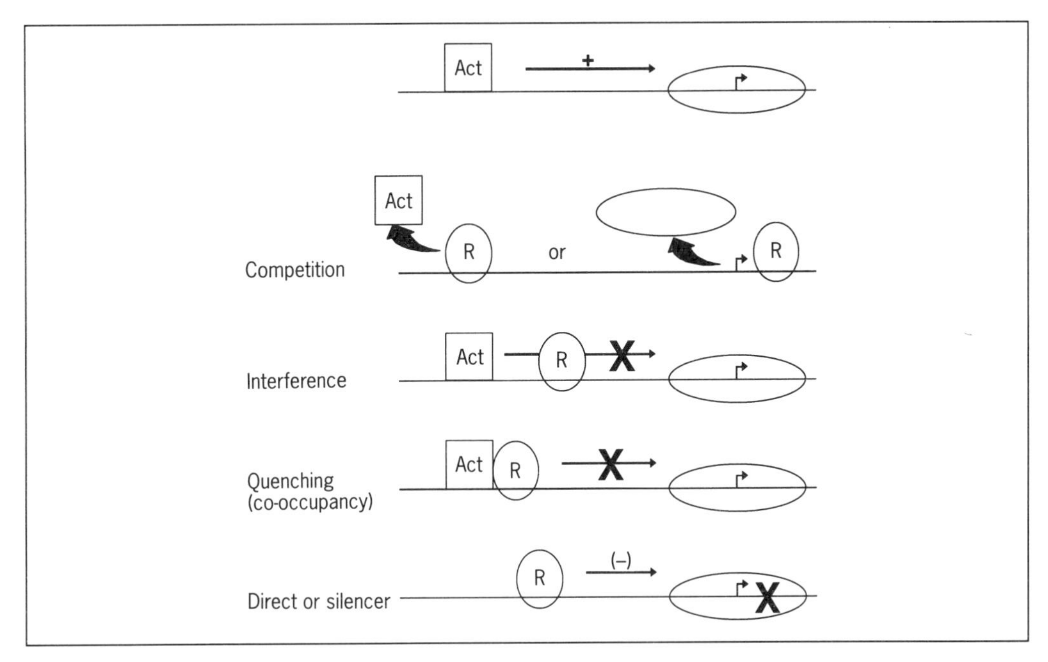

Fig. 1 Repression resulting from repressor (receptor) binding to DNA. The diagrams illustrate the respective roles of activator (Act) of transcription and repressor (R) in different mechanisms of repression. In each of these models, R binds to a specific DNA sequence which constitutes at least part of the negative hormone response element

transcriptional repression (5). Repression may result from binding of the repressor to DNA (Fig. 1) or to an activator of transcription without making direct contact with DNA (Fig. 2). For the purpose of this review, the repressor is a ligand-activated nuclear receptor that is devoid of activity is absence of hormone. Up till now, this framework has been adequate to describe the many actions of nuclear receptors such as GR. However, other receptors like TR appear to have the capacity to be transcriptionally active in both their liganded and unliganded state, thus providing other mechanisms for repression; these will also be considered below.

2.1 Competition for overlapping binding sites

Promoter occlusion or factor displacement is the classical mechanism of repression, as it was first described for the *lac* operon in which repressor displaces an activating protein from overlapping binding sites on the DNA (6). This model has been proposed to account for repression by many nuclear receptors, but recent work has indicated that this mechanism does not appear to be involved in most cases.

2.2 Null binding site

As nuclear receptors are potent activators of transcription, their action as DNA-bound transcriptional repressors requires in most cases that their positive *trans*-activation functions be silent in this context. This can be achieved by interaction with a specific DNA sequence that binds the receptor efficiently but which does not allow positive *trans*-activation in whatever context. Thus, such a null binding site would behave as a true negative HRE (nHRE), because repression would be the only property of this site. The nGRE identified in the proximal promoter of the pro-opiomelanocortin (POMC) gene provides an example of a null binding site (7). Some HREs, like the oestrogen response element (ERE), could behave as a positive response element for one receptor (ER in this case) and as a null binding site for another receptor, like TR (8).

2.3 Silencer or direct repressor element

This mechanism of repression is the mirror image of the positive HRE. Elements acting in this way have been called silencers because they are in themselves sufficient to repress transcription in a position- and orientation-independent manner. A direct repressor element would repress transcription as a result of repressor binding to the element independently of other proteins and by a direct effect on the transcription machinery. An example of this mechanism is provided by the *Escherichia coli* regulatory protein *mer*R which represses transcription by tight binding with operator DNA: this interaction prevents formation of open complexes as a result of constraint on DNA conformation (9). In this case, binding of mercury to *mer*R results in a loosening of the interaction with DNA, and as a result the protein which behaved as a repressor now acts as an activator of transcription. As for nuclear receptors, the best documented example of dominant repressor element and factors is the TRE and unliganded TR or its oncogenic mutant, the v-*erb*A gene product (10).

2.4 Repression by interference

Repression by interference would result from binding of repressor to an element that is not a binding site for other factors and which is not sufficient on its own to confer repression, unlike a silencer. The DNA-bound repressor would interfere with the interactions of other proteins or factors bound to DNA upstream and/or downstream of itself. For example, it could hinder interactions between upstream promoter-bound factors and basic transcription complex (Fig. 1).

2.5 Repression by quenching

Another mechanism of repression involves the binding of an activator protein and a repressor to contiguous binding sites. Interaction between the DNA-bound proteins results in dampening or 'quenching' of the *trans*-activation function of the activator. The activation of yeast-specific genes by the MCM1 transcription factor appears to be repressed by the mating type protein α_2 by a squelching mechanism (11). Indeed, the two proteins bind co-operatively to neighbouring sites but the presence of α_2 prevents *trans*-activation by MCM1. This mechanism has been proposed to account for glucocorticoid repression of the proliferin gene (12).

2.6 Repression by squelching

Squelching was originally described as an artefact of gene transfer experiments in which activators were overexpressed (13). Overexpression of one activator resulted in repression of a target gene for a second activator and this effect did not require the DNA-binding domain of the first overexpressed activator. Thus, it was proposed that interactions between the two proteins titrated cellular activators away from their DNA targets—hence the term 'squelching' (Fig. 2). This mechanism requires only the intact interaction interfaces of the two proteins and does not require the repressor to bind to DNA. This model has been proposed to explain the cross-talk between various signal transduction pathways involving nuclear

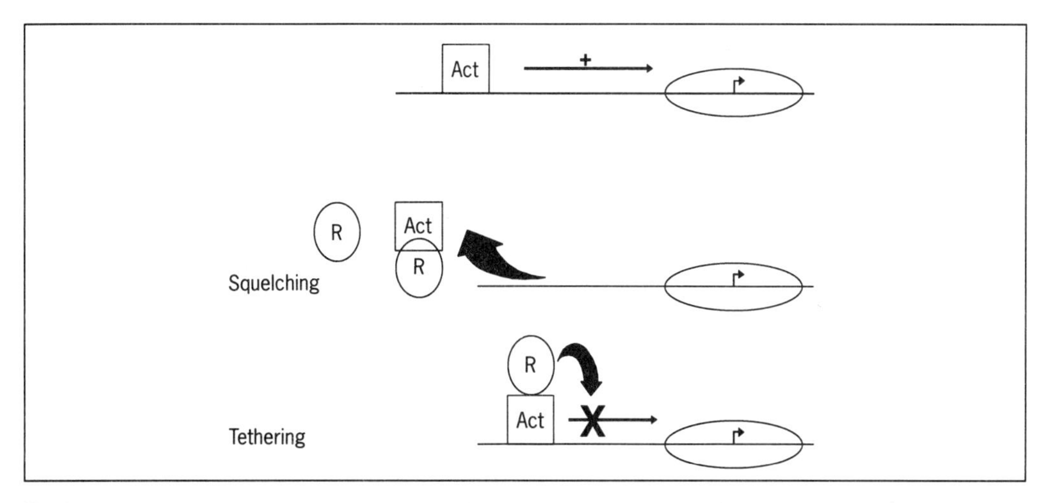

Fig. 2 Repression without direct DNA binding of repressor. Act, activator of transcription; R, repressor

receptors and members of the leucine zipper transcription factor family, such as AP-1 (14, 15).

2.7 Repression by tethering

Another mechanism was also proposed recently to explain repression by nuclear receptors in the absence of direct interaction between receptor and DNA. In this 'tethering' mechanism, the receptor binds an activator which is itself bound to a target DNA sequence. Binding of the repressing receptor hinders the *trans*-activation function of the activator. This mechanism was proposed for glucocorticoid repression of the collagenase gene (16, 17).

3. Negative hormone response elements

Attempts to derive consensus nHREs for any nuclear receptor have not been very successful. This is due to the fact that a relatively small number of repressed genes have been studied in detail and that a variety of mechanisms appear to be involved in repression of these genes. Thus, for any mechanism and receptor, only a few examples are available to develop a clear consensus. Be that as it may, different DNA sequences have been proposed as candidates for nuclear receptor-mediated repression of transcription: these include binding sites for nuclear receptors and binding sites for transcription factors of other structural families, notably the AP-1 family of leucine zipper transcription factors. This latter class of nHREs is discussed below in Section 9 and is not discussed further here because receptor-dependent repression does not involve direct interaction of receptor with DNA.

3.1 Negative glucocorticoid response elements

3.1.1 Pro-opiomelanocortin

Glucocorticoids are synthesized in the adrenals in response to adrenocorticotrophic hormone (ACTH) which derives from processing of POMC. Glucocorticoids exert a negative feedback effect on anterior pituitary POMC transcription rate (18–20). A GR-binding site has been identified in the proximal region of the POMC promoter, and repression of this promoter by glucocorticoids appears to depend on the presence of this site (21, 22). The affinity of GR for this site appears to be similar to that of receptor for positive GREs despite limited homology between the two sites. The sequence of this site is compared to the consensus GRE in Figure 3. Detailed analysis of the interaction between purified rat liver GR and nGRE indicated that three moieties of the receptor form a complex with the POMC nGRE (7), in contrast with the GRE which binds a GR homodimer (23–25). Indeed, each half site of the GRE palindrome interacts with a GR moiety so that all interactions between GR homodimer and GRE take place in two consecutive major grooves accessible on the same side of the double helix. In the nGRE complex, a GR homodimer has a similar

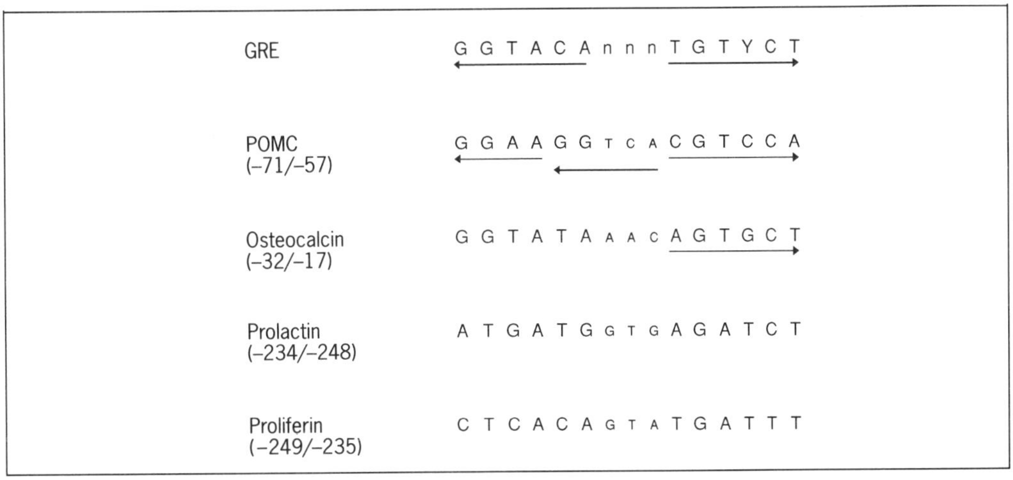

GRE	G G T A C A n n n T G T Y C T
POMC (−71/−57)	G G A A G G t c a C G T C C A
Osteocalcin (−32/−17)	G G T A T A a a c A G T G C T
Prolactin (−234/−248)	A T G A T G g t g A G A T C T
Proliferin (−249/−235)	C T C A C A g t a T G A T T T

Fig. 3 Negative glucocorticoid response elements (nGRE). The sequence of various glucocorticoid receptor binding sites identified in nGREs is compared with the consensus GRE. The following nGRE sequences are shown: pro-opiomelanocortin (POMC) (7, 21), osteocalcin (28), prolactin (29), and proliferin (12, 30)

association with one side of the sequence and a third GR molecule binds to the opposite side of the DNA in the major groove, which is between the sites of interaction of the first two GR molecules. Thus, the POMC nGRE is sandwiched by three GR molecules which do not constitute a true trimer (Fig. 4). Interestingly, a short 85-amino-acid fragment of GR, consisting only of the zinc-finger region (26), discriminates similarly hGRE from GRE, such that it forms tri-molecular complexes with nGRE and homodimeric complexes with GRE (7). The ability to form unique complexes with the POMC nGRE is therefore an intrinsic property of the DNA-binding domain. Thus, the nGRE is a high-affinity site for GR, which prevents the activity of receptor *trans*-activation domains and behaves as a null binding site as it appeared to be insufficient on its own to confer either glucocorticoid inducibility or repression.

3.1.2 Osteocalcin

Glucocorticoids repress vitamin D-induced osteocalcin gene expression (27). A putative nGRE has been identified in the promoter of this gene which overlaps the TATA box (28). Like the POMC nGRE, this GR-binding site has a relatively high affinity for GR, despite its limited homology to consensus GREs (see Fig. 3). The stereochemistry of this interaction and the transcriptional properties of these sequencees have not yet been fully assessed. However, it has been proposed that binding of GR to this site might impede function of the overlapping TATA box.

3.1.3 Prolactin and proliferin

Other GR-binding sites that have been implicated in negative response to gluco-

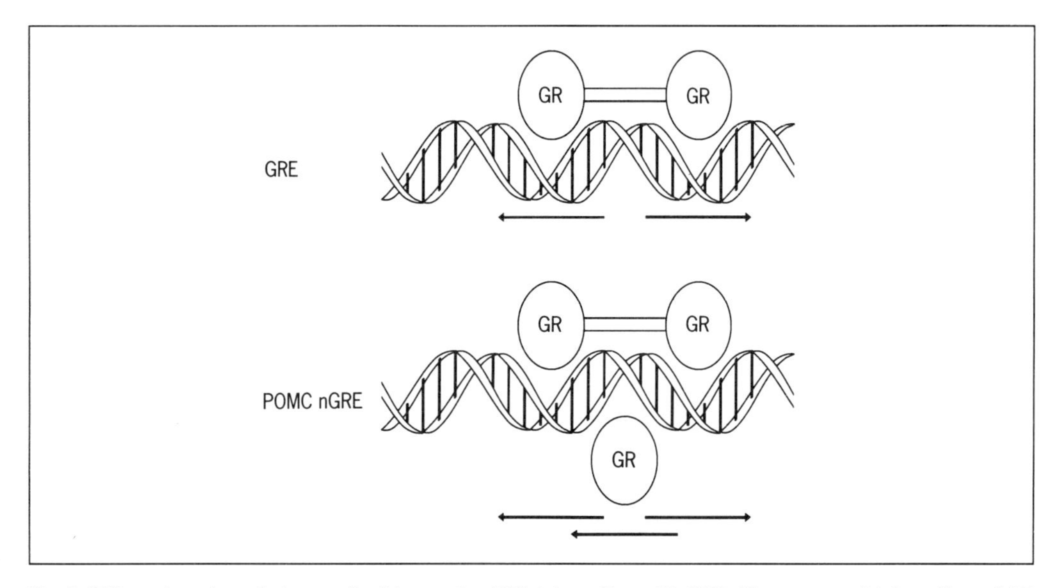

Fig. 4 Different modes of glucocorticoid receptor (GR) interaction with DNA. The proposed interaction of GR homodimers with glucocorticoid response element (GRE) is contrasted with the proposed formation of a unique complex with the pro-opiomelanocortin (POMC) negative GRE which contains three molecules of GR per binding site (7). Arrows indicate sequences related to the hexanucleotide TGTYCT which constitutes each half of the GRE palindrome

corticoids are even more divergent from the consensus GRE than the POMC and osteocalcin nGREs described above. They include GR-binding sites present in the bovine prolactin gene promoter (29) and a binding site in the proliferin gene (12, 30). In both cases, the affinity of GR for its target sequence is significantly lower than for GRE, and these sites have skeletal homology with consensus GRE—so much so that, in the absence of dimethylsulfate (DMS) interference data, it is difficult to be sure of the exact position of the binding site within the regulatory element. It has been proposed that glucocorticoids repress prolactin transcription by displacement of a positive factor that binds to GR binding site 3 (29), whereas the action of GR on the proliferin nGRE appears to involve interaction with other factors leading to the quenching of their activity (12).

3.1.4 Glycoprotein hormone α-subunit

It has been proposed that glucocorticoids repress transcription of the glycoprotein hormone α-subunit (α-CG or α-chorionic gonzadotropin) gene by receptor interaction with GR-binding sites present in the promoter of that gene and displacement of the cyclic AMP-response element binding protein (CREB) transcription factor (31). However, more recent work has indicated that receptor interaction with DNA might not be required for repression in this system and that repression might be due to direct interactions between GR and CREB by squelching or tethering mechanisms (32, 33).

3.2 Negative thyroid hormone response elements

Various putative nTREs have been identified in thyroid hormone-repressed genes, like those encoding the α- and β-subunits of thyroid-stimulating hormone (TSH). These nTRE elements are all *in vitro* binding sites for TR; however, they contain different arrangements of the hexanucleotide motif GGGTCA, so that it is impossible to derive any kind of consensus nTRE. In contrast to GREs, a similar situation exits for positive TREs, which are found in many different arrangements in the promoters of positively regulated genes (4). This diversity of target DNA sequences might reflect the complexity of the TR family of receptors, as well as the complexity of other receptor families such as retinoid X receptor (RXR) which can form heterodimers with TR (34–40). Given that many heterodimers appear to have higher affinity for DNA than TR monomers or homodimers, it is possible that different heterodimers might exhibit binding preference for some target sequences. Thus, a specific nTRE might become active only when its cognate receptors are expressed. This hypothesis awaits experimental confirmation but it could explain the wide array of TRE and nTRE sequences.

3.2.1 TSH β-subunit

TSH, which is formed by the association of its α- and β-subunits, stimulates thyroid hormone release by the thyroid gland. Thyroid hormone exerts a negative feedback to repress transcription of the TSH α- and β-subunit genes in thyrotrope cells of the pituitary (41, 42). The TSH-β promoter contains *in vitro* binding sites for TR in the vicinity of the site of transcription initiation and this region of the TSH-β gene appears to confer nTRE activity (43–47) (Fig. 5). The nTRE of mouse TSH-β appears to be sufficient to confer hormone-dependent repression as its placement upstream of the thymidine kinase (TK) promoter is sufficient to render this viral

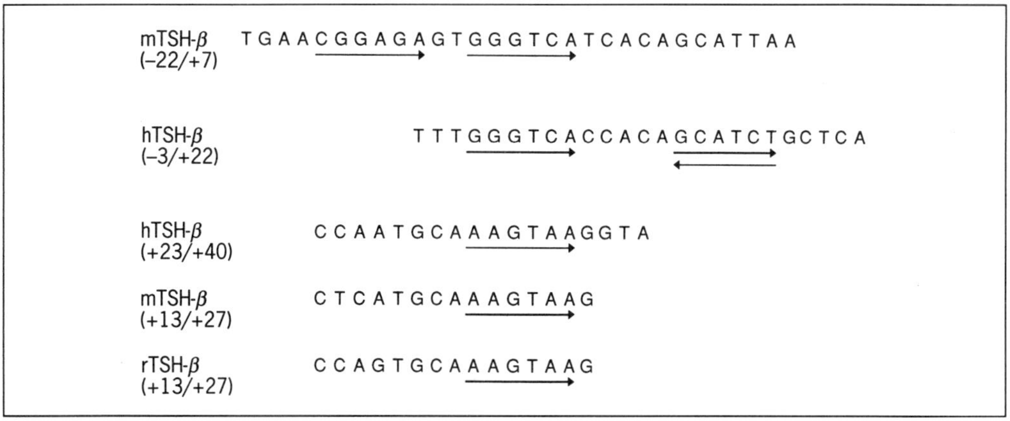

mTSH-β (−22/+7) TGAACGGAGAGTGGGTCATCACAGCATTAA

hTSH-β (−3/+22) TTTGGGTCACCACAGCATCTGCTCA

hTSH-β (+23/+40) CCAATGCAAAGTAAGGTA

mTSH-β (+13/+27) CTCATGCAAAGTAAG

rTSH-β (+13/+27) CCAGTGCAAAGTAAG

Fig. 5 Proposed negative thyroid hormone response element of thyroid-stimulating hormone (TSH) β-subunit genes from mouse (m), rat (r), and human (h). Data are from references 43, 46, and 47

promoter hormone sensitive (47). The nTRE activity has been ascribed to slightly different sequences within the TSH-β promoter of various species. Näär *et al.* (47) proposed that a DR-2 binding site (direct repeats of sequences related to GGGTCA spaced by two base pairs (bp)) present in the mouse TSH-β promoter (−22 to +8 bp) was responsible for activity (in the original publication, this element was labelled DR-0 because a repeat of 8 bp was considered rather than 6 bp). Bodenner *et al.* (46) pointed out that the upstream half of this repeat is not conserved in human and rat TSH-β promoters, and suggested that sequences, including the second GGGTCA motif of the mouse TSH-β promoter and downstream sequences, were the most important for nTRE activity. These sequences, which could be viewed as inverted or direct repeats with 5-bp spacing, are relatively well conserved in human, mouse, and rat TSH-β promoters (Fig. 5), and they were included in functional studies with all three nTREs (44, 46, 47). The human TSH-β promoter was also shown to require other nTRE sequences that are present further downstream within transcribed sequences (46), and it is interesting to note that these sequences are also relatively well conserved in the three species (Fig. 5).

3.2.2 TSH α-subunit

The human glycoprotein hormone α-subunit gene also contains a putative nTRE and TR-binding site in the region of the TATA box (48). It has been proposed that this element is an imperfect palindrome of the IR-0 or Pal type (inverted repeat without spacing between the two halves), but it may also be possible to view it as a DR-2 element. However, recent evidence has raised doubts about the role of receptor interaction with this DNA sequence in the mechanism of thyroid hormone repression (49).

It is also interesting to note that the α-CG promoter is a target for oestrogen repression in the pituitary, although in this case gonadotrophic cells are more likely to be involved than thyrotropes. Indeed, a short α-CG promoter fragment containing only 314-bp 5′-flanking sequences was sufficient to confer oestrogen repression specifically in the pituitaries of transgenic mice (50). This response does not appear to depend on ER binding to the α-CG promoter, although the exact mechanism of repression is not clear and may even involve competition for limiting factors (51).

3.2.3 Oestrogen response element

The oestrogen response element (ERE) can also be viewed as a nTRE although the physiological relevance of this remains to be established. Indeed, it was shown that TR binds EREs with high affinity but that it is incapable of activating transcription from this element (8). Thus, the ERE behaves as a null binding site for TR. In this way, TR appears to reverse oestrogen- and ER-dependent activation and thus repress transcription. A similar interaction was also documented for the palindromic TRE, which behaves as an imperfect ERE. In this case, only the unliganded TR behaves as repressor of oestrogen-dependent ER activation (52).

4. Dominant negative forms of nuclear receptors

Whereas some genes may be targeted for nuclear receptor-dependent repression because of the presence of a negative regulatory element as discussed above, other genes containing positive HREs may be subject to transcriptional repression because of the expression or activation of a dominant negative nuclear receptor in target cells. The oncogenic mutant of TR-α, v-*erb*A, is a good example of this. Indeed, the v-*erb*A gene product has been shown to bind TREs and to act as a transcriptional repressor at those sites (53, 54). v-*erb*A represses constitutive expression from both complete and minimal promoters (55) and this property is shared with its normal cellular counterpart, TR-α. In contrast to TR-α which responds to ligand by relieving this repressor activity, v-*erb*A does not bind thyroid hormone and thus behaves only as a dominant negative repressor of thyroid hormone action. Thus, overexpression of v-*erb*A leads to repression of TR-α and ligand-dependent transcription, and the oncogenic variant of the receptor behaves as an antagonist or dominant negative form of the receptor (53, 54). The repressor or silencing function of v-*erb*A has been localized, and it was found that similar activities are present in the C-terminal portion of the TR and RAR (10).

A splicing variant of the human and rat TR-α called TR-α_2 also behaves as a dominant negative form of this receptor (56–58). This variant differs from the TR-α_1 gene product in its C-terminus, and its transcriptional properties are somewhat similar to that of the v-*erb*A gene product. It may act by a similar mechanism, although its properties have not been extensively studied. It has been proposed that this hormone-independent splicing variant might serve to modulate thyroid hormone responsiveness in tissues where it is expressed, by competing with the action of the functiuonal TR-α_1.

Recent evidence has suggested that the ovalbumin upstream promoter (COUP) nuclear receptor and related orphan receptors, like apolipoprotein AI regulatory protein (ARP1), act as transcriptional repressors. Of course, this repressor activity might reflect properties of these orphan receptors only in the absence of ligand, as by definition we do not know the ligand involved. Irrespective of this caveat, it was shown that COUP transcription factor (TF) and the related v-*erb*A-related (EAR2) and ARP1 can repress ligand-induced transcription mediated by the RXR, retinoic acid receptor (RAR), TR, and vitamin D receptor (VDR) in some conditions (59–61). The repressor activity is not general and depends on the specific combination of receptors and response element. For example, COUP-TF represses retinoic acid (RA)-dependent RXR-α-mediated transcription but not RAR-α-induced transcription from a DR1 response element (59). In contrast, COUP-TF represses RA and RAR-α-dependent transcription from a DR5 response element (61). COUP-TF and ARP1 were also shown to antagonize the positive activity of another member of the steroid receptor superfamily, HNF4 by their action on a regulatory site of the apolipoprotein CIII gene (62). Whereas the predominant activity observed with synthetic promoters has been the repression of ligand-induced transcription, ARP1 or its related receptors might modulate ligand responsiveness in a more subtle way on natural promoters like the apolipoprotein AI promoter. Indeed, ARP1 expression sensitizes this promoter to the action of RA and RXR-α because it appears to have a larger inhibitory effect in the absence than in the presence of RA (60). Thus, the simple interactions documented with simple synthetic promoters may be modulated by other regulatory elements found in natural promoters. The mechanisms by which COUP and its related orphan receptors repress transcription are not yet clear, although it was shown that they readily form heterodimers with RXR-α (59, 60).

5. Receptor heterodimers as repressors

Although heterodimerization of receptors has been proposed as a general mechanism for increasing the regulatory potential of receptors, formal proof of the involvement of heterodimerization is still available mostly for positively regulated systems such as the synergism between RXR, TR, RAR, and VDR (34–40). It is not unlikely that specific target DNA sequences and specific receptor heterodimers might function as nHRE and repressor. However, antagonism between receptors could also result from competitive binding of homodimers to common or overlapping DNA targets. Alternatively, antagonism between receptors could also result from competition for limiting transcription factors (51). Thus, conclusive formal proof of heterodimerization as a mechanism for repression is not available in most cases.

The role of receptor heterodimers as repressor of transcription was suggested for the RAR-α and RXR-α isoforms that readily form heterodimers in the presence and absence of DNA (39). Indeed, it was shown that RAR-α antagonizes the RXR-mediated induction of transcription from the RX response element (RXRE) identified in the cellular retinoic acid binding protein type II (CRABPII) gene (63). Despite the fact that both of these receptors bind the same ligand, retinoic acid, the RAR does not activate transcription from the CRABPII element, and at equal amounts of expression vectors for RAR and RXR it suppresses RXR-dependent *trans*-activation by 90%. The DNA sequence of the CRABPII RXRE appears determinant in directing antagonism between the two receptors, as the same two receptors were shown to act in synergy on another target sequence, the DR5 retinoic acid response element (RARE) (39). Thus, for this pair of receptors, the target DNA sequence may direct either synergism or antagonism. Similarly, it was shown that the RAR-γ_1 isoform antagonizes the activity of other RARs at the β-RARE target sequence, whereas this same receptor isoform is stimulatory on a palindromic TRE (64).

As discussed above, the COUP and ARP1 orphan receptors repress transcription of many genes by antagonizing the action of positively acting receptors. It appears that both COUP and ARP1 may achieve this effect by forming heterodimers with RXR-α, because heterodimers of these receptors bind some target sequences with an *in vitro* affinity that is almost ten times greater than that of either receptor alone (59, 60).

In contrast to the examples outlined above, there is no evidence for the formation of heterodimers involving GR, PR, AR, and ER.

6. Silencer or direct repressor activity

The only well-characterized nuclear receptor-dependent direct repressor or silencer element is the TRE. Indeed, it was shown that both the palindromic TRE or the inverted palindromic sequence found in the chicken lysozyme silencer behave as silencer element in the presence of unliganded TR or the oncogenic variant of this receptor, v-*erb*A (10, 53, 55, 65). With regard to this activity, the unliganded natural receptor behaves very much like the oncogenic viral variant. The TRE exhibits true silencer activity, as repressor function is independent of position or orientation, and it is even active on a minimal TATA-containing promoter (55). A silencing domain responsible for this activity has been localized within the C-terminal domain of TR, which is conserved in the v-*erb*A oncogene product and in RAR (10); in the case of the natural receptors, it appears that ligand binding reverses the effect of the silencing domain (Fig. 6). v-*erb*A has lost this property and it does not bind ligand.

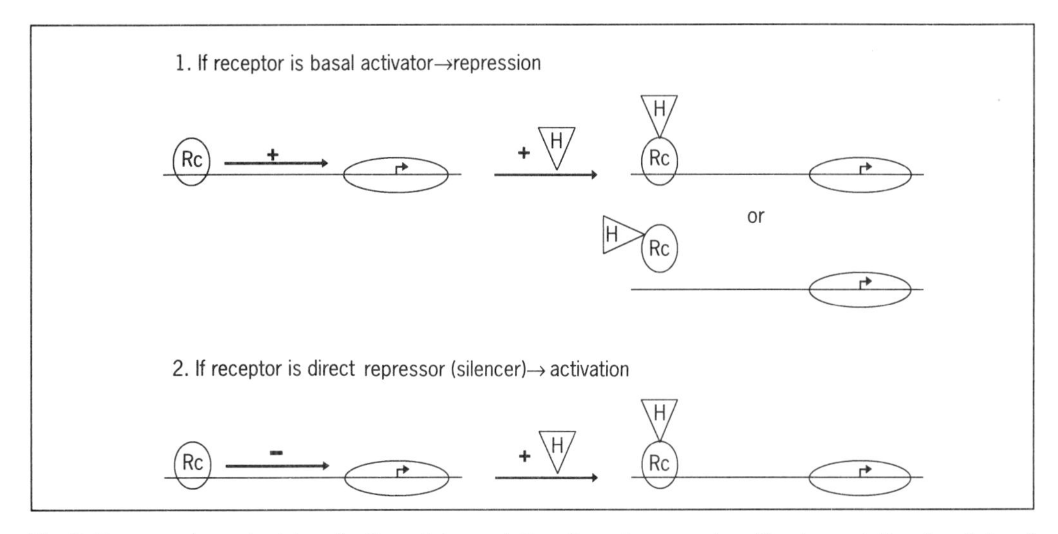

Fig. 6 Hormone-dependent inactivation of transcriptionally active receptor. The transcriptional activity of receptor (Rc) is indicated on the diagram before and after binding of hormone (H)

Most other negative HREs do not exhibit this type of dominant repressor activity as they require either specific promoter context or overlapping binding activities to exert repressor activity.

7. Competition for overlapping binding sites

The classical mechanism of repression was initially proposed to account for repression in various steroid-dependent systems. However, in many cases, further scrutiny has raised questions about the viability of this model, as outlined below.

7.1 Competition between receptors

As most nuclear receptors have the capacity to act as positive *trans*-activators, their action as transcriptional repressor is usually constrained by specific requirements, such as a target DNA sequence that behaves as a null binding site or the absence of ligand for TR. Except for the dominant negative receptor isoforms which are devoid of activation function, the repressor activity of a receptor is often the property of a unique combination of receptor(s) and target DNA sequence. This is best exemplified by the antagonism that ER and TR exert on each other's positive response element. Indeed, whereas TR acts as a strong positive activator by binding to a TRE in which the two palindromic halves are directly abutting each other, the same receptor also binds efficiently to an ERE which contains a 3-bp spacing between the same palindromic halves, but it is unable to activate transcription in this context and will even compete for ER-mediated activation of transcription (8). Whereas the biological relevance of this antagonism remains to be defined, there is more evidence to suggest that the interplay between various members of the TR–RAR–RXR subfamily of receptors may modulate transcription at various target sequences in significant ways as a result of changes in the ratio of receptor expression. Indeed, modulation of receptor gene expression may lead to a greater abundance of a receptor with repressor activity, such as COUP which will displace receptors already present on a subset of response elements. Expression of receptor genes has been shown to be regulated by a variety of hormones and ligands including autoregulation and cross-regulation between receptors and ligand of the same family, as well as during development. For such mechanisms to be operative, the affinity of the new receptor complex, whether homo- or heterodimer, has to be higher than that of the previous complex in order to displace it from their common target sequence, or the relative abundance of the receptors has to be modified to achieve significant displacement. For example, it has been shown that heterodimers of the RXR-α and ARP1 receptors have an affinity that appears to be about ten times greater than that of RXR-α homodimers for a common target sequence (60).

The complex picture that is currently emerging from analysis of receptor isoforms and related receptor subfamilies which have the potential to act on common target sequences reveals the potential for very subtle regulation of different target genes that contain related regulatory DNA sequences. It would not be surprising to find, as we learn more about these receptors and their target sequences, that they exhibit within the living cell even greater selectivity than suggested by *in vitro* affinities.

7.2 Competition for binding sites of other factors

Most HREs that contain *in vitro* binding sites for nuclear receptors and other transcription factors were proposed to be the site of antagonism between these two, such that mutually exclusive binding would result in hormone-dependent repression of transcription. Interestingly, formal proof of this competition is still lacking for all these systems, and in some cases it now appears that other mechanisms of repression are implicated. Although displacement of a positively acting factor may formally be the easiest mechanism for repression, as is often observed in bacterial systems, involvement of nuclear receptors in the regulation of many target genes, including positively regulated ones, may impose constraints on the affinity and/or abundance of nuclear receptors such that this mechanism is hard to reconcile with the complex biology of hormone response in mammalian cells. The current status of various systems in which factor displacement has been proposed as a mechanism for repression by nuclear receptors is outlined below.

7.2.1 Prolactin

Transcription of the bovine prolactin gene is repressed by glucocorticoids. Seven *in vitro* binding sites for GR have been identified in the promoter of this gene and, for at least one of these, it has been shown that overlapping sequences enhance promoter activity in the absence of glucocorticoids or the GR (29). This element is active on a heterologous promoter whether it is placed upstream or downstream of that promoter. Thus, it has been proposed that the bovine prolactin nGRE acts by conferring basal promoter activity via the action of an unidentified positive transcription factor, and GR was postulated to repress this activity by displacement of the factor (29).

7.2.2 Glycoprotein α-subunit

A similar mechanism has been proposed for glucocorticoid repression of the glycoprotein hormone α-subunit gene (31). In this case, three closely spaced *in vitro* binding sites for GR were mapped in the promoter of the human α-CG gene, and these binding sites overlap two cyclic AMP (cAMP) response elements (CREs). It had been postulated that GR acts by displacing the CREB transcription factor which is responsive to cAMP and binds the CREs. However, more recent work has indicated that glucocorticoid repression of this gene does not depend on GR–DNA interactions but rather seems to be mediated via protein–protein interactions, possibly by a squelching or tethering mechanism involving CREB (32, 33).

7.2.3 POMC

We have also proposed that glucocorticoid repression of the POMC gene might result from displacement of a positive transcription factor that binds sequences overlapping the nGRE; indeed, the POMC nGRE binds the COUP factor *in vitro* (21, 66). However, when this hypothesis was tested by linker scanning or replacement mutagenesis of the COUP *in vitro* binding site which is within nGRE, it was found that this mutation does not affect basal POMC promoter activity (67). Surprisingly, the region in and around the nGRE is almost the only part of the promoter that does not appear to contain any element for basal promoter activity, whereas the upstream 400 bp have numerous binding sites for various transcription factors (either ubiquitous or cell-specific factors). In agreement with these observations, oligonucleotides containing the POMC nGRE do not enhance basal promoter activity of various heterologous promoters (7, 21). Thus, there is no basis for believing that GR binding to the nGRE might displace any basal transcription factor, and other mechanisms of repression must be considered. It is also difficult to envisage a quenching mechanism because the closest regulatory element is at least 25 bp away from the nGRE. At this juncture, the simplest hypothesis to explain nGRE-mediated repression would involve GR-dependent interference of promoter function. This interference could result from the presence of DNA-bound GR in the proximal region of the promoter, which would in this way interfere in protein–protein interactions between factors bound to upstream DNA sequences and downstream promoter elements; for example, the interaction between factors bound upstream of nGRE and basic transcription machinery could be disrupted. Alternatively, formation of the three-subunit GR–nGRE complexes might hinder promoter function by changing or restricting DNA conformation of the promoter. Indeed, binding of GR on both sides of the nGRE double helix might freeze that sequence in a conformation that impedes promoter function. It is striking that a short GR DNA-binding domain polypeptide expressed in bacteria is sufficient to form the three-subunit complexes with nGRE, considering that the sequences involved in the binding of third GR molecule do not usually bind GR on their own as they represent a half ERE sequence (7). The unique feature of the POMC nGRE has provided clues as to why this sequence behaves a null binding site, but the exact mechanism by which nGRE-bound GR might repress transcription remains to be elucidated.

7.2.4 α-Fetoprotein, osteocalcin, and interleukin 6

The α-fetoprotein gene is repressed by glucocorticoids in the developing liver. A nGRE which binds GR *in vitro* was localized in the promoter of this gene (68) and found to overlap with an AP1 binding site (69). It was proposed that GR and AP1 could either synergize or antagonize each other at this unusual GR binding site by mutually exclusive binding and/or by protein–protein interactions.

It has been proposed that glucocorticoid repression of the human osteocalcin gene might result from GR-dependent displacement of the TATA binding factor, as an *in vitro* GR binding site was found to overlap this important basic promoter element (28). Similarly, it was proposed that glucocorticoid repression of the interleukin 6 gene might depend on occlusion of various promoter elements including enhancer, TATA box, and RNA start sites (70).

7.2.5 TSH subunits

In thyrotrophic cells of the pituitary, transcription of the TSH-β and TSH-α (common α-subunit of the glycoprotein hormones) gene is repressed by thyroid hormones (41, 42). Various investigators have studied TR repression of these promoters and identified putative nTRE elements (described above in section 3.2.1). The nTREs of the TSH-β promoter increase basal expression of TK, mouse and human TSH-β omoters in the presence of TR but in the absence of hormone; thyroid hormone reverses this increase of basal activity, and thus represses transcription (45–47). It would then appear that, despite its position within the promoter, this nTRE might not repress transcription by interference with the function of the TATA box or initiation but rather by working as a hormone-repressed positive element (Fig. 6). In the case of the mouse TSH-β promoter, it has been shown that the positive activity observed in the absence of hormone is dependent on TR (47). It is not known whether this dependence is also true for the human TSH-β promoter. However, it has been proposed that binding of TR downstream of the initiation site might interfere either with the action of positive factors (other than TR) binding in this region or with the action of basic transcription machinery (45, 46). Clearly, more work is needed to determine whether displacement of factors is involved in repression of the TSH-β promoter or whether hormone repression of the basal positive activity of TR is sufficient to confer hormone sensitivity.

It has been similarly proposed that TR represses transcription of the glycoprotein α-subunit gene by binding a putative TRE which has been identified adjacent to the TATA box (48). Although it was initially proposed that TR-dependent promoter occlusion might provide a mechanism for thyroid hormone repression of this promoter, recent evidence has suggested that thyroid hormone binding to this site is not required for repression (49).

8. Steroid repression by quenching mechanism

This mechanism of repression has mostly been defined by analysis of the proliferin gene. Proliferin is induced by mitogens, serum, and phorbol esters like tetradecanoyl phorbol acetate (TPA). The transcriptional effect of these agents appear to be mediated via an element of the proliferin gene which is a binding site for transcription factor AP1 (30). Interestingly, glucocorticoids repress transcription of the proliferin gene, and this transcriptional response was mapped to the same sequences that mediate phorbol ester stimulation. The composite proliferin gene regulatory element contains neighbouring binding sites for AP1 and GR (12, 30). Accordingly, this element confers responsiveness to both agents when transferred to a heterologous promoter.

Investigation of the mechanism by which AP1 and GR might interact revealed that the response to glucocorticoids is highly dependent on the predominant form of AP1 activity. AP1 is a generic name for factors acting as dimers on the TPA response element (TPARE); depending on the cells from which it is isolated or their growth status, it may be mostly constituted of c-*jun* homodimers or of c-*jun*/c-*fos* heterodimers (71, 72). Thus, in the presence of c-*fos* and c-*jun*, glucocorticoids were found to repress transcription by a receptor-dependent mechanism, whereas in the presence of c-*jun* only (which presumably forms homodimers) glucocorticoids were found to induce transcription from promoters containing the composite proliferin element (12). It was proposed that both GR and AP1, be it *jun* homodimers or *fos*/*jun* heterodimers, co-occupy the regulatory element and that protein–protein interactions in addition to protein–DNA interactions are responsible for determining whether GR will activate in synergy with AP1 or whether it will quench the activity of the *fos/jun* heterodimers (12, 73). This is an interesting model because it increases greatly the regulatory potential of simple response elements. However, the formation of complexes containing GR, AP1 and DNA has not yet been demonstrated directly, although it has been shown that GR and AP1 independently bind the regulatory element. In addition, formal genetic and functional proof of the requirement for co-occupancy of the DNA element is still lacking and may require more detailed biochemical analysis of the putative GR–AP1–DNA complexes.

9. Repression without direct DNA binding of receptor

Novel mechanisms for transcriptional repression have been proposed as a result of the analysis of glucocorticoid repression of genes induced in response to inflammation. In particular, analysis of the collagenase gene, which is induced by inflammation but also by tumour promoters and mitogenic agents, indicated that the same element confers responsiveness to these agents and sensitivity to glucocorticoids (14–17). Indeed, it was found that the TPARE, which mediates inductive signal via binding of the AP1 transcription factor, is also the target for GR-dependent repression. In contrast to the proliferin gene composite element, the TPARE does not bind GR, and thus a mechanism involving protein–protein interactions between GR and AP1 must be envisaged to explain repression in this system. Two studies have provided evidence for direct interaction between GR and c-*jun* and/or c-*fos* which constitute the AP1 activity (14, 16). In overexpression experiments, GR, *jun*, or c-*jun*/c-*fos* exhibit mutual antagonism as if progressive increases in one titrated the activity of the other. Taken together, these observations support models in which direct interactions between GR and c-*jun* or c-*fos* modulate the ability of AP1 to act at the collagenase TPARE. For reasons that are still unclear, divergent observations have been made on the effect of this protein–protein interaction on the ability of GR and/or AP1 to bind their respective target DNA sequences. Yang-Yen *et al.* (14) and Schüle *et al.* (15) found that the ability of both GR and *jun* to bind their respective target DNA sequences *in vitro* was inhibited by the other protein. Thus, these observations led the authors to propose that GR and AP1 antagonize each other's action by a squelching mechanism. In contrast, Jonat *et al.* (16) observed that the TPARE binding activity extracted from HeLa cell nuclei after dexamethasone TPA + (DEX) treatment was no less than in TPA-treated cells. They also provided evidence for co-immunoprecipitation in GR and c-*jun*. These observations suggested that GR interferes with AP1 activity by a tethering mechanism in which GR binds to TPARE-bound AP1 to prevent its transcriptional activity. More recently, *in vivo* footprinting data have provided support for the latter model as they did not document any change in the occupancy of the TPARE after glucocorticoid treatment (74). A decrease in TPARE-bound AP1 would have been expected if GR antagonizes AP1 activity by a squelching mechanism.

A similar antagonism has been observed between AP1 and TR (75, 76). Interestingly, the oncogenic mutant of TR, v-*erb*A does not share that property and even acts as a dominant negative factor by overcoming the repression of AP1 activity by TR (76). The RAR also antagonizes AP1 responsive genes, including collagenase and stromelysin, by a similar antagonism (77, 78). Antagonism between AP1 and ER has also been documented (79, 80). However, in this case, the action may not be reciprocal. Previous work had suggested the existence of complex interactions between ER and the *fos/jun* complex, which are independent of DNA binding by ER (81). In all likelihood, nature is much more complex and subtle than the simple model presented above, and an analysis of the interaction between AP1 components and GR, PR, and AR using different reporters and different cell lines, has confirmed interactions between nuclear receptor and AP1 regulatory pathways, but added a new order of complexity by showing that interactions between the two systems are highly dependent on receptor, promoter, cell type, and specific member of the AP1 family (82). Clearly, much work will be needed to understand the significance in the living cell of all the putative protein–protein interactions that have been described so far using mostly *in vitro* binding analysis and overexpression studies, which may produce disproportionate levels of expression.

Similar interactions have been proposed between GR and a transcription factor related to AP1, CREB, to account for glucocorticoid repression of the α-CG gene (32, 33).

10. Conclusion

Although it is hoped that this review will help the reader by defining a formal framework within which repression by nuclear receptors can be approached, it also illustrates the limited understanding of this regulatory process. In most cases, repression requires much more than receptor binding to nHRE, and protein–protein interactions appear to play a major role in the mechanism of transcriptional repression by nuclear receptors.

References

1. Beato, M. (1987) Gene regulation by steroid hormones. *J. Steroid Biochem.*, **27**, 9.
2. Evans, R. M. (1988) The steroid and thyroid hormone receptor superfamily. *Science*, **240**, 889.
3. Lucas, P. C. and Granner, D. K. (1992) Hormone response domains in gene transcription. *Annu. Rev. Biochem.*, **61**, 1131.
4. Glass, C. K. and Holloway, J. M. (1990) Regulation of gene expression by the thyroid hormone receptor. *Biochem. Biophys. Acta*, **1032**, 157.
5. Levine, M. and Manley, J. L. (1989) Transcriptional repression of eukaryotic promoters. *Cell*, **59**, 405.
6. Ptashne, M. (1987) *A genetic switch*. Cell Press and Blackwell Scientific Publications, Palo Alto, CA.
7. Drouin, J., Sun, Y. L., Chamberland, M., Gauthier, Y., De Léan, A., Nemer, M., and

Schmidt, T. J. (1993) Novel glucocorticoid receptor complex with DNA element of the hormone-repressed POMC gene. *EMBO J.*, **12,** 145.

8. Glass, C. K., Holloway, J. M., Devary, O. V., and Rosenfeld, M. G. (1988) The thyroid hormone receptor binds with opposite transcriptional effects to a common sequence motif in thyroid hormone and estrogen response elements. *Cell*, **54,** 313.
9. O'Halloran, T. V., Frants, B., Shin, M. K., Ralston, D. M., and Wright, J. G. (1989) The *mer*R heavy metal receptor mediates positive activation in a topologically novel transcription complex. *Cell*, **56,** 119.
10. Baniahmad, A., Köhne, A. C., and Renkawitz, R. (1992) A transferable silencing domain is present in the thyroid hormone receptor, in the v-*erb*A oncogene product and in the retinoic acid receptor. *EMBO J.*, **11,** 1015.
11. Keleher, C. A., Goutte, C., and Johnson, A. D. (1988) The yeast cell-type-specific repressor alpha$_2$ acts cooperatively with a non-cell-type-specific protein. *Cell*, **53,** 927.
12. Diamond, M. I., Miner, J. N., Yoshinaga, S. K., and Yamamoto, K. R. (1990) Transcription factor interactions: selectors of positive or negative regulation from a single DNA element. *Science*, **249,** 1266.
13. Gill, G. and Ptashne, M. (1988) Negative effect of the transcriptional activator GAL4. *Nature*, **334,** 721.
14. Yang-Yen, H. S., Chambard, J. C., Sun, Y. L., Smeal, T., Schmidt, T. J., Drouin, J., and Karin, M. (1990) Transcriptional interference between c-*jun* and glucocorticoid receptor due to mutual inhibition of DNA-binding activity. *Cell*, **62,** 1205.
15. Schüle, R., Rangarajan, P., Kliewer, S., Ransone, L. J., Bolado, J., Yang, N., Verma, I. M., and Evans, R. M. (1990) Functional antagonism between oncoprotein c-*jun* and the glucocorticoid receptor. *Cell*, **62,** 1217.
16. Jonat, C., Rahmsdorf, H. J., Park, K.-K., Cato, A. C. B., Gebel, S., Ponta, H., and Herrlich, P. (1990) Antitumor promotion and antiinflammation: down-modulation of AP-1 (*fos/jun*) activity by glucocorticoid hormone. *Cell*, **62,** 1189.
17. Lucibello, F. C., Slater, E. P., Jooss, K. U., Beato, M., and Müller, R. (1990) Mutual transrepression of *fos* and the glucocorticoid receptor: involvement of a functional domain in *fos* which is absent in *fos*B. *EMBO J.*, **9,** 2827.
18. Gagner, J.-P. and Drouin, J. (1985) Opposite regulation of pro-opiomelanocortin gene transcription by glucocorticoids and CRH. *Mol. Cell. Endocrinol.*, **40,** 25.
19. Gagner, J.-P. and Drouin, J. (1987) Tissue-specific regulation of pituitary pro-opiomelanocortin gene transcription by corticotropin-releasing hormone, 3′,5′-cyclic adenosine monophosphate, and glucocorticoids. *Mol. Endocrinol.*, **1,** 677.
20. Birnberg, N. C., Lissitzky, J. C., Hinman, M., and Herbert, E. (1983) Glucocorticoids regulate pro-opiomelanocortin gene expression *in vivo* at the levels of transcription and secretion. *Proc. Natl Acad. Sci. USA*, **80,** 6982.
21. Drouin, J., Trifiro, M. A., Plante, R. K., Nemer, M., Eriksson, P., and Wrange, Ö. (1989) Glucocorticoid receptor binding to a specific DNA sequence is required for hormone-dependent repression of pro-opiomelanocortin gene transcription. *Mol. Cell. Biol.*, **9,** 5305.
22. Riegel, A. T., Lu, Y., Remenick, J., Wolford, R. G., Berard, D. S., and Hager, G. L. (1991) Proopiomelanocortin gene promoter elements required for constitutive and glucocorticoid-repressed transcription. *Mol. Endocrinol.*, **5,** 1973.
23. Scheidereit, C. and Beato, M. (1984) Contacts between hormone receptor and DNA double helix within a glucocorticoid regulatory element of mouse mammary tumor virus. *Proc. Natl Acad. Sci. USA*, **81,** 3029.

24. Tsai, S. Y., Carlstedt-Duke, J., Weigel, N. L., Dahlman, K., Gustafsson, J.-Å., Tsai, M.-J., and O'Malley, B. W. (1988) Molecular interactions of steroid hormone receptor with its enhancer element: evidence for receptor dimer formation. *Cell*, **55**, 361.
25. Drouin, J., Sun, Y. L., Tremblay, S., Lavender, P., Schmidt, T. J., De Léan, A., and Nemer, M. (1992) Homodimer formation is rate-limiting for high affinity DNA binding by glucocorticoid receptor. *Mol. Endocrinol.*, **6**, 1299.
26. Freedman, L. P., Luisi, B. F., Korszun, Z. R., Basavappa, R., Sigler, P. J., and Yamamoto, K. R. (1988) The function and structure of the metal coordination sites within the glucocorticoid receptor DNA binding domain. *Nature*, **334**, 543.
27. Morrison, N. A., Shine, J., Fragonas, J.-C., Verkest, V., McMenemy, M. L., and Eisman, J. A. (1989) 1,25-Dihydroxyvitamin D-responsive element and glucocorticoid repression in the osteocalcin gene. *Science*, **246**, 1158.
28. Strömstedt, P.-E., Poellinger, L., Gustafsson, J.-Å., and Carlstedt-Duke, J. (1991) The glucocorticoid receptor binds to a sequence overlapping the TATA box of the human osteocalcin promoter: a potential mechanism for negative regulation. *Mol. Cell. Biol.*, **11**, 3379.
29. Sakai, D. D., Helms, S., Carlstedt-Duke, J., Gustafsson, J.-Å., Rottman, F. M., and Yamamoto, K. R. (1988) Hormone-mediated repression: a negative glucocorticoid response element from the bovine prolactin gene. *Genes Dev.*, **2**, 1144.
30. Mordacq, J. C. and Linzer, D. I. H. (1989) Co-localization of elements required for phorbol ester stimulation and glucocorticoid repression of proliferin gene expression. *Genes Dev.*, **3**, 760.
31. Akerblom, I. E., Slater, E. P., Beato, M., Baxter, J. D., and Mellon, P. L. (1988) Negative regulation by glucocorticoids through interference with a cAMP responsive enhancer. *Science*, **241**, 350.
32. Chatterjee, V. K. K., Madison, L. D., Mayo, S., and Jameson, J. L. (1991) Repression of the human glycoprotein hormone alpha-subunit gene by glucocorticoids: evidence for receptor interactions with limiting transcriptional activators. *Mol. Endocrinol.*, **5**, 100.
33. Stauber, C., Altschmied, J., Akerblom, I., Marron, J., and Mellon, P. L. (1991) Glucocorticoid receptor represses transcription of the human glycoprotein alpha-subunit gene in a DNA-binding independent manner. *J. Cell. Biochem.*, **15B**, 277.
34. Darling, D. S., Beebe, J. S., Burnside, J., Winslow, E. R., and Chin, W. W. (1991) 3,5,3'-Triiodothyronine (T_3) receptor–auxiliary proten (TRAP) binds DNA and forms heterodimers with the T_3 receptor. *Mol. Endocrinol.*, **5**, 73.
35. Marks, M. S., Hallenbeck, P. L., Nagata, T., Segars, J. H., Appella, E., Nikodem, V. M., and Ozato, K. (1992) H-2RIIBP (RXR_{beta}) heterodimerization provides a mechanism for combinatorial diversity in the regulation of retinoic acid and thyroid hormone responsive genes. *EMBO J.*, **11**, 1419.
36. Zhang, X.-K., Hoffmann, B., Tran, P. B.-V., Graupner, G., and Pfahl, M. (1992) Retinoid X receptor is an auxiliary protein for thyroid hormone and retinoic acid receptors. *Nature*, **355**, 441.
37. Bugge, T. H., Pohl, J., Lonnoy, O., and Stunnenberg, H. G. (1992) RXR_{alpha}, a promiscuous partner of retinoic acid and thyroid hormone receptors. *EMBO J.*, **11**, 1409.
38. Yu, V. C., Delsert, C., Andersen, B., Holloway, J. M., Devary, O. V., Näär, A. M., Kim, S. Y., Boutin, J.-M., Glass, C. K., and Rosenfeld, M. G. (1991) RXR_{beta}: a coregulator that enhances binding of retinoic acid, thyroid hormone, and vitamin D receptors to their cognate response elements. *Cell*, **67**, 1251.
39. Kliewer, S. A., Umesono, K., Mangelsdorf, D. J., and Evans, R. M. (1992) Retinoid X

receptor interacts with nuclear receptors in retinoic acid, thyroid hormone and vitamin D_3 signalling. *Nature,* **355,** 446.

40. Leid, M., Kastner, P., Lyons, R., Nakshatri, H., Saunders, M., Zacharewski, T., Chen, J.-Y., Staub, A., Garnier, J.-M., Mader, S., and Chambon, P. (1992) Purification, cloning, and RXR identity of the HeLa cell factor with which RAR or TR heterodimerizes to bind target sequences efficiently. *Cell,* **68,** 377.
41. Shupnik, M. A., Chin, W. W., Habener, J. F., and Ridgway, E. C. (1985) Transcriptional regulation of the thyrotropin subunit genes by thyroid hormone. *J. Biol. Chem.,* **260,** 2900.
42. Shupnik, M. A., Ardisson, L. J., Meskell, M. J., Bornstein, J., and Ridgway, E. C. (1986) Triiodothyronine (T_3) regulation of thyrotropin subunit gene transcription is proportional to T_3 nuclear receptor occupancy. *Endocrinology,* **118,** 367.
43. Darling, D. S., Burnside, J., and Chin, W. W. (1989) Binding of thyroid hormone receptors to the rat thyrotropin-beta gene. *Mol. Endocrinol.,* **3,** 1359.
44. Carr, F. E., Burnside, J., and Chin, W. W. (1989) Thyroid hormones regulate rat thyrotropin beta gene promoter activity expressed in GH3 Cells. *Mol. Endocrinol.,* **3,** 709.
45. Wondisford, F. E., Farr, E. A., Radovick, S., Steinfelder, H. J., Moates, J. M., McClaskey, J., and Weintraub, B. D. (1989) Thyroid hormone inhibition of human thyrotropin beta-subunit gene expression is mediated by a *cis*-acting element located in the first exon. *J. Biol. Chem.,* **264,** 14 601.
46. Bodenner, D. L., Mroczynski, M. A., Weintraub, B. D., Radovick, S., and Wondisford, F. E. (1991) A detailed functional and structural analysis of a major thyroid hormone inhibitory element in the human thyrotropin beta-subunit gene. *J. Biol. Chem.,* **266,** 21 666.
47. Näär, A. M., Boutin, J.-M., Lipkin, S. M., Yu, V. C., Holloway, J. M., Glass, C. K., and Rosenfeld, M. G. (1991) The orientation and spacing of core DNA-binding motifs dictate selective transcriptional responses to three nuclear receptors. *Cell,* **65,** 1267.
48. Chatterjee, V. K. K., Lee, J.-K., Rentoumis, A., and Jameson, J. L. (1989) Negative regulation of the thyroid-stimulating hormone alpha gene by thyroid hormone: receptor interaction adjacent to the TATA box. *Proc. Natl Acad. Sci. USA,* **86,** 9114.
49. Jameson, J. L., Madison, L. D., Datta, S., Krishna, B., Chatterjee, K., and Nagaya, T. (1992) Transcriptional repression of the thyroid stimulating hormone alpha gene by thyroid hormone. *J. Cell. Biochem.,* **16C,** 12.
50. Keri, R. A., Andersen, B., Kennedy, G. C., Hamernik, D. L., Clay, C. M., Brace, A. D., Nett, T. M., Notides, A. C., and Nilson, J. H. (1991) Estradiol inhibits transcription of the human glycoprotein hormone alpha-subunit gene despite the absence of a high affinity binding site for estrogen receptor. *Mol. Endocrinol.,* **5,** 725.
51. Meyer, M.-E., Gronemeyer, H., Turcotte, B., Bocquel, M.-T., Tasset, D., and Chambon, P. (1989) Steroid hormone receptors compete for factors that mediate their enhancer function. *Cell,* **57,** 433.
52. Graupner, G., Zhang, X.-K., Tzukerman, M., Wills, K., Hermann, T., and Pfahl, M. (1991) Thyroid hormone receptors repress estrogen receptor activation of a TRE. *Mol. Endocrinol.,* **5,** 365.
53. Damm, K., Thompson, C. C., and Evans, R. M. (1989) Protein encoded by v-*erb*A functions as a thyroid-hormone receptor antagonist. *Nature,* **339,** 593.
54. Sap, J., Muñoz, A., Schmitt, J., Stunnenberg, H., and Vennström, B. (1989) Repression of transcription mediated at a thyroid hormone response element by the v-*erb*A oncogene product. *Nature,* **340,** 242.

55. Baniahmad, A., Steiner, C., Köhne, A. C., and Renkawitz, R. (1990) Modular structure of a chicken lysozyme silencer: involvement of an unusual thyroid hormone receptor binding site. *Cell,* **61,** 505.
56. Koenig, R. J., Lazar, M. A., Hodin, R. A., Brent, G. A., Larsen, P. R., Chin, W. W., and Moore, D. D. (1989) Inhibition of thyroid hormone action by a non-hormone binding c-*erb*A protein generated by alternative mRNA splicing. *Nature,* **337,** 659.
57. Lazar, M. A., Hodin, R. A., and Chin, W. W. (1989) Human carboxyl-terminal variant of alpha-type c-*erb*A inhibits *trans*-activation by thyroid hormone receptors without binding thyroid hormone. *Proc. Natl Acad. Sci. USA,* **86,** 7771.
58. Rentoumis, A., Chatterjee, V. K. K., Madison, L. D., Datta, S., Gallagher, G. D., Degroot, L. J., and Jameson, J. L. (1990) Negative and positive transcriptional regulation by thyroid hormone receptor isoforms. *Mol. Endocrinol.,* **4,** 1522.
59. Kliewer, S. A., Umesono, K., Heyman, R. A., Mangelsdorf, D. J., Dyck, J. A., and Evans, R. M. (1992) Retinoid X receptor–COUP-TF interactions modulate retinoic acid signaling. *Proc. Natl Acad. Sci. USA,* **89,** 1448.
60. Widom, R. L., Rhee, M., and Karathanasis, S. K. (1992) Repression by ARP-1 sensitizes apolipoprotein A1 gene responsiveness to RXR_{alpha} and retinoic acid. *Mol. Cell. Biol.,* **2,** 3380.
61. Cooney, A. J., Tsai, S. Y., O'Malley, B., and Tsai, M.-J. (1992) Chicken ovalbumin upstream promoter transcription factor (COUP-TF) dimers bind to different GGTCA response elements, allowing COUP-TF to repress hormonal induction of the vitamin D_3, thyroid hormone, and retinoic acid receptors. *Mol. cell.* **12,** 4153.
62. Mietus-Snyder, M., Sladek, F. M., Ginsburg, G. S., Kuo, C. F., Ladias, J. A. A., Darnell, J. E. J. R., and Karathanasis, S. K. (1992) Antagonism between apolipoprotein A1 regulatory protein 1, Ear3/COUP-TF, and hepatocyte nuclear factor 4 modulates apolipoprotein CIII gene expression in liver and intestinal cells. *Mol. Cell. Biol.,* **12,** 1708.
63. Mangelsdorf, D. J., Umesono, K., Kliewer, S. A., Borgmeyer, U., Ong, E. S., and Evans, R. M. (1991) A direct repeat in the cellular retinol-binding protein type II gene confers differential regulation by RXR and RAR. *Cell,* **66,** 555.
64. Husmann, M., Lehmann, J., Hoffmann, B., Hermann, T., Tzukerman, M., and Pfahl, M. (1991) Antagonism between retinoic acid receptors. *Mol. Cell. Biol.,* **11,** 4097.
65. Graupner, G., Wills, K. N., Tzukerman, M., Zhang, X.-K., and Pfahl, M. (1989) Dual regulatory role for thyroid-hormone receptors allows control of retinoic-acid receptor activity. *Nature,* **340,** 653.
66. Drouin, J., Sun, Y. L., and Nemer, M. (1990) Regulatory elements of the pro-opiomelanocortin gene. Pituitary specificity and glucocorticoid repression. *Trends Endocrinol. Metab.,* **1,** 219.
67. Therrien, M. and Drouin, J. (1991) Pituitary POMC expression requires synergistic interactions of several regulatory elements. *Mol. Cell. Biol.,* **11,** 3492.
68. Guertin, M., Larue, H., Bernier, D., Wrange, Ö., Chevrette, M., Gingras, M.-C., and Bélanger, L. (1988) Enhancer and promoter elements directing activation and glucocorticoid repression of the $alpha_1$-fetoprotein gene in hepatocytes. *Mol. Cell. Biol.,* **8,** 1398.
69. Zhang, X.-K., Dong, J.-M., and Chiu, J.-F. (1991) Regulation of alpha-fetoprotein gene expression by antagonism between AP-1 and the glucocorticoid receptor at their overlapping binding site. *J. Biol. Chem.,* **266,** 8248.
70. Ray, A., LaForge, K. S., and Sehgal, P. B. (1990) On the mechanism for efficient

repression of the interleukin-6 promoter by glucocorticoids: enhancer, TATA box, and RNA start site (Inr motif) occlusion. *Mol. Cell. Biol.*, **10,** 5736.

71. Chiu, R., Boyle, W. J., Meek, J., Smeal, T., Hunter, T., and Karin, M. (1988) The c-*fos* protein interacts with c-*jun*/AP-1 to stimulate transcription from AP-1 responsive genes. *Cell*, **54,** 541.
72. Rauscher, F. J. III, Cohen, D. R., Curran, T., Bos, T. J., Vogt, P. K., Bohmann, D., Tjian, R., and Franza, B. R. Jr. (1988) *Fos*-associated protein p39 is the product of the *jun* proto-oncogene. *Science*, **240,** 1010.
73. Miner, J. N., Diamond, M. I., and Yamamoto, K. R. (1991) Joints in the regulatory lattice: composite regulation by steroid receptor-AP1 complexes. *Cell Growth Differ.*, **2,** 525.
74. König, H., Ponta, H., Rahmsdorf, H. J., and Herrlich, P. (1992) Interference between pathway-specific transcription factors: glucocorticoids antagonize phorbol ester-induced AP-1 activity without altering AP-1 site occupation *in vivo*. *EMBO J.*, **11,** 2241.
75. Zhang, X.-K., Wills, K. N., Husmann, M., Hermann, T., and Pfahl, M. (1991) Novel pathway for thyroid hormone receptor action through interaction with *jun* and *fos* oncogene activities. *Mol. Cell. Biol.*, **11,** 6016.
76. Desbois, C., Aubert, D., Legrand, C., Pain, B., and Samarut, J. (1991) A novel mechanism of action for v-*erb*A: abrogation of the inactivation of transcription factor AP-1 by retinoic acid and thyroid hormone receptors. *Cell*, **67,** 731.
77. Nicholson, R. C., Mader, S., Nagpal, S., Leik, M., Rochette-Egly, C., and Chambon, P. (1990) Negative regulation of the rat stromelysin gene promoter by retinoic acid is mediated by an AP1 binding site. *EMBO J.*, **9,** 4443.
78. Yang-Yen, H. F., Zhang, X.-K., Graupner, G., Tzukerman, M., Sakamoto, B., Karin, M., and Pfahl, M. (1991) Antagonism between retinoic acid receptors and AP-1: implications for tumor promotion and inflammation. *New Biol.*, **3,** 1.
79. Doucas, V., Spyrou, G., and Yaniv, M. (1991) Unregulated expression of c-*jun* or c-*fos* proteins but not *jun*D inhibits oestrogen receptor activity in human breast cancer derived cells. *EMBO J.*, **10,** 2237.
80. Tzukerman, M., Zhang, X.-K., and Pfahl, M. (1991) Inhibition of estrogen receptor activity by the tumor promoter 12-O-tetradeconylphorbol-13-acetate: a molecular analysis. *Mol. Endocrinol.*, **5,** 1983.
81. Gaub, M.-P., Bellard, M., Scheuer, I., Chambon, P., and Sassone-Corsi, P. (1990) Activation of the ovalbumin gene by the estrogen receptor involves the *fos–jun* complex. *Cell*, **64,** 1267.
82. Shemshedini, L., Knauthe, R., Sassone-Corsi, P., Pornon, A., and Gronemeyer, H. (1991) Cell-specific inhibitory and stimulatory effects of *fos* and *jun* on transcription activation by nuclear receptors. *EMBO J.*, **10,** 3839.

7 Structure and function of the steroid receptor zinc finger region

LEONARD P. FREEDMAN

1. Introduction

Steroid hormone receptors are ligand-inducible transcriptional regulatory proteins. Like many eukaryotic transcription factors, nuclear hormone receptors appear to be organized into relatively discrete functional domains (1), and it is this common domain organization that groups these receptors into a superfamily of functionally and most likely structurally related proteins (reviewed in Chapter 1). Because of this modularity, domains for ligand binding (located in the carboxy-terminal region), DNA binding (located toward the central or amino terminus), nuclear localization (within both the DNA and ligand-binding domains), and transcriptional modulation (localized to more variable regions of the receptors, including the *N*-terminus) (Fig. 1), can all confer their specific functions when linked to unrelated, non-receptor proteins (2–7). This has led to the efficacy of examining in detail the properties of a particular receptor functional domain, such as the DNA-binding domain, and of avoiding, at least initially, the potential physical/chemical complexities of studying a full-length native receptor.

Along these lines, this chapter focuses exclusively on the structure and function of the steroid receptor DNA-binding domain. Among the superfamily of nuclear hormone receptors, this domain is by far the most highly conserved (Fig. 1), and it was by exploiting this fact that many receptor-encoding complementary DNAs (cDNAs) were cloned. Several amino acids in this domain are invariant throughout the family, including eight cysteines that were shown, in the case of the glucocorticoid receptor (GR), to co-ordinate tetrahedrally two Zn^{2+} ions (8) in an arrangement reminiscent of the 'zinc finger' co-ordination scheme originally proposed for transcription factor IIIA of *Xenopus* (9, 10). Mutagenesis experiments indicated that specific residues within the zinc finger region of the glucocorticoid receptor were critical for DNA-binding specificity (11–13), DNA-dependent dimerization (13, 14), and positive control of transcription (15). These observations have recently been confirmed and expanded by three-dimensional structural analysis (16–18). Thus,

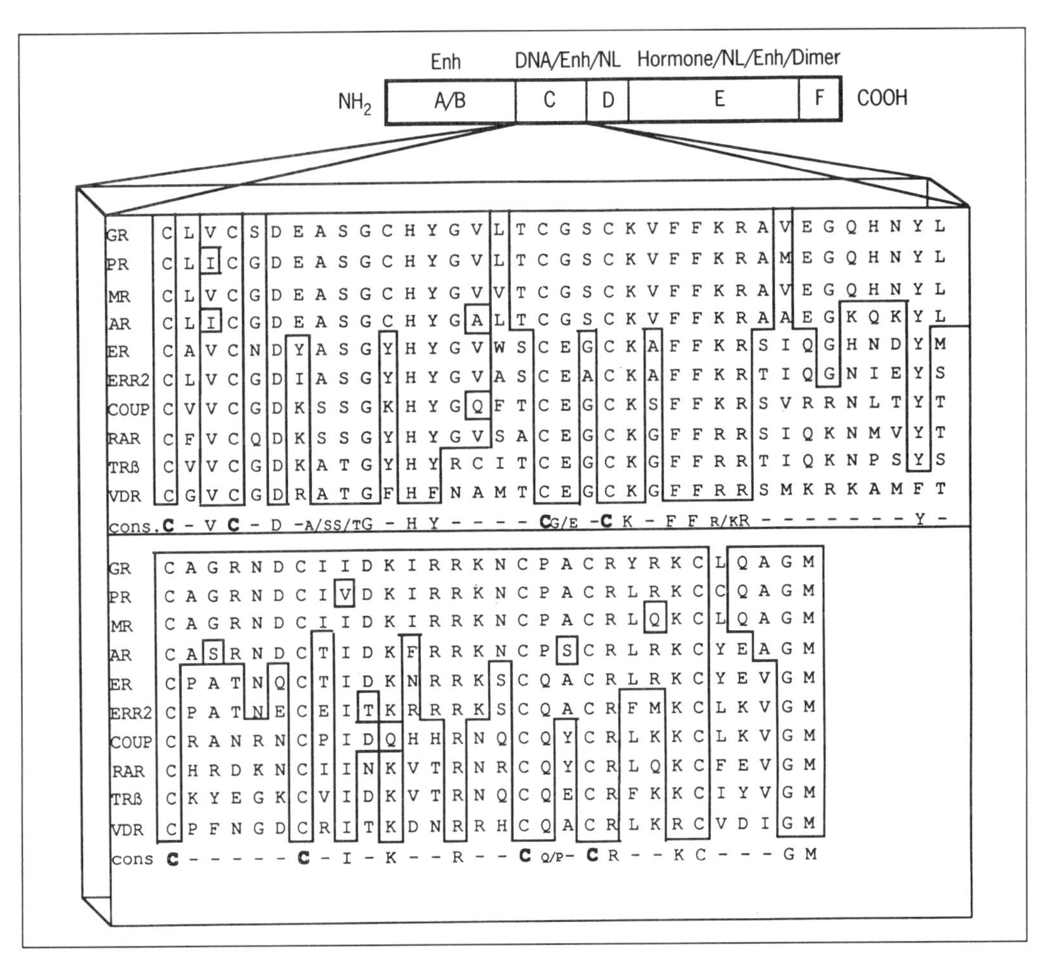

Fig. 1 Domain organization of steroid receptors, divided into regions A–F, as per the convention defined for the oestrogen receptor. 'Enh' refers to regions that confer transcriptional enhancement activity; 'DNA' and 'Hormone' denote the DNA- and ligand-binding domains, respectively; 'NL' refers to two portions of the GR sufficient for nuclear localization; and 'Dimer' refers to a portion of the ligand-binding domain believed to mediate homo- and/or hetero-dimerization. Below is an alignment of amino acid sequences of DNA-binding domains of the members of the nuclear receptor superfamily. Conserved residues are boxed; outlined cysteines are those demonstrated to be involved in zinc co-ordination. See text for appropriate references

the DNA-binding domain, a small yet self-contained region within nuclear receptors, is itself subdivided into substructures of distinct conformation and function. This review attempts to dissect these functions within the context of the various genetic, biochemical, and structural experiments that have to date been described primarily for the DNA-binding domains of the GR and the oestrogen receptor (ER). Given the strong sequence conservation of this region, several generalities can be made based on these existing structures. Where appropriate, however, differences that might be important to the function of other receptor DNA-binding domains will be pointed out.

2. Hormone response elements

To begin an examination of how nuclear hormone receptors recognize and bind specific DNA elements, an understanding of the characteristics of the *cis* elements themselves is essential (see Chapter 4). Historically, much more was known about such hormone response elements (HREs) well before a comprehensive study of the corresponding *trans* binding proteins was begun. In the early 1980s, it was shown that the glucocorticoid-stimulated expression of genes encoding the mouse mammary tumour virus (MMTV) was mediated by short sequences localized upstream of its promoter (reviewed in reference 19). These segments could confer hormonal control to heterologous genes in a distance- and orientation-independent manner (20), and correlated perfectly to regions of the MMTV long-terminal repeat that had been identified *in vitro* as high-affinity binding sites for partially purified rat liver GR (21, 22). Thus, glucocorticoid response elements (GREs), and later other HREs, were defined as hormone-inducible enhancers, and the cognate *trans* factor—the receptor—as an inducible transcriptional regulatory protein.

The original DNA-binding studies on the MMTV long-terminal repeat, utilizing DNase I and methylation protection assays, generated a 15 base pairs (bp) consensus sequence that was partially palindromic, consisting of two 'half-sites' separated by 3 bp (23). Several genes also identified as glucocorticoid inducible, such as those encoding human metallothioneine IIA, chicken lysozyme, and rat tyrosine aminotransferase (24 and references therein), were subsequently found to have similar sequences in their promoter regions. The functional GRE consensus sequence is 5′-GGTACA*nnn*TGTTCT-3′ (24), where the three '*n*' bases can be any nucleotide but, as discussed below, the spacing of the three is invariant. Strong, functional GREs can be designed as perfect palindromes, again provided that the spacer is 3 bp in length (25) (Fig. 2). Interestingly, this and similar GREs can function virtually as efficiently as progesterone (PRE), androgen (ARE), and mineralocorticoid (MRE) response elements, and in fact are typically used as PREs, AREs, and MREs in transcriptional regulatory studies involving these respective receptors (26–30).

Saturation mutagenesis experiments of GREs indicate that changes at positions G+4, T+5, C+7, C−4, and A−5 (defining the middle n as +1 and moving 5′ → 3′ for positive numbering, 3′ → 5′ for negative numbering; see Fig. 2) are not tolerated for either DNA binding of GR *in vitro* (31) or glucocorticoid inducibility of a reporter gene *in vivo* (32), while changes in other bases are tolerated, to differing degrees. These results are consistent with methylation protection and interference data, which show that GR contacts guanines at positions +4, +7, and −4, −7 (31, 33, 34). Thymine at position +5 appears to be critical for target-site discrimination (31, 35, 36), and a functional consensus oestrogen response element (ERE) differs from a GRE by only 2 bp in each half-site (nnnTGTTCT → nnnTGACCT; Fig. 2) (37). While methylation and ethylation interference assays indicate that ERs and PRs (and presumably GRs) make equivalent contacts with guanines and the phosphate backbone when bound to a GRE or ERE (38), analysis of DNA binding by

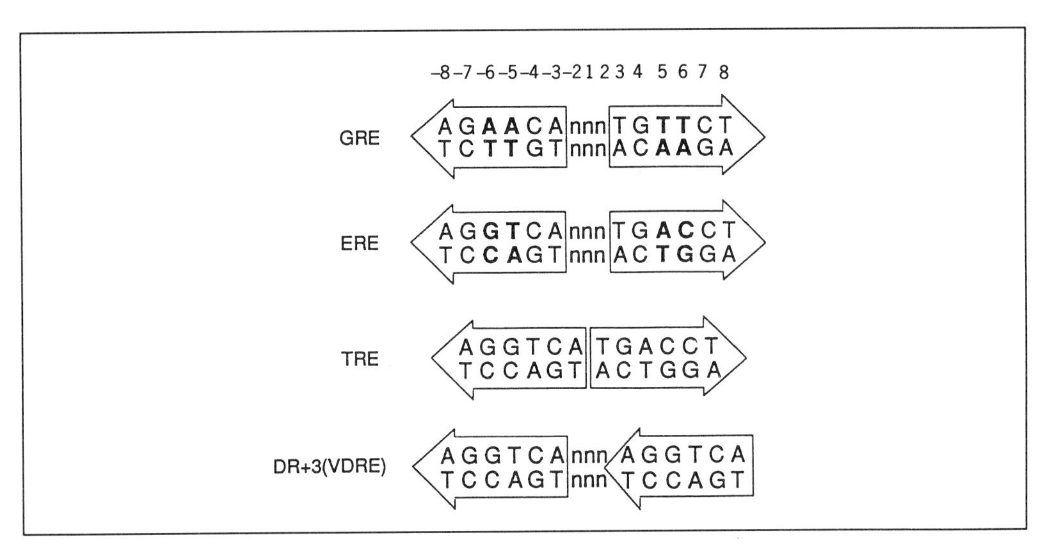

Fig. 2 Idealized hormone response elements are organized as two half-sites. Arrows indicate the directionality of the half-sites

potassium permanganate interference demonstrates that both GRs and PRs contact the 5-methyl group of T(+5), but that the ER does not (38).

The facts that bona fide GREs consist of two imperfect palindromes separated by 3 bp, and that both mutagenesis and DNA-binding data have revealed an approximate twofold symmetry between affected nucleotides in each half-site, strongly suggest that the GR binds a GRE as a head-to-head dimer. This indeed appears to be the case for the steroid receptors (see sections 3.2 and 4.3), but a subfamily of receptors that also appear to homo- or hetero-dimerize in solution are able to bind and activate from response elements irrespective of half-site orientation. The thyroid hormone (TR) and retinoic acid (RAR) receptors can recognize the same half-site element found in a functional ERE—AGGTCA—as an inverted repeat with no half-site spacing (39, 40; Fig. 2) or as a direct repeat with a spacer of 4–5 bp, respectively (25, 41). The vitamin D_3 receptor (VDR) can bind and activate from the same element if the half-site spacing is 3 bp (41) (DR+3; Fig. 2). Moreover, this direct repeat with no intervening bases appears to confer constitutive transcriptional activation by TR in the absence, and ligand-dependent repression in the presence, of hormone (25). Thus, for this subclass of receptors, the orientation and spacing of the same core DNA sequence element can potentially influence both binding discrimination and how a given receptor regulates the transcription of a linked gene (i.e. whether it represses or activates); the latter in turn appears to be modulated by the status of ligand occupancy of the receptor. Taken together, these results imply that the structures of VDR, TR, and RAR, and of other putative members of this subclass of nuclear receptors, permit a much more flexible dimer association than appears to be the case for steroid receptors upon DNA binding. Unlike steroid receptors, these receptor dimers can recognize a specific sequence regardless of its orientation; nevertheless, they too are sensitive to constraints imposed by half-site spacing. Whether or not the '3, 4, 5 rule' (41) or the 'orientation and spacing code' (25), as they have been called, represent a biologically relevant means of conferring specific hormone receptor responsiveness to genuine target genes is not yet clear. These observations certainly suggest, however, that binding site selectivity for this group of receptors may be influenced more by the organization of the response element than particular nucleotide differences *per se*, as appears to be true for GR and ER (see below).

3. Functional dissection of the steroid receptor DNA-binding domain

3.1 Mutagenesis of the DNA-binding domain

The DNA-binding domain was originally inferred to be located somewhere within a 40-KD *C*-terminal fragment of the purified rat GR by limited proteolysis of the protein (42, 43). In the mid 1980s, the isolation and cloning of cDNAs encoding steroid receptors permitted the rapid delineation of multiple functional domains. The generation and analysis of nested sets of *C*- and *N*-terminal deletion mutants of the rat and human GR cDNA clones indicated that a subregion of approximately 86 amino acids was sufficient to confer specific DNA binding (44–47; see Chapter 5). In addition, it appeared that this small fragment, defined by residues 440–525 (numbering convention is that of the rat GR), also contained sequences that could confer weak constitutive (i.e. hormone independent) transcriptional activation to GRE-driven promoters (48, 49), as well as a strong nuclear localization signal (50). Finally, sequences sufficient to mediate the negative regulation of the bovine prolactin gene (46) map to the DNA-binding domain. Thus, at least four apparently distinct activities were found to co-localize within this small region.

The cloning and sequencing of several steroid receptor cDNAs revealed a highly conserved, cysteine-rich region within the delineated DNA-binding domain (Fig. 1). These sequences were proposed (47, 51, 52) to fold into two zinc-stabilized peptide loop structures similar to what had previously been suggested as a folding scheme for a series of nine tandemly repeated sequences found in transcription factor IIIA (TFIIIA) (9, 10). The TFIIIA motif is characterized by pairs of cysteines and histidines that co-ordinate tetrahedrally a zinc ion (53, 54), resulting in a peptide loop of 12 amino acids (consensus sequence Cys–$X_{2,4}$–Cys–X_3–Phe–X_5–Leu–X_2–His–X_3–His) (reviewed in reference 55). In contrast, the nuclear receptor DNA-binding domain was proposed to utilize four cysteines in each of two fingers to co-ordinate zinc, yielding two non-equivalent zinc fingers (Cys–X_2–Cys–X_{13}–Cys–X_2–Cys–X_{15}–Cys–X_5–Cys–X_9–Cys–X_2–Cys–X_4–Cys) (Fig. 3). Klug and Rhodes (57) proposed that TFIIIA-type proteins bind DNA via finger loop–major groove interactions, and although this model was generalized to include the nuclear receptors, mutagenesis data soon suggested that the receptor fingers mediate DNA binding through a distinct mechanism.

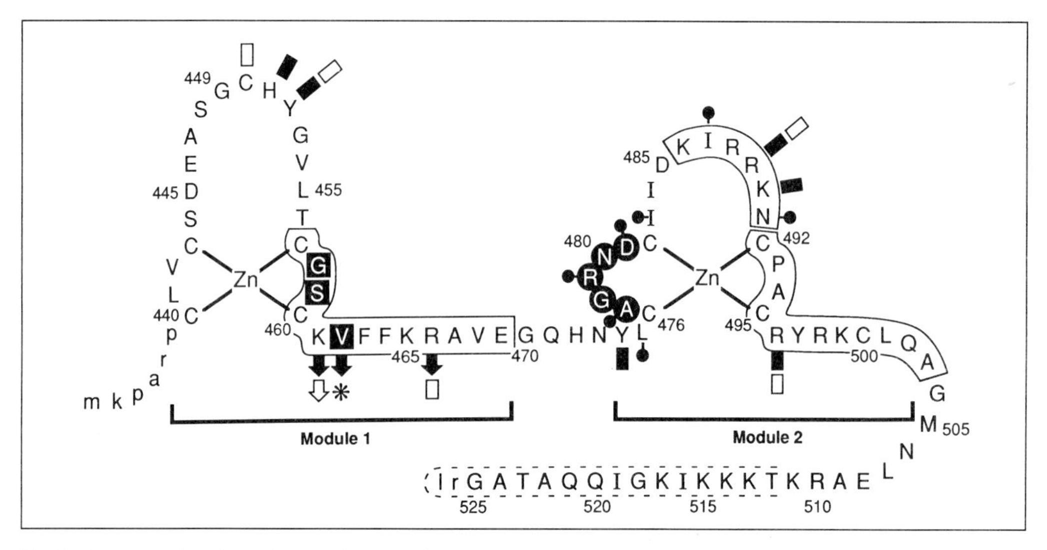

Fig. 3 Zinc co-ordination scheme of the rat GR DNA-binding domain. Numbering scheme is based on the full-length receptor. Indicated residues and regions are based on the crystal structure of the protein bound to a GRE (18). Residues making specific phosphate backbone contacts are indicated by solid rectangles; those making non-specific contacts are indicated by open rectangles. Residues making direct contacts with bases are depicted by solid arrows. Amino acids involved in dimer interface interactions are indicated by a solid dot. Three amino acids (residues 458, 459, 462) that confer specificity in mutagenesis experiments (11–13) are shown in solid boxes; those demonstrated to confer half-site spacing requirements (residues 477–481; 13, 14, 56) are shown in solid circles. α-helical regions are enclosed by solid lines. A disordered section at the *C*-terminus is enclosed by dashed lines. Amino acids as lower-case letters derive from the vector sequence of the overexpression plasmid. Reprinted by permission from *Nature*, **352,** 497–505. © 1991 Macmillan Magazines Ltd

While several groups proceeded to use site-directed mutagenesis to begin defining residues important for DNA binding (see below) Schena *et al.* (15) took the novel approach of using yeast to screen a large number of chemically generated *random* point mutations within the GR DNA-binding domain to dissect these overlapping functions. They and others had found that many mammalian transcription factors, including steroid receptors, functioned in yeast, (58, 59), and so devised a strategy whereby a GR cDNA fragment encoding the DNA-binding domain alone was chemically mutagenized *in vitro*, and then inserted back into a wild-type GR recipient present on a yeast expression plasmid. This pool was then used to transform a yeast strain containing an integrated GRE-dependent *lacZ* reporter gene. Mutant candidates were then initially screened by a colour phenotype on x-gal indicator plates.

Two classes of mutants were isolated in this way that were subsequently shown to be defective in *in vitro* GRE binding: (1) those that mapped to predicted co-ordinating cysteines; and (2) those that mapped to residues immediately *C*-terminal to both finger loops. Several mutations occurring in residues within the finger loops themselves appeared to have little or no effect on function, implying that the two fingers do not necessarily play a direct role in mediating DNA interaction, as had been proposed for the TFIIIA-type of Cys_2His_2 group finger proteins

(57). A third class of isolated mutants were wild-type for DNA binding but were unable to enhance transcription from a GRE-linked promoter in yeast or mammalian cells. Interestingly, these positive control (*pc*)-like mutants all cluster around the 'tip' of the second finger, and Schena *et al.* (15) proposed that this region may define an area of the GR, which they called *enh*1, that mediates protein–protein contact with other factors involved in the process of transcription initiation.

Like TFIIIA and the TFIIIA-class of zinc finger proteins, each steroid receptor zinc finger is encoded by separate exons (60, 61). Unlike the tandemly repeating TFIIIA-type fingers, however, it is almost immediately apparent that the receptor fingers are functionally distinct, since there are numerous differences in the number and kinds of amino acids in each finger (Fig. 3). One difference between the two receptor zinc fingers is the number of amino acids in each peptide loop (13 versus 9). While the first finger contains several hydrophobic residues, the second is more basic. The first finger contains two amino acids between each pair of four coordinating cysteines; the second has five residues between the first two coordinating cysteines (Fig. 3). Taken together, it is unlikely that the two receptor zinc fingers arose from a simple duplication event during evolution, as is likely for the multiple, essentially superimposable TFIIIA-type fingers, but rather that they represent more or less independent substructures within the DNA-binding domain.

That the steroid receptor zinc finger region is necessary and sufficient for specific DNA binding was unequivocally demonstrated by the construction of a number of chimeric GR–ER derivatives. The first such chimera consisted of the full-length ER in which its own DNA-binding domain was substituted for that of the GR. This ER derivative activated a GRE-dependent—but not ERE-dependent—CAT reporter gene in response to oestradiol (62). By swapping individual ER fingers for those of GR at the exon–intron border, Green *et al.* (63) demonstrated that residues within the first finger were primarily responsible for determining target specificity, although both fingers, including a basic region immediately *C*-terminal to the second finger (and important for nuclear localization (50)), were required for DNA binding (see Chapter 5). Indeed, when the GR DNA-binding domain alone was overexpressed and purified from *Escherichia coli*, it could recognize and bind to GRE elements with relatively high affinity, but a trypsin-generated subfragment that included all of the first finger but only half of the second was unable to bind DNA at all, indicating that both fingers and perhaps the *C*-terminal basic region were required for DNA binding (8). Thus, the *N*-terminal finger is the principal determinant for specificity, but it requires the contribution of the *C*-terminal finger (and beyond) for overall binding affinity.

By focusing on the first finger, individual residues that appear to mediate DNA-binding specificity have been identiied. Combining the results of CAT transactivating assays from three laboratories (11–13), it became apparent that three amino acids, located at the *C*-terminal side of the first finger at positions 458, 459, and 462 (Fig. 3, residues in solid boxes), play a key role in distinguishing between a GRE and ERE. A substitution of two of these amino acids in the context of full-length GR to the corresponding ER residues (Gly–Ser → Glu–Gly) switches the receptor specificity from a GRE to an ERE, but not to the level of wild-type ER, as assayed by the induction of CAT enzyme expressed from reporter constructs regulated by a GRE or ERE (11, 13). Moving from ER towards GR, Mader *et al.* (12) demonstrated that a three-amino-acid change of Glu–Gly – Ala (EGA) to Gly–Ser – Val (GSV) completely changed specificity so that this mutant ER transactivated strongly from a GRE-driven reporter and not at all from an ERE.

Thus it is in this region, beginning at the *C*-terminal base of the first finger in GR, ER, and probably all nuclear receptors, that DNA-sequence discrimination occurs. This stretch of amino acids (457–469 in GR) was predicted to be α-helical and making direct contact with bases (64); from the structures, this indeed seems to be the case (see sections 4.2 and 4.3). Several residues are absolutely conserved within this region, and some are conserved within subgroups of receptors, which can in turn be used as a predictor for the 'core' binding sequence that a given nuclear receptor will recognize: receptors carrying the GSV motif (i.e. GR, PR, mineralocorticoid receptor (MR) and androgen receptor (AR)) all recognize a GRE with high affinity; receptors carrying the EGA or EGG motif (i.e. ER VDR, TR, RAR, and ecdysone receptor) all appear to bind to the ERE core (AGGTCA) with high affinity. Within the second group, specificity is further conferred by the spacing and the relative orientation of the two half-sites (see below). In cases where specificity between receptors appears identical and yet the proteins are presumably regulating distinct sets of genes (e.g. the first group), we can only speculate that differential modifications or interactions with specific auxillary factors confer subtle but decisive changes in affinity and specificity to target sites.

3.2 Role of dimerization in DNA binding

The palindromic nature of many GREs and other steroid response elements suggested that hormone receptors can bind to DNA as symmetrical dimers. The full-length GR has been detected as both a monomer and homodimer in solution (i.e. in the absence of DNA); however, it appears to bind exclusively as a dimer to GREs (65). This suggests that an equilibrium exists between GR monomer and dimer in solution, and either that a preformed GR dimer binds to a GRE, or that GR binds as a monomer to one half-site and strong positive co-operativity results in the binding of the second monomer, which is essentially undetectable (65).

Kumar and Chambon (66) first showed that the principal ER dimerization domain is within the ligand-binding domain. By deletion and site-directed mutagenesis, this region has been sublocalized to ER residues between 500 and 540 (68). Interestingly, this region, when aligned with other members of nuclear receptor superfamily, contains a heptad repeat of hydrophobic amino acids that could form a dimerization interface with other conserved residues in between them (67, 68). Conceivably, coiled-coil interactions could take place within this region that would confer both homo- and hetero-dimerization. Indeed, this region appears to mediate hetero-dimerization between several nuclear receptors and the retinoid X receptor. This interaction potentiates receptor DNA binding and transactivation, presumably by forming a more stable dimer than the homodimeric species (69–74).

Recent work also suggests that the GR can trimerize functionally with *jun–fos* and *jun–jun* complexes, which will in turn determine whether transcription of a given gene will be enhanced or repressed by the receptor (56, 75, 77). The regulatory consequences of such combinations of complexes are clearly profound, and go beyond the scope of this review.

It is within the DNA-binding domain itself, however, that the influence of dimerization on specific DNA binding has been most clearly demonstrated. Mixing DNA-binding domain derivatives of different sizes with oligonucleotide binding sites, investigators have been able to show specific shifts in electrophoretic mobility consistent with two protein monomers bound to each half-site (31, 78). Similar experiments using extracts from transfected cells show that the ER DNA-binding domain binds as a dimer to an ERE, but with a lower apparent affinity than derivatives containing the hormone-binding domain (66). Likewise, the GR DNA-binding domain purified from *E. coli* binds GREs with approximately 80-fold lower affinity than full-length GR purified from rat liver (L. P. Freedman, unpublished results). The ability to use both the GR and ER DNA-binding domains to examine GR–GRE and ER–ERE interactions, however, has been advantageous, since the proteins, even at millimolar concentrations, are monomeric in solution (16, 17, 79) but, as mentioned above, bind as a dimers. DNA-dependent dimerization has been studied quantitatively, and it is clearly co-operative in nature (31, 78, 80). Thus the interaction of one GR or ER DNA-binding domain monomer with one of two GRE or ERE half-sites facilitates the binding of a second monomer to the second half-site. Independently, LaBaer and Yamamoto (31) and Hard *et al.* (80) have shown that the intrinsic association constant GR DNA-binding domain for a specific site relative to a non-specific site differs by only two orders of magnitude (approximately 10^8 M^{-1} versus 10^6 M^{-1}, respectively). This differential between specific and non-specific sites is considerably less than that of other sequence-specific DNA-binding proteins. This could be due in part to the use of the DNA-binding domain alone, minus the stabilizing influence of the additional domains of the receptor, and/or because nuclear receptors simply have lower intrinsic binding affinities than, for example, helix-turn-helix proteins. Since, generally speaking, natural GREs consist of one 'good' half-site (i.e. close to the TGTTCT consensus) and a second half-site that deviates considerably from the consensus, binding to the low-affinity half-site would be dependent on the initial association of GR to the first, high-affinity half-site. Quantitative gel shift analysis using the TGTTCT half-site alone versus the full GRE indicates that the contribution due to co-operative binding is approximately two orders of magnitude (31, 81, 82). Presumably, regulation of a given gene by glucocorticoids could relate in part to how far or close GRE half-sites deviate from the consensus sequence, in which graded affinities could be achieved by changes in sequences in both half-sites. Perhaps an intrinsically low-affinity site could attract GR in association with other factors; this would represent yet another level of regulatory complexity by virtue of protein–protein and protein–DNA interaction.

The fact that the GR and ER DNA-binding domains alone cannot dimerize but require association with a palindromic GRE suggests that the architecture of both the protein and DNA are important for dimerization. The sequence composition of a GRE can vary considerably, but the spacing between imperfect palindromes appears invariant, strongly implying that dimer binding is somehow constrained by the distance between each half-site. Given the fact that GREs are more or less palindromic, monomers sitting on each half-site should be facing each other in a symmetrically head-to-head orientation. In this way, symmetrical pairs of contacts between residues in each monomer could take place that would facilitate dimerization. It is clear that half-site spacing of 3 bp is essential for both co-operative dimer binding by GR and induction of a GRE-dependent reporter gene: increasing or decreasing the spacing completely abolishes transactivation of a responsive gene *in vivo* (32, 81), and results in non-cooperative binding of the GR DNA-binding domain *in vitro*. Here, the monomeric species predominates, and the dimer is detected only at relatively high protein concentrations (14, 82). When residues at the base of the second GR finger, between co-ordinating Cys476 and Cys482, were replaced by the corresponding amino acids from the TR, a GR derivative that had already been mutagenized such that it recognized an ERE now transactivated from a TRE (13). This functional TRE is identical in sequence to an ERE but lacks the 3-bp spacer. Umesono and Evans (13) concluded that by changing these GR residues to those corresponding to TR, the half-site spacing restriction normally in place was relaxed. Using *in vitro* gel mobility shift assays, other investigators subsequently demonstrated that similarly mutated proteins no longer bound co-operatively to GRE oligonucleotides, although their intrinsic binding affinities for a GRE half-site were not affected (14, 82). Thus this five-amino-acid stretch between the first two co-ordinating cysteines of the second finger (residues 477–481), termed the 'D box' (13) was predicted to mediate spacing requirements critical for co-operative dimer binding to palindromic HREs, presumably through a dimer interface involving those residues in each monomer. With the solution of the three-dimensional structure of the GR DNA-binding domain complexed to a GRE, this prediction has been borne out, as discussed in section 4.3.

4. Structure of the steroid receptor zinc finger region

4.1 Biochemistry of zinc co-ordination

Until the last few years, structural studies of steroid receptors have been hampered by the limited availability of pure protein. The receptor is present in very small amounts in animal tissues and expressed at relatively low levels by cloned receptor cDNAs in tissue culture cells. Although several overexpression systems have been employed, severe limitations on yield of full-length receptors due to insolubility or instability have forced investigators to focus their efforts on individual functional domains. The steroid receptor DNA-binding domain has lent itself very well to this approach. Milligram quantities of both the GR and ER DNA-binding domains overexpressed from a cDNA subcloned fragment in *E. coli* have been obtained and purified to homogeneity with relative ease (8, 17, 82–84). One such GR derivative, encompassing sequences 407–556 (T7X556; 23), and a smaller version corresponding to residues 440–525 (16, 18, 85) have been used extensively for structural and functional studies. Importantly, these receptor subfragments bind DNA with the same specificity as rat liver purified GR, albeit with a lower affinity (see section 3.2). As predicted, the T7X556 protein was shown to contain two zinc atoms for each protein molecule, as assessed by atomic absorption spectroscopy (8). Dialysis at low pH in the presence of chelating agents released the metal from the protein. Zinc binding is quite strong ($K_{app} = 4 \times 10^{12}\ M^{-1}$ at pH 7.4 (79)) and is resistant to known chelators such as ethylenediamine tetra-acetic acid and 1,10-phenanthroline (L. P. Freedman, unpublished results). Chelation of metal by low pH dialysis yields an apoprotein that is unable to bind to DNA and is protease sensitive. DNA-binding activity and protease resistance could be restored, however, if the apoT7X556 protein was preincubated with either Zn(II) or Cd(II) (8), demonstrating that the metal ion is an essential co-factor for specific DNA binding, apparently by maintaining the protein in its native, active form.

4.2 Solution structure by two-dimensional nuclear magnetic resonance

The accumulation of functional data from mutagenesis and biochemical experiments led to a concentrated effort to solve the three-dimensional structure of the steroid receptor DNA-binding domain. The first structures solved were by two-dimensional nuclear magnetic resonance (2-D NMR). The method, in which molecules can be studied in solution at or near physiological conditions, utilizes the magnetic effects of adjacent or covalently connected atoms. Structural information can be obtained from the method because these spectrascopic effects are dependent on distances between atoms and their chemical connectivities. Distance geometry algorithms are then applied to generate initial structures based on the spectroscopic data and polypeptide structure constraints derived from the amino acid sequence (reviewed in references 86 and 87). The major limitations of 2-D NMR, however, are that it requires tens of milligrams of homogeneous material, and that the molecule studied be preferably smaller than 20 kD. The protein also has to be stable in solution. Because large amounts of highly pure and soluble 13-kD GR DNA-binding domain were available and had already been characterized extensively (8, 15), as described in preceding sections, it was a very attractive candidate for structure determination by 2-D NMR. The ER DNA-binding domain also turned out to highly expressable and easily purified (17, 78), and consequently was also suitable for 2-D NMR analysis.

Using two overlapping GR DNA-binding domain fragments (residues 440–525 from rat and 394–500 from human), Hard and co-workers (16, 88) determined the structure of the zinc finger region based on a set of 470 non-redundant connectivities found by NMR. These data allowed the identification of several secondary structure elements from Cys440–Ile519. They include two α-helices C-terminal to each finger, stretching from Ser459 to Glu469 and from Pro493 to Gly504. Thus, the overall pattern is that of a finger-helix-extended region repeated twice. Other structural elements include a type I reverse turn between Arg479 and Cys482, and a type II reverse turn between Leu475 and Gly478, both of which are within the D-box, a short anti-parallel β-sheet in the first finger involving residues Cys440–Leu441 and Leu455–Cys457 (88). A very similar pattern is seen with the ER DNA-binding domain (17). The data did not reveal any α-helicity within the fingers themselves, although the crystal structure of the GR DNA-binding domain clearly shows a distorted helix within the second finger (18; see below). Besides the β-sheet, several other long-range interactions were detected between, for example, regions within the first and second α-helices, between the first finger and the *C*-terminal extended region, and between a cluster of aromatic and hydrophobic residues in both α-helices and into the two extended regions (88).

By refinement of distance geometry, the tertiary structures of the GR and ER DNA-binding domains in solution were determined independently (16, 17). Their overall structures are quite similar. As depicted in Figure 4, the ER DNA-binding domain has an oblate shape, with a short axis of 20 Å (2 nm) and a long axis of 35 Å (3.5 nm), and with one flat face and one conical face. Interestingly, the two α-helices are oriented perpendicularly to each other, somewhat reminiscent of the helix-turn-helix motif found in many other DNA-binding proteins (89). Together with the extended regions that follow them, the two α-helices form a rather compact, hydrophobic core with a diameter of approximately 20 Å (2 nm). In the GR, several conserved hydrophobic and aromatic residues (Phe463, Phe464,

Fig. 4 2-D NMR structure of the ER DNA-binding domain. Stereo-view of an overlay of 49 confomers (shown by Cα atoms). Each confomer was calculated independently using a set of 351 inter-residue distance constraints and 23 × 1 angle constraints derived from 2-D NMR data. The reader has a 'DNA's eye view' of the domain, with the DNA-recognition helix oriented vertically. The *N*-terminus of the domain is at the bottom of the figure and the *C*-terminus on the left-hand side. The region to the right in the figure is poorly structured in solution but becomes ordered when the domain binds DNA. Figure courtesy of Dr J. Schwabe, Medical Research Council, Cambridge, UK. See also reference 17

Ala467, Val468, Ala494, Tyr497, Leu501, Leu507) make up this core, suggesting that the hydrophobic core is a critical and common structural component to all receptor DNA-binding domains. Indeed, Schwabe *et al.* (17) described a virtually identical interhelical hydrophobic core for the ER DNA-binding domain, involving Tyr19, Val21, Phe30, Phe31, Ile35, Tyr41, Leu64, Cys67, Tyr68, Met72, and Lys74. In the GR, the two zinc atoms are located outside the core, 13 Å (1.3 nm) apart from each other. The first finger is folded on top of the core, making several contacts with the two α-helices and the *C*-terminal extended region. The second finger, which is less defined than the first in both the GR and ER NMR structures (16, 17), appears to form two loops as it projects out from the hydrophobic core, involving Ala477–Asp481 and Ile483–Asn491. The two fingers make several contacts, principally between Val442–Cys443 and Asp485–Ile487.

Based on this structure and the previously described genetic and biochemical data, Hard *et al.* (16) proposed a model for how the GR DNA-binding domain forms a specific complex with a GRE. A similar model has been proposed by Schwabe and colleagues (17) for the ER DNA-binding domain. In the model, two monomers would bind to each HRE half-site in a head-to-head, symmetrical orientation, mediated in part by protein–protein contacts between residues within the D-box at the base of the second finger. The orientation is such that the first α-helix (Ser459–Glu469 in the GR) presents itself to bases in the major groove of each half-site. GR residues shown to be critical for specificity (Gly458, Ser459, Val462) are close to the discriminating T–T bases in each GRE half-site, and would presumably be able to make direct side-chain contacts. In addition, conserved basic residues within the 'recognition helix', Lys461, Lys465, and R466, would also be able to make close contacts with bases within the major groove. On the other hand, the fingers themselves are oriented away from the DNA, in contrast to the role of the Cys_2–His_2 type fingers, in which an α-helix *within* the *C*-terminal side of the finger loop itself appears to make critical contacts with DNA (90–92). In the NMR structure of the ER DNA-binding domain, the second finger loop is in a position such that two conserved basic residues, Arg56 and Lys57, can make minor groove interactions with the phosphate backbone (17). Consistent with this, the corresponding residues in GR, Lys489 and Arg490, are found to make phosphate backbone contacts in the X-ray structure of the GR DNA-binding domain–GRE complex, and are in fact part of a distorted α-helix that begins and ends within the finger loop (18). Since the distorted helix is not seen in the NMR structures, the presence of this helix may be a dimer-induced structural change that would be present only in the protein–DNA complex of the crystal structure, since the protein is a monomer in the absence of DNA.

4.3 Crystallographic analysis of the GR zinc finger–DNA complex

Crystal structures of the GR DNA-binding domain were determined by Luisi and colleagues (18) using the same fragment, GR440–525, that was used in the NMR structural analysis, but in this case complexed to DNA. Thus, the crystal structures obtained, while to a large extent confirming the 2-D NMR structures of GR and ER and the predictions they made concerning the mechanism of DNA binding, have provided substantially more mechanistic detail, since they were crystallized as protein–DNA complexes. The structures, solved by multiple isomorphous replacement, are of the same protein bound to two different GRE oligonucleotides. Both target sites were conceived as idealized GREs, organized as symmetrical inverted repeats (Fig. 2), but differing in the number of bases between the two half-sites. GRE_s3 has the typical spacing of three nucleotides; GRE_s4 has a spacing of four nucleotides. The protein–GRE_s3 complex has been resolved to 4.0 Å (0.4 nm); the protein GRE_s4 complex to 2.9 Å (0.29 nm).

The crystal and 2-D NMR studies have yielded roughly the same tertiary structure, although neither the GR nor ER NMR structures has revealed a distorted helix present in the second finger, as mentioned above, nor a short region of anti-parallel β structure at the tip of the first finger that appears in crystal structure. Nevertheless, the ER 2-D NMR and GR crystal structures are strikingly similar and the closest of the three. The principal difference is in the conformation of the second finger, in which the GR 2-D NMR structure deviates significantly from the other two structures. This suggests that this portion of the second GR finger was incorrectly resolved in the 2-D NMR structure.

In the crystal structure, the β-sheet at the tip of the first finger appears to orient His451 and Tyr452 to make specific contacts with the DNA phosphate backbone. His451 is also part of a hydrogen-bonded network involving residues in the tip of the first finger. In this network, the Nε of His451 accepts a hydrogen bond from the OH group of Ser448, and Ser448 is in turn an acceptor for the amide group of Gly458. Interestingly, His451 is absolutely conserved in the nuclear receptor superfamily, and either Ser or Thr is present at position 448. Further, Asp445 and Ser459 accept hydrogen bonds from Arg489, another conserved residue that interacts with the phosphate backbone. Arg489 also donates a hydrogen bond to Asn480 at the base of the second finger. Thus, the extensive interactions of Arg489, both with amino acids in each finger and with the phosphate backbone of DNA, imply that it is required to stabilize the overall fold of the DNA-binding domain, both structurally and functionally. Finally, the crystal structure also reveals that while the geometry of the amino-terminal Zn–S centre is in the *S*-configuration, typical of many other metal-binding proteins (91, 93), the Zn co-ordination of the carboxy-terminal finger assumes the *R*-configuration. Luisi *et al.* (18) speculated that the large number of amino acids (five) between the first and second co-ordinating cysteines in the second finger may allow for a greater conformational flexibility, resulting in a chirality opposite that of the first finger and most other metal co-ordination sites.

As was proposed by modelling of the GR (16) and ER (17) 2-D NMR structures, and from methylation protection and interference experiments (31–35), two monomers of GR440–525 in the crystal structure (18) bind to one face of the target GRE oligonucleotide as a symmetrical dimer, making contacts with the major groove only. Two residues of the amino-terminal finger loop, His451 and Tyr452,

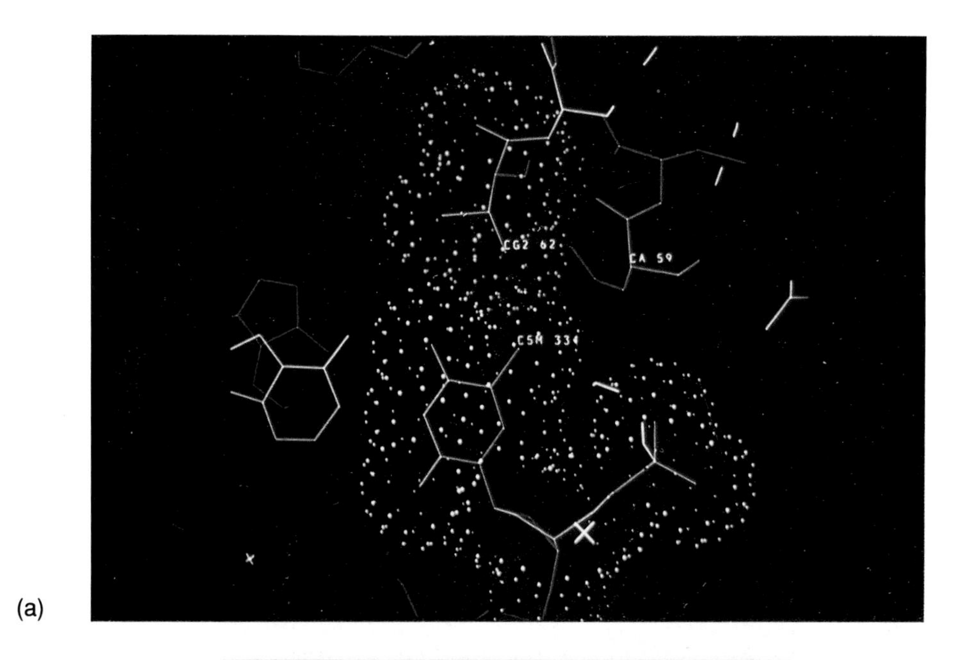

(a)

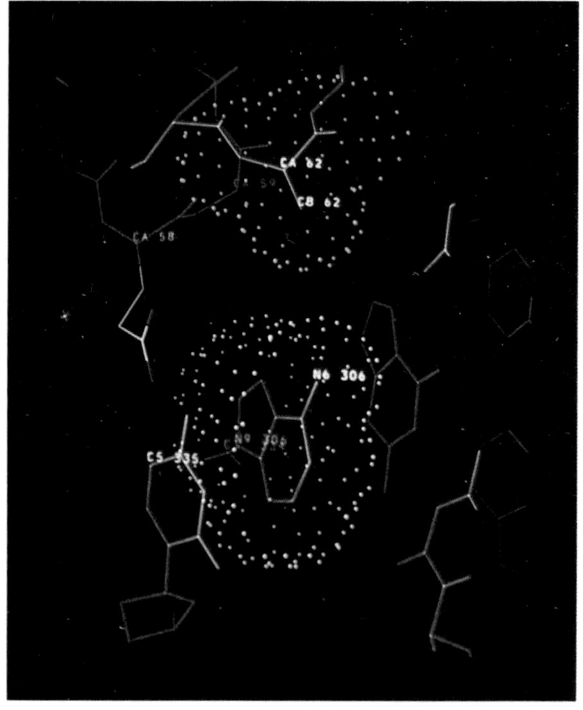

(b)

Fig. 6 Specific side-chain contacts with bases in the GRE or ERE. (a) The methyl group of GR Val 462 ('CG2 62)' makes a van der Waals' contact with the 5-methyl group of T(5) of a GRE ('C5M 334'); shown are the van der Waals' surfaces. (b) Proposed interaction of ER Glu203 carboxyl group (red ends of 'CA 58') with 4-amino group of C(6) of an ERE ('C5 335'). Note that ER Ala207 ('CB 62') is distal to both the ERE-specific A(5) and C(6), suggesting that this residue is not making a specifying side-chain contact.

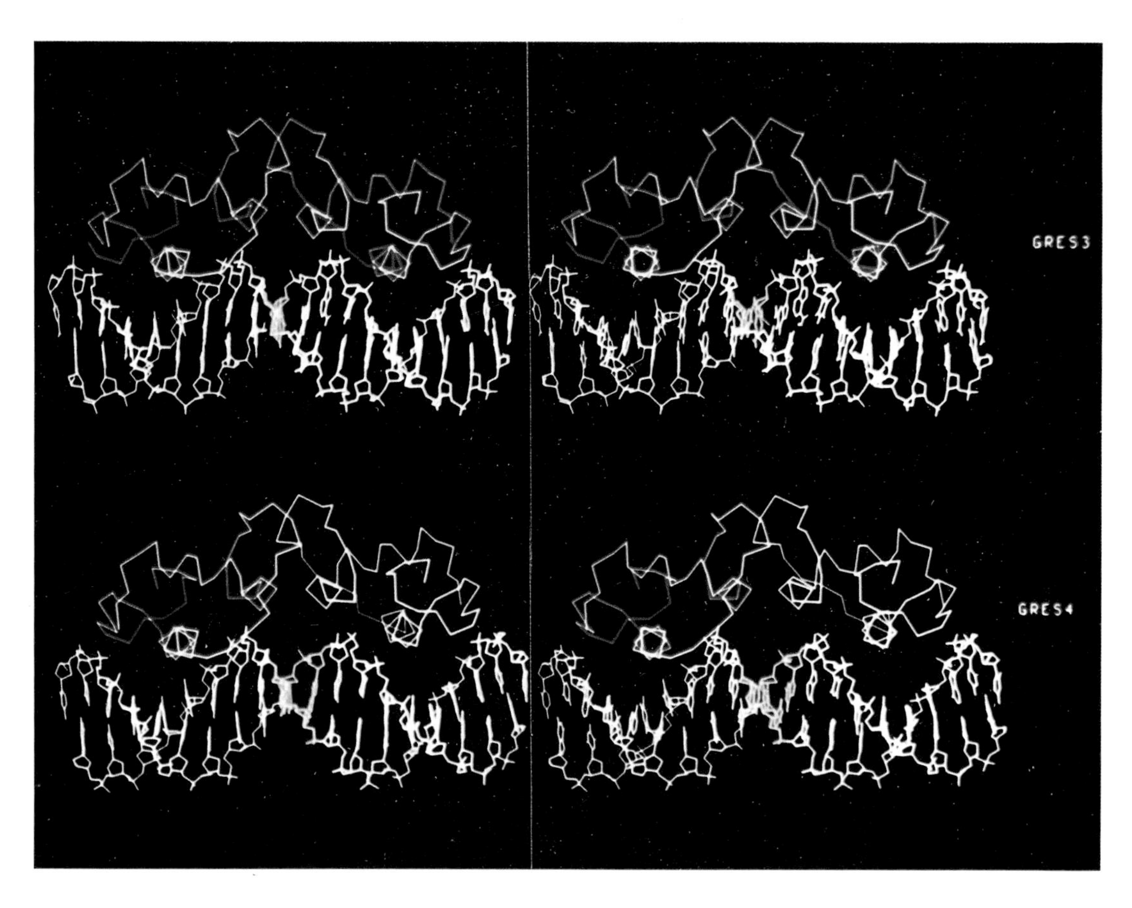

Fig. 7 Stereoscopic view of the GR DNA-binding domain—GRE complex derived from two crystal structures. Top: Dimeric protein binding to a GRE_s3. Note that the dyads of the protein and the DNA are aligned. Bottom: Dimeric protein binding to a GRE_s4, where the axis of the dimer is displaced by approximately one-half base pair relative to the dyad axis of the GRE (depicted in yellow). The non-specific monomer is shown in blue; the subunit making specific contacts is orange. Reprinted by permission from *Nature*, **352,** 497–505. © 1991 Macmillan Magazines Ltd.

make phosphate contacts with the DNA strand at positions furthest from the dyad that separates the two half-sites (at A(−8) and G(−7), respectively); two residues in the carboxy-terminal finger, Arg489 and Lys491, make phosphate contacts very close to and within the spacer, respectively (G(+4) and N(+1)). Only three residues, all within the amino-terminal α-helix, make direct side-chain contacts with bases. G(+4), which appears to be a conserved feature of steroid response elements, interacts via two hydrogen bonds with Arg466 (Fig. 5), an amino acid that is absolutely conserved throughout the nuclear receptor superfamily. Not surprisingly, mutation of this residue to Lys or Gly renders the GR non-functional both *in vivo* and *in vitro* (5, 15). Another completely conserved amino acid in the α-helix, Lys461, donates a hydrogen bond directly to G(−7) and, via a water molecule to G(−7) and T(+6) (Fig. 5). G(−7) is also found in most HREs. The third residue, Val462, makes a van der Waals' contact via two methyl side-chain groups with the 5-methyl group of T(+5) (Fig. 5 and Fig. 6a). In an ERE, position 5 is occupied by an adenine, and the amino acid that corresponds to Val462 in the ER is Ala (or Gly or Ser in several other nuclear receptors). Thus, at the current level of resolution of

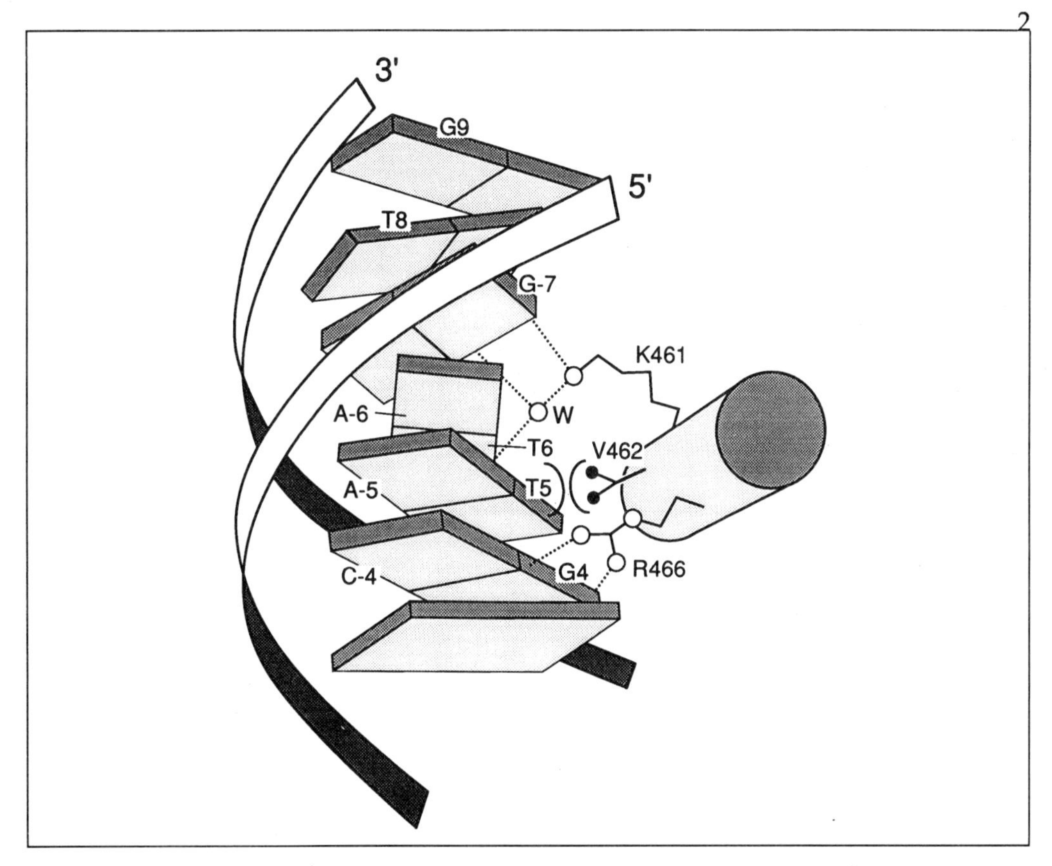

Fig. 5 Direct side-chain contacts by GR amino acids of the first α-helix to bases within the major groove of one GRE half-site. Reprinted by permission from *Nature*, **352**, 497–505. © 1991 Macmillan Magazines Ltd

the crystal structure, it is the interaction of Val462 with T(+5) that confers much of the protein's target site specificity. But from mutagenesis data (12–14, 94; see section 3.2), two additional residues within the α-helix, which like valine are not conserved among members of the nuclear receptor superfamily (but like Val462, are conserved within a subgroup consisting of the PR, AR, and MR), appear to be critical for discrimination: Gly458 and Ser459. Swapping these amino acids for the corresponding amino acids present in ER, Glu, and Gly is sufficient to change binding specificity to an ERE, even though the specifying GR valine is still present (94). On the other hand, making a single substitution at Val462 with the corresponding residue from ER, alanine, does not convert the protein to an ERE-binding species. The conclusion one might draw from these mutants is that while residues within the first α-helix of both the GR and ER DNA-binding domains are clearly critical for discrimination, their positions are not equivalent. That is to say, while valine at position 462 within GR is necessary for GRE recognition, the corresponding alanine in ER is not the determining residue for ERE recognition. Rather, alanine may be contributing to target site discrimination not by increasing affinity to specific bases, but by providing effective *repulsion* to a non-target site, such as a GRE. It has been proposed (94) that the ER-specific glutamic acid, five amino acids more *N*-terminal than the alanine, is what specifies protein binding to an ERE. This could be done via a direct side-chain contact of Glu to the amino group of C(+6) (Fig. 6b) in the major groove, one of two bases within an ERE half-site that distinguishes it from a GRE.

What of the roles of Gly458 and Ser459, which, while within the amino terminal α-helix, do not appear from the GR crystal structure to contact the DNA and yet functionally appear to be important for specificity? It is possible that their effect is indirect. For example, Ser459 interacts with the conserved Arg489, which in turn contacts phosphates within the GRE DNA (see above). This interaction may not be critical for target site discrimination *per se*, but may affect affinity. The presence of glycine either at the second (for GR) or third (for ER) position from the beginning of the helix (in both cases between two of four Zn^{2+} co-ordinating cysteines) might be important in allowing the α-helix to fit optimally into the major groove, owing to glycine's small size, or as an initiator of the α-helix due to its relative flexibility (64). Interestingly, a glycine is present at one of these two positions in every member of the nuclear receptor superfamily.

As mentioned above, two GRE-binding sites were used in crystal structure determination, differing only in their half-site spacing (18). The rationale behind the use of the GRE_s4 oligonucleotide duplex was to provide a perfectly symmetrical target site, i.e. the dyad of symmetry is *between* the second and third base pairs of the spacer, rather than *through* the second base pair of the spacer in GRE_s3. Ultimately, a comparison of each of the crystal structures (18) revealed a critical role for the spacer between each half-site: monomeric GR DNA-binding domains dimerize on DNA such that they interact specifically with each half-site only if those sites are separated by 3 bp. The complex solved with GRE_s4 revealed that one monomer associated specifically with one half-site, and the second monomer bound nonspecifically. From the structure, it was clear that residues within the second finger form critical contacts between monomers that restrict this interaction to the equivalent distance of three nucleotides. That is, the monomers need to be in complementary register with the binding sites. This registry is skewed in the structure of the GR DNA-binding domain–GRE_s4 complex. As depicted in the stereoscopic view in Figure 7, the dyad axis passing through bases −1 and +1 (e.g. in the middle of the spacer) does not coincide with the dyad formed by the dimeric protein, which is displaced by the equivalent of approximately 0.5 bp towards the monomer (in orange) making specific contacts with DNA. In contrast, when the half-site spacing is three nucleotides, the protein and DNA dyads align, and residues in each monomer make equivalent, specifying contacts with bases and phosphates (Fig. 7). Because this dimer interface is stabilized by DNA binding, any deviation from 3 bp would result in a loss of the positively co-operative binding that the GR DNA-binding domain exhibits upon association with a GRE. This is in fact what is observed experimentally: GR cannot transactivate from a binding site like GRE_s4 (14, 42), and GR and ER DNA-binding domains bind non-cooperatively to GRE_s4 and ERE_s4 oligonucleotides *in vitro*, respectively (56, 76; L. P. Freedman unpublished results; J. Schwabe, unpublished results) (see section 3.3).

The dimer interface is mediated exclusively by residues within the second zinc finger. Three of these residues — Ala477, Arg479, and Asp481 — are located at the *N*-terminal base of the second finger, between co-ordinating Cys476 and Cys482, that was originally inferred by mutagenesis to be an important determinant for half-site spacing — the D-box (section 3.3). The crystal structure shows that the residues corresponding to the D-box form a reverse β turn; this conformation appears to be maintained by the *R*-configuration of the second finger zinc atom. Within the *D*-box, Arg479 and Asp481 of each monomer form a symmetrical pair of inter-subunit salt bridges. The carbonyl oxygen of Ala477 in each monomer hydrogen bonds with the NH of Ile483, a residue outside the D-box. In fact, the structure indicates that the dimer interface is not restricted to residues within the D-box alone: a symmetrical pair of hydrogen bonds is also formed between the carbonyl O of Asn491 in one monomer and the NH_2 side-chain of Asn491 of the other, and a hydrophobic contact occurs between Ile487 and Leu475.

Both the GR and ER DNA-binding domains are monomeric in solution (15, 16, 74), indicating that the dimer interface within the second zinc finger is relatively weak. On the other hand, GRE_s4 crystal structure indicates that one monomer is forced to interact with a non-specific sequence (18), suggesting that the stability of the interface is greater than any gain in stability conferred by the monomer interacting specifically with bases of that half-site. Otherwise, the monomers would be exposed to pull apart and bind specifically but independently to each of the two half-sites. Thus, Luisi *et al.* (18) view the target DNA as an allosteric effector of GR recognition, in that it provides the correct stereospecificity for dimerization, and in doing so stabilizes the correct interactions required to provide a strong dimer interface. Whether or not native, full-length steroid receptors, which appear to dimerize in solution, require the same stabilizing contribution by

DNA is unclear, but the fact that they too require properly spaced half-sites from which to activate transcription is at least consistent with the notion that this same mechanistic constraint is imposed on the full-length proteins.

A similar binding strategy was recently described for the yeast zinc finger transcription factor GAL4 (95). The recognition element for GAL4, like that of steroid receptors, is palindromic, and the GAL4 DNA-binding domain, normally monomeric in solution, dimerizes symmetrically upon binding DNA. In contrast to the GR structure, however, the crystal structure of this domain bound to its recognition site reveals that an α-helix interacts with major groove bases of each half-site that are one and a half helical turns apart, placing monomers on opposite faces of DNA. In addition, GAL4 binds two Zn(II) ions through six cysteines (rather than two pairs of four), in which two of the cysteines are bridging ligands (85). In this way, the GAL4 DNA-binding domain more closely resembles metallothioneine than the Cys_2–His_2 TFIIA-like fingers or the Cys_4 steroid receptor fingers. Nevertheless, in all three variations of the zinc finger motif, metal coordination is used to expose an α-helix that slips into the major groove and interacts with specific bases.

5. Stereochemical basis of nuclear receptor dysfunction in disease

The crystal structure of the GR–DNA complex provides a useful reference for understanding the molecular basis of nuclear receptor dysfunctions occurring in certain inherited diseases (Chapter 9). Many such diseases have been found to be caused by mutations within the coding regions of nuclear receptors, and several have been mapped to the DNA-binding domain. For instance, in four cases of vitamin D-resistant inherited rickets, defects have been traced to mutations in the VDR zinc finger region that alter amino acids conserved in the superfamily. One mutation occurs at position 47 of VDR (residue 466 in the GR), where a conserved Arg is replaced by Gln (96). From the GR crystal structure, this mutation lies in the middle of the recognition helix, where Arg makes a direct side-chain contact with G(4), a base that is generally present in all HREs. A Gln at this position could not provide the same interaction. Another conserved Arg in the second finger (residue 77 in VDR; residue 496 in GR), is also substituted by Gln in some resistant patients (97). Gln could not make the same critical phosphate backbone contact that Arg does in the GR structure. Members of a third family with inherited vitamin D-resistant rickets carry yet another Arg → Gln mutation in the second finger of the VDR at amino acid 70 (98), analogous to position 489 in GR. Arg489 in the GR structure donates a hydrogen bond to Asn480 at the base of the second finger, as well as to several other residues in the first finger (see section 4.3). The extensive interactions of Arg489, both with amino acids in each finger and with the phosphate backbone of DNA, imply that it is required to stabilize the overall fold of the DNA-binding domain, both structurally and functionally. Gln at this position cannot provide the same hydrogen bonding pattern as Arg. Finally, members of a fourth family suffering from rickets were found to have a mutation at Gly 40 at the tip of the first finger of VDR (equivalent to position 449 in GR) (98). Gly at this position, which is conserved throughout the nuclear receptor superfamily, ensures a special peptide-backbone conformation that is not favoured for any other amino acid. This conformation is required to orient the nearby conserved His (His31 in VDR), which contacts the DNA phosphate backbone by donating a hydrogen bond (B. F. Luisi, personal communication). An analogous mutation occurs at the same position in the androgen receptor (Gly559) from an individual with partial androgen resistance, except that the change is to Val. Two other patients suffering from complete androgen insensitivity syndrome carry mutations in one of the four co-ordinating cysteines of the first finger (Y. T. Chang and T. Brown, unpublished results). Finally, two children with partial androgen resistance carry a Leu → Arg substitution at position 616 (A. De Bellis and F. French, personal communication) (residue 497 in GR). An Arg at this positiuon would conceivably disrupt the hydrophobic core that stabilizes the packing of the two α-helices of the finger domain.

6. Conclusions

Of the various functional domains within nuclear receptors, the DNA-binding domain is the most highly conserved. Those conserved amino acids appear to play critical roles in both the structure and function of the domain. Thus, the structures of the zinc finger regions of many, if not all, of the members of the nuclear receptor superfamily are likely to be similar to those described for the GR and ER by 2-D NMR and crystallographic analyses. In addition, these structures are probably a fairly close reflection of a similar domain within an ancient progenitor receptor that gave rise to the superfamily (99, 100). On the other hand, some of the mechanistic details of the interaction between the various receptors and specific DNA target sites are likely to vary among the subgroups of receptors. For example, steroid receptors recognize and bind response elements organized as inverted repeats. As discussed in this review, the structure of the steroid receptor DNA-binding domain is such that zinc co-ordination stabilizes both an α-helix at the carboxy-terminal end of the first finger that is the principal determinant for specificity, and a dimer interface in the second finger that insures a twofold symmetrical orientation of receptor monomers in register with properly spaced half-sites. Other receptors, however, such as those for vitamin D_3, thyroid hormone, and retinoic acid, can bind to and activate transcription from half-sites oriented as direct repeats, differing only in the length of the spacer sequence between the half-sites. This suggests that this class of receptors bind DNA as dimers without twofold symmetry, and that spacing may confer receptor specificity by allowing or prohibiting dimer binding to DNA, rather than through positive co-operativity. If this is indeed the case, then subtle but critical differences should be found in the structures of the DNA-binding domains of these receptors that distinguish them from the details gleaned from the three-dimensional structures of the GR and ER DNA-binding domains. Finally, the functional and structural relationships between the DNA-binding domain and other receptor domains, such as that for ligand binding, will be of keen interest from the vantage points of both the effects on the structure of the ligand itself, and how the association of other nuclear receptors (68–74, 101–103) — or non-receptor nuclear factors (56, 75–77) — affect affinity and selectivity of DNA binding, as well as the modulation of transcriptional initiation. The future awaits such structural descriptions.

References

1. Frankel, A. D. and Kim, P. S. (1991) Modular structure of transcription factors. *Cell,* **65,** 717.
2. Picard, D., Salser, S. J., and Yamamoto, K. R. (1988) A movable and regulable inactivation function within the steroid binding domain of the glucocorticoid receptor. *Cell,* **54,** 1073.
3. Godowski, P. J., Picard, D., and Yamamoto, K. R. (1988) Signal transduction and transcriptional regulation by glucocorticoid receptor–LexA fusion proteins. *Science,* **241,** 812.
4. Webster, N. J. G., Green, S., Jin, J. R., and Chambon, P. (1988) The hormone-binding domains of the estrogen and glucocorticoid receptors contain an inducible transcription activation function. *Cell,* **54,** 199.
5. Hollenberg, S. M. and Evans, R. M. (1988) Multiple and cooperative trans-activation domains of the human glucocorticoid receptor. *Cell,* **55,** 899.
6. Eilers, M., Picard, D., Yamamoto, K. R., and Bishop, J. M. (1989) Chimeras of *myc* oncoprotein and steroid receptors cause hormone-dependent transformation of cells. *Nature,* **340,** 66.
7. Umek, R. M., Friedman, A. D., and McKnight, S. J. (1991) CCAAT-enhancer binding protein: a component of a differentiation switch. *Science,* **251,** 288.
8. Freedman, L. P., Luisi, B. F., Korszun, Z. R., Basavappa, R., Sigler, P. J., and Yamamoto, K. R. (1988) The function and structure of the metal coordination sites within the glucocorticoid receptor DNA binding domain. *Nature,* **334,** 543.
9. Miller, J., McClachlin, A. D., and Klug, A. (1985) Repetitive zinc-binding domains in the protein transcription factor IIIA from *Xenopus* oocytes. *EMBO J.,* **4,** 1609.
10. Brown, R. S., Sander, C., and Argos, P. (1985) The primary structure of transcription factor TFIIIA has 23 consecutive repeats. *FEBS Lett.,* **186,** 271.
11. Danielson, M., Hinck, L., and Ringold, G. M. (1989) Two amino acids within the knuckle of the first zinc finger specify response element activation by the glucocorticoid receptor. *Cell,* **57,** 1131.
12. Mader, S., Kumar, V., deVereneuil, H., and Chambon, P. (1989) Three amino acids of the oestrogen receptor are essential to its ability to distinguish an oestrogen from a glucocorticoid-responsive receptor. *Nature,* **338,** 271.
13. Umesono, K. and Evans, R. M. (1989) Determinants of target gene specificity for steroid/thyroid hormone receptors. *Cell,* **57,** 1139.
14. Dahlman-Wright, K., Wright, A., Gustafsson, J. A., and Carstedt-Duke, J. (1991) Interaction of the glucocorticoid receptor DNA-binding domain with DNA as a dimer is mediated by a short segment of five amino acids. *J. Biol. Chem.,* **266,** 3107.
15. Schena, M., Freedman, L. P., and Yamamoto, K. R. (1989) Mutations in the glucocorti-

coid receptor zinc finger region that distinguish interdigitated DNA binding and transcription enhancement activities. *genes Dev.*, **3**, 1590.
16. Hard, T., Kellenbach, E., Boelens, R., Maler, B. A., Dahlman, K., Freedman, L. P., Carlstedt-Duke, J., Yamamoto, K. R., Gustafsson, J. A., and Kaptein, R. (1990) Solution structure of the glucocorticoid receptor DNA binding domain. *Science*, **249**, 157.
17. Schwabe, J. W. R., Neuhaus, D., and Rhodes, D. (1990) Solution structure of the DNA binding domain of the oestrogen receptor. *Nature*, **348**, 458.
18. Luisi, B. F., Xu, W. X., Otwinoski, Z., Freedman, L. P., Yamamoto, K. R., and Sigler, P. B. (1991) Crystallographic analysis of the interaction of the glucocorticoid receptor with DNA. *Nature*, **352**, 497.
19. Yamamoto, K. R. (1985) Steroid receptor regulated transcription of specific genes and gene networks. *Ann. Rev. Genet.*, **19**, 209.
20. Chandler, V. L., Maler, B. A., and Yamamoto, K. R. (1983) DNA sequences bound specifically by glucocorticoid receptor *in vitro* render a heterologous promoter hormone responsive *in vivo*. *Cell*, **33**, 489.
21. Payvar, F., DeFranco, D., Firestone, G. L., Edgar, B., Wrange, O., Okret, S., Gustafsson, J. A., and Yamamoto, K. R. (1983) Sequence-specific binding of the glucocorticoid receptor to MTV-DNA at sites within and upstream of the transcribed region. *Cell*, **35**, 81.
22. Scheidereit, C., Geisse, S., Westphal, H. M., and Beato, M. (1983) The glucocorticoid receptor binds to defined nucleotide sequences near the promoter of mouse mammary tumour virus. *Nature*, **304**, 749.
23. Beato, M., Chalepakis, G., Schauer, M., and Slater, E. P. (1989) DNA regulatory elements for steroid hormones. *J. Steroid Biochem.*, **32**, 737.
24. Beato, M. (1989) Gene regulation by steroid hormones. *Cell*, **56**, 335.
25. Naar, A. M., Boutin, J. M., Lipkin, S. M., Yu, V. C., Holloway, J. M., Glass, C. K., and Rosenfeld, M. G. (1991) The orientation and spacing of core DNA-binding motifs dictate selective transcriptional responses to three nuclear receptors. *Cell*, **65**, 1267.
26. Strahle, U., Klock, G., and Schutz, G. (1987) A DNA sequence of 15 base pairs is sufficient to mediate both glucocorticoid and progesterone induction of gene expression. *Proc. Natl Acad. Sci. USA*, **84**, 7871.
27. Darbre, P., Page, M., and King, R. J. B. (1986) Androgen regulation by the long terminal repeat of mouse mammary tumour virus. *Mol. Cell. Biol.*, **6**, 2847.
28. Cato, A. C. B., Henderson, D., and Ponta, H. (1987) The hormone response element of the mouse mammary tumour virus DNA mediates the progestin and androgen induction of transcription in the proviral long terminal repeat region. *EMBO J.*, **6**, 363.
29. Ham, J., Thomson, A., Neddham, M., Webb, P., and Parker, M. (1988) Characterization of response elements for androgens, glucocorticoids and progestins in mouse mammary tumour virus.*Nucleic Acids Res.*, **16**, 5263.
30. Cato, A. C. B. and Weinmann, J. (1988) Mineralocorticoid regulation of transfected mouse mammary tumour virus DNA in cultured kidney cells. *J. Cell Biol.*, **106**, 2119.
31. LaBaer, J. (1989) A detailed analysis of the DNA-binding affinity and sequence specificity of the glucocorticoid receptor DNA-binding domain. PhD thesis, University of California, San Francisco.
32. Nordeen, S. K., Suh, B. J., Kuhnel, B., and Hutchison, C. A. (1990) Structural determinants of a glucocorticoid receptor recognition element. *Mol. Endocrinol.*, **4**, 1866.
33. Scheidereit, C. and Beato, M. (1984) Contacts between receptor and DNA double helix within a glucocorticoid regulatory element of mouse mammary tumor virus. *Proc. Natl Acad. Sci. USA*, **81**, 3029.

34. Scheidereit, C., Westphal, H. M., Carlson, C., Boshard, H., and Beato, M. (1986) Molecular model of the interaction between the glucocorticoid receptor and the regulatory elements of inducible genes. *DNA*, **5**, 383.
35. Truss, M., Chalepakis, G., and Beato, M. (1990) Contacts between steroid hormone receptors and thymines in DNA: a new interference method. *Proc. Natl Acad. Sci. USA*, **87**, 7180.
36. Cairns, C., Gustafsson, J. A., and Carlstedt-Duke, J. (1991) Identification of protein contact sites within the glucocorticoid/progestin response element. *Mol. Endocrinol.*, **5**, 598.
37. Klock, G., Strahle, U., and Schutz, G. (1987) Oestrogen and glucocorticoid responsive elements are closely related but distinct. *Nature*, **329**, 734.
38. Truss, M., Chalepakis, G., Slater, E. P., Mader, S., and Beato, M. (1991) Functional interaction of hybrid response elements with wild type and mutant steroid hormone receptors. *Mol. Cell. Biol.*, **11**, 3247.
39. Umesono, K., Giguere, V., Glass, C. K., Rosenfeld, M. G., and Evans, R. M. (1988) Retinoic acid and thyroid hormone nduce gene expression through a common responsive element. *Nature*, **336**, 262.
40. Glass, C. K., Holloway, J. M., Devary, O. V., and Rosenfeld, M. G. (1988) The thyroid hormone receptor binds with opposite transcriptional effects to common sequence motif in thyroid hormone and estrogen response elements. *Cell*, **54**, 313.
41. Umesono, K., Murakami, K. K., Thompson, C. C., and Evans, R. M. (1991) Direct repeats as selective response elements for the thyroid hormone, retinoic acid, and vitamin D_3 receptors. *Cell*, **65**, 1255.
42. Wrange, O. and Gustafsson, J. A. (1978) Separation of the hormone- and DNA binding sites of the hepatic glucocorticoid receptor by means of proteolysis. *J. Biol. Chem.*, **253**, 856.
43. Carlstedt-Duke, J., Stromstedt, P. E., Wrange, O., Bergman, T., Gustafsson, J. A., and Jornvall, H. (1987) Domain structure of the glucocorticoid receptor. *Proc. Natl Acad. Sci. USA*, **84**, 4437.
44. Rusconi, S. and Yamamoto, K. R. (1987) Functional dissection of the hormone and DNA binding domains of the glucocorticoid receptor. *EMBO J.*, **6**, 1309.
45. Danielson, M., Northrop, J. P., Jonklaas, J., and Ringold, G. M. (1987) Domains of the glucocorticoid receptor involved in specific and nonspecific deoxyribonucleic acid binding, hormone activation, and transcriptional enhancement. *Mol. Endocrinol.*, **1**, 816.
46. Sakai, D. D., Helms, S., Carlstedt-Duke, J., Gustafsson, J. A., Rottman, F. M., and Yamamoto, K. R. (1988) Hormone-mediated repression of transcription: a negative glucocorticoid response element from the bovine prolactin gene. *Genes Dev.*, **2**, 1144.
47. Weinberger, C., Hollenberg, S. M., Rosenfeld, M. G., and Evans, R. M. (1985) Domain structure of human glucocorticoid receptor and its relationship to the v-*erb*A oncogene product. *Nature*, **318**, 670.
48. Miesfeld, R., Godowski, P. J., Maler, B. A., and Yamamoto, K. R. (1987) Glucocorticoid receptor mutants that define a small region sufficient for enhancer activation. *Science*, **236**, 423.
49. Hollenberg, S. M., Giguere, V., Segui, P., and Evans, R. M. (1987) Colocalization of DNA binding and transcriptional activation functions in the human glucocorticoid receptor. *Cell*, **49**, 39.
50. Picard, D. and Yamamoto, K. R. (1987) Two signals mediate hormone-dependent nuclear localization of the glucocorticoid receptor. *EMBO J.*, **6**, 3333.

51. Danielson, M., Northrop, J. P., and Ringold, G. M. (1986) The mouse glucocorticoid receptor: mapping of functional domains by cloning, sequencing and expression of wild-type and mutant receptor proteins. *EMBO J.*, **5,** 2513.
52. Evans, R. M. and Hollenberg, S. M. (1988) Zinc fingers: gilt by association. *Cell,* **52,** 1.
53. Diakun, G. P., Fairall, L., and Klug, A. (1986) EXAFS study of the zinc-binding sites in the protein transcription factor IIIA. *Nature,* **324,** 698.
54. Frankel, A. D., Berg, J. M., and Pabo, C. (1987) Metal-dependent folding of a simple zinc finger from transcription factor IIIA. *Proc. Natl Acad. Sci. USA,* **84,** 4841.
55. Berg, J. (1990) Zinc fingers and other metal-binding domains. *J. Biol. Chem.,* **265,** 6513.
56. Yang-Yen, H. F., Chambard, J. C., Sun, Y. L., Smeal, T., Schmidt, T. J., Drouin, J., and Karin, M. (1990) Transcriptional interference between c-*jun* and the glucocorticoid receptor: mutual inhibition of DNA binding due to direct protein–protein interaction. *Cell,* **62,** 1205.
57. Klug, A. and Rhodes, D. (1987) 'Zinc fingers': a novel protein motif for nucleic acid recognition. *TIBS,* **12,** 464.
58. Metzger, D., White, J. H., and Chambon, P. (1988) The human oestrogen receptor functions in yeast. *Nature,* **334,** 31.
59. Schena, M. and Yamamoto, K. R. (1988) Mammalian glucocorticoid receptor derivatives enhance transcription in yeast. *Science,* **241,** 965.
60. Ponglikimongkol, M., Green, S., and Chambon, P. (1988) Genomic organization of the human oestrogen receptor gene. *EMBO J.,* **7,** 3385.
61. Jacobson, M. (1990) The structure and regulation of the rat glucocorticoid receptor gene. PhD thesis, University of California, San Francisco.
62. Green, S. and Chambon, P. (1987) Oestradiol induction of a glucocorticoid-response gene by a chimaeric receptor. *Nature,* **325,** 75.
63. Green, S., Kumar, V., Theulaz, I., Whali, W., and Chambon, P. (1988) The *N*-terminal DNA binding zinc finger of the oestrogen and glucocorticoid receptors determines target gene specificity. *EMBO J.,* **7,** 3037.
64. Berg, J. (1989) DNA binding specificity of steroid receptors. *Cell,* **57,** 1065.
65. Perlmann, T., Eriksson, P., and Wrange, O. (1990) Quantitative analysis of the glucocorticoid receptor–DNA interaction at the mouse mammary tumor virus glucocorticoid response element. *J. Biol. Chem.,* **265,** 17 222.
66. Kumar, V. and Chambon, P. (1988) The estrogen receptor binds tightly to its responsive element as a ligand-induced homodimer. *Cell,* **55,** 145.
67. Forman, B. M., Yang, C. R., Au, M., Cassanova, J., Ghysdael, J., and Samuels, H. H. (1989) A domain containing leucine zipper-like motifs mediate novel *in vivo* interactions between the thyroid hormone and retinoic acid receptors. *Mol. Endocrinol.,* **3,** 1610.
68. Fawell, S. E., Lees, J. A., White, R., and Parker, M. G. (1990) Characterization and localization of steroid binding and dimerization activities in the mouse estrogen receptor. *Cell,* **60,** 953.
69. Yu, V. C., DElsert, C., Andersen, B., Holloway, J. M., Devary, O. V., Naar, A. M., Kim, S. Y., Boutin, J. M., Glass, C. K., and Rosenfeld, M. G. (1991) RXRβ: a coregulator that enhances binding of retinoic acid thyroid hormone and vitamin D receptors to their cognate response elements. *Cell,* **67,** 1251.
70. Leid, M., Kastner, P., Lyons, R., Nakshatri, H., Saunders, M., Zacharewski, T., Chen, J. A., Staub, A., Garnier, J. M., Mader, S., and Chambon, P. (1992) Purification,

cloning, and RXR identity of the HeLa cell factor with which RAR or TR heterodimerizes to bind target sequences. *Cell*, **68,** 377.

71. Zhang, X. K., Hoffmann, B., Tran, P. B.-V., Graupner, G., and Pfahl, M. (1992) Retinoid X receptor is an auxillary protein for thyroid hormone and retinoic acid receptors. *Nature*, **355,** 441.
72. Kliewer, S. A., Umesono, K., Mangelsdorf, D. J., and Evans, R. M. (1992) Retinoic X receptor interacts with nuclear receptors in retinoic acid, thyroid hormone, and vitamin D_3 signalling. *Nature*, **355,** 446.
73. Bugge, T. H., Pohl, J., Lonnoy, O., and Stunnenberg, H. (1992) RXRα, a promiscuous partner of retinoic acid and thyroid hormone receptors. *EMBO J.*, **11,** 1409.
74. Marks, M. S., Hallenbeck, P. L., Nagata, T., Segars, J. H., Appella, E., Nikodem, V. M., and Ozato, K. (1992) H-2RIIBP (RXRβ) heterodimerization provides a mechanism for combinatorial diversity in the regulation of retinoic acid and thyroid hormone responsive genes. *EMBO J.*, **11,** 1419.
75. Diamond, M. I., Miner, J. N., Yoshinaga, S. K., and Yamamoto, K. R. (1990) Transcription factor interactions: selectors for positive or negative regulation from a single DNA element. *Science*, **249,** 1266.
76. Jonat, C., Rahmsdorf, H. J., Park, K. K., Cato, A. C. B., Gebel, S., Ponta, H., and Herrlich, P. (1990) Antitumor promotion and anti-inflammation: down-modulation of AP-1 (*fos/jun*) activity by glucocorticoid hormone. *Cell*, **62,** 1189.
77. Schule, R., Rangarajan, P., Kliewer, S., Ransone, L. J., Bolado, J., Yang, N., Verma, I. M., and Evans, R. M. (1990) Functional antagonism between oncoprotein c-*jun* and the glucocorticoid receptor. *Cell*, **62,** 1217.
78. Tsai, S. Y., Carlstedt-Duke, J., Weigel, N. L., Dahlman, K., Gustafsson, J. A., Tsai, M. J., and O'Malley, B. W. (1988) Molecular interactions of steroid hormone receptor with its enhancer element: evidence for receptor dimer formation. *Cell*, **55,** 361.
79. Freedman, L. P., Yamamoto, K. R., Luisi, B. F., and Sigler, P. J. (1988) More fingers in hand. *Cell*, **54,** 444.
80. Hard, T., Dahlman, K., Carlstedt-Duke, J., Gustafsson, J. A., and Rigler, R. (1990) Cooperativity and specificity in the interactions between DNA and the glucocorticoid receptor DBA-binding domain. *Biochemistry*, **29,** 5358.
81. Dahlman-Wright, K., Siltata-Roos, H., Carlstedt-Duke, J., and Gustafsson, J. A. (1990) Protein–protein interactions facilitate DNA binding by the glucocorticoid receptor DNA binding domain. *J. Biol. Chem.*, **265,** 14 030.
82. Freedman, L. P. and Towers, T. (1991) DNA binding properties of the vitamin D_3 receptor zinc finger region. *Mol. Endocrinol.*, **5,** 1815.
83. Dahlman, K., Stromsedt, P. E., Rae, C., Jornvall, J. I., Carlstedt-Duke, J., and Gustafsson, J. A. (1989) High level expression in *Escherichia coli* of the DNA binding domain of the glucocorticoid receptor in a functional form utilizing domain-specific cleavage of a fusion protein. *J. Biol. Chem.*, **264,** 804.
84. Nardulli, A. M., Lew, D., Erijman, L., and Shapiro, D. J. (1991) Purified estrogen receptor DNA binding domain expressed in *Escherichia coli* activates transcription of an estrogen responsive promoter in cultured cells. *J. Biol. Chem.*, **266,** 24070.
85. Pan, T., Freedman, L. P., and Coleman, J. E. (1990) ^{113}Cd NMR studies of the DNA binding domain of the mammalian glucocorticoid receptor. *Biochemistry*, **29,** 9015.
86. Wuthrich, K. (1989) Protein structure determination in solution by nuclear magnetic resonance spectroscopy. *Science*, **243,** 45.

87. Wright, P. (1989) What can two-dimensional NMR tell us about proteins? *TIBS*, **14**, 255.
88. Hard, T., Kellenbach, E., Boelens, R., Dahlman, K., Carlstedt-Duke, J., Freedman, L. P., Maler, B. A., Hyde, E. I., Yamamoto, K. R., Gustafsson, J. A., and Kaptein, R. (1990) ^{1}H-NMR studies of the glucocorticoid receptor DNA-binding domain: sequential assignments and identification of secondary structure elements. *Biochemistry*, **29**, 9015.
89. Harrison, S. C. (1991) A structural taxonomy of DNA-binding domains. *Nature*, **353**, 715.
90. Parraga, G., Horvath, S. J., Eisen, A., Taylor, W. E., Hood, L., Young, E. T., and Klevit, R. E. (1988) Zinc-dependent structure of a single-finger domain of yeast ADR1. *Science*, **241**, 1489.
91. Lee, M. S., Gippert, G. P., Soman, K. V., Case, D. A., and Wright, P. E. (1989) Three-dimensional solution structure of a single zinc finger DNA-binding domain. *Science*, **245**, 635.
92. Pavletich, N. P. and Pabo, C. O. (1991) Zinc finger–DNA recognition: crystal structure of a Zif268–DNA complex at 2.1 Å. *Science*, **252**, 809.
93. Berg, J. M. (1988) Proposed structure for the zinc-binding domains from transcription factor IIIA and related proteins. *Proc. Natl Acad. Sci. USA*, **85**, 99.
94. Alroy, I. and Freedman, L. P. (1992) DNA binding analysis of glucocorticoid receptor specificity mutants. *Nucleic Acids Res.*, **20**, 1045.
95. Marmomstein, R., Carey, M., Ptashne, M., and Harrison, S. C. (1992) DNA recognition by GAL4: structure of a protein–DNA complex. *Nature*, **356**, 408.
96. Sone, T., Kerner, S. A., Saijo, T., Takeda, E., and Pike, J. W. (1991) Mutations in the DNA binding domain of the vitamin D receptor associated with hereditary resistance of 1,25-dihydroxyvitamin D_3. In *Vitamin D: Gene Regulation, Structure–Function Analaysis, and Clinical Applications*, Norman, A. W., Bouillon, R., Thomasset, M. (eds). Walter de Gruyter, New York, p. 84.
97. Sone, T., Marx, S. J., Liberman, U. A., and Pike, J. W. (1990) A unique point mutation in human vitamin D receptor chromosomal gene confers hereditary resistance of 1,25-dihydroxyvitamin D_3. *Mol. Endocrinol.*, **4**, 623.
98. Hughes, M. R., Malloy, P. J., Kieback, D. G., Kesterson, R. A., Pike, J. W., Feldman, D., and O'Mally, B. W. (1989) Point mutations in the human vitamin D receptor gene associated with hypocalcemic rickets. *Science*, **242**, 1702.
99. Amero, S. A., Kretsinger, R. H., Moncrief, N. D., Yamamoto, K. R., and Pearson, W. R. (1992) The origin of nuclear receptor proteins: a single precursor distinct from other transcription factors. *Mol. Endocrinol.*, **6**, 3.
100. Laudet, V., Hanni, C., Coll, J., Catzflis, F., and Stehelin, D. (1992) Evolution of the nuclear receptor gene superfamily. *EMBO J.*, **11**, 1003.
101. Glass, C. K., Lipkin, S. M., Devary, O., and Rosenfeld, M. G. (1989) Positive and negative regulation of gene transcription by a retinoic acid–thyroid hormone heterodimer. *Cell*, **59**, 697.
102. Beebe, J. S., Darling, D. S., and Chin, W. W. (1991) 3,5,3′-Triiodothyronine (thyroid hormone) receptor–auxillary protein (TRAP) enhances receptor binding by interactions within the thyroid hormone response element. *Mol. Endocrinol.*, **5**, 85.
103. Sone, T., Ozono, K., and Pike, J. W. (1991) A 55-kilodalton accessory factor facilitates vitamin D receptor DNA binding. *Mol. Endocrinol.*, **5**, 1578.

8 Mechanism of action of hormone antagonists

S. DAUVOIS and M. G. PARKER

1. Introduction

Steroid hormone antagonists are compounds that compete with steroid hormones for receptor binding but fail to trigger the normal response and thereby inhibit the action of the hormone. They have a number of important clinical applications, particularly in the control of fertility and for the treatment of hormone-dependent cancer, and are also useful for elucidating the role of steroid hormones in the regulation of a wide range of physiological responses. The advantage of using anti-hormones to suppress hormone action compared with other approaches such as endocrine gland ablation is their greater degree of specificity: in general, their effects are restricted to target tissues for the hormone. Thus, while endocrine gland ablation is being used successfully to treat a number of clinical syndromes, specific hormone antagonists are now available which in certain circumstances are used as the preferred treatment.

This chapter outlines the current understanding of the molecular mechanism of action of hormone antagonists, focusing mainly on anti-oestrogens, but also summarizes the actions of antiandrogens and antiprogestins. Although the general pharmacology and clinical uses of steroid hormone antagonists are mentioned, the reader should consult other reviews for a more complete discussion (1–6).

2. Antioestrogens

2.1 General classification and structure

The first synthetic antioestrogen to be described was a triphenolic compound called MER25 (7). Although this antioestrogen was unsuitable as a drug, owing to its toxicity, it led to the synthesis of clomiphene (Fig. 1), which can be used to induce ovulation in subfertile women (8). Numerous antioestrogens were subsequently identified and analysed for their activity as contraceptives, but as it emerged that these compounds were unsuitable for this purpose they were investigated instead for their potential in the treatment of advanced breast cancer. Currently the main interest for the development and use of antioestrogens stems from

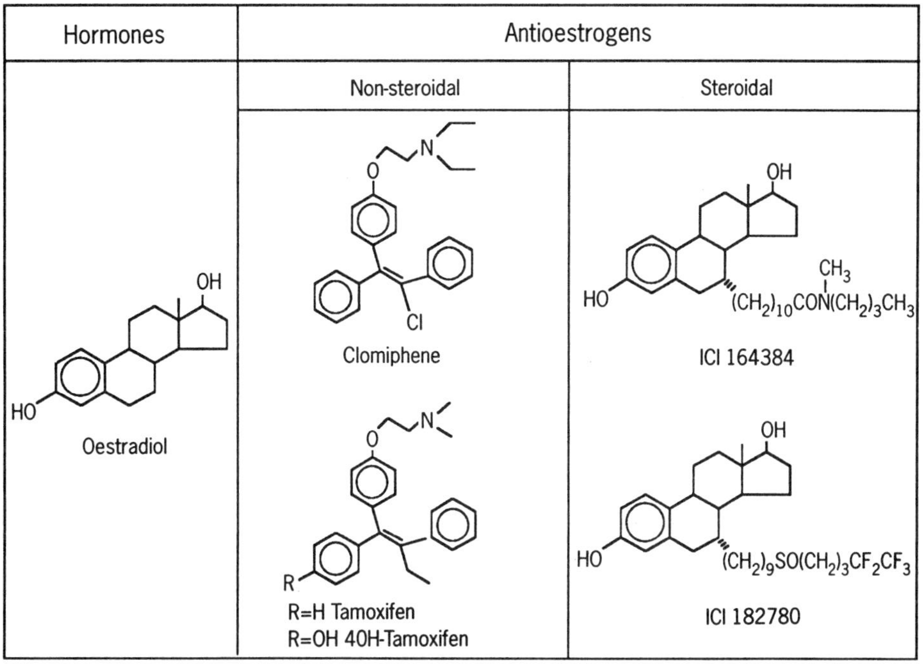

Hormones	Antioestrogens	
	Non-steroidal	Steroidal
Oestradiol	Clomiphene	ICI 164384 $(CH_2)_{10}CON(CH_3)(CH_2)_3CH_3$
	R=H Tamoxifen R=OH 4OH-Tamoxifen	ICI 182780 $(CH_2)_9SO(CH_2)_3CF_2CF_3$

Fig. 1 Structure of antioestrogens

the observation that oestrogen is a mitogen in breast cancer and that approximately one-third of patients will respond to endocrine therapy (9).

The main therapeutic antioestrogen that has been used to date is tamoxifen. Tamoxifen has a triphenyl-ethylene structure with two rings corresponding to the A and D rings of oestradiol. While the *trans* form of tamoxifen is an antagonist, the *cis* form is a weak agonist. Tamoxifen is converted, mainly in the liver, to 4-hydroxytamoxifen, and this conversion increases its affinity for receptor by approximately 100-fold and consequently its potency (10). Another important feature of the molecule includes the alkylamide side-chain, which is essential for its antioestrogenic activity, although the composition can be varied to some extent (see reference 1 for a review).

Although tamoxifen has turned out to be an effective drug for the treatment of hormone-dependent breast cancer it was realized during testing that it was actually a partial agonist. In other words, while tamoxifen is able to inhibit the action of oestrogen it exhibits some oestrogenic activity when it is examined by itself. The relative antagonistic and agonistic activities depend on the physiological response. This property, which is illustrated in an idealized fashion in Figure 2, has led to the term 'partial agonist' or 'mixed agonist–antagonist' to describe compounds such as tamoxifen. In view of this agonist activity, attempts were made to synthesize antioestrogens devoid of any oestrogenic activity. The steroidal antioestrogen

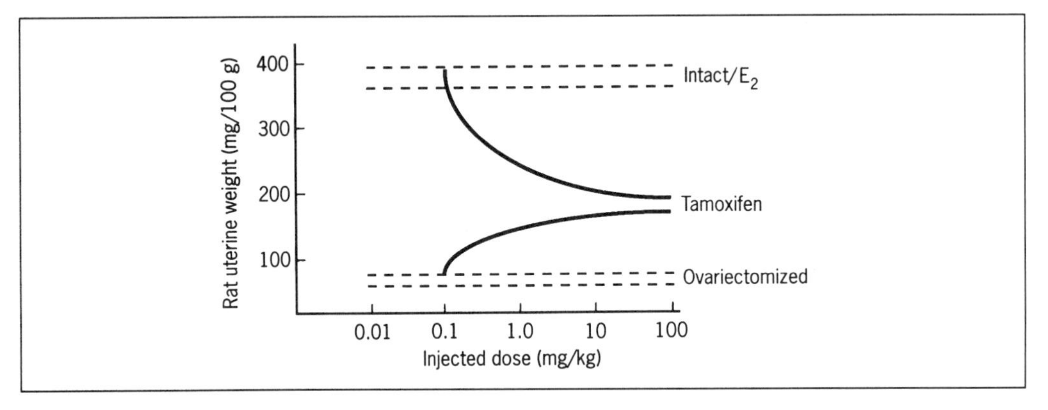

Fig. 2 Typical curves showing the effect of increasing doses of tamoxifen on basal or oestradiol-stimulated uterine weight in rats

ICI 164384 completely inhibits the stimulatory effects of oestradiol on uterine weight and, since it appeared to lack agonist activity, this compound is referred to as a pure antioestrogen. Its effectiveness as an antioestrogen depends on an alkylamide side-chain at the 7α-position of the B ring in the steroid, which has an optimum length of 16–18 carbon atoms (11). The hydrophobic nature of ICI 164384 limits its use as an effective drug but solubility has been increased in ICI 182780 by the introduction of fluorine atoms in the side-chain; this is also results in a slight increase in affinity for the receptor (12).

2.2 Clinical uses

Antioestrogens were first developed as potential contraceptives, but in fact they are now used as profertility agents. Clomiphene effectively suppresses the negative feedback of oestrogen on the hypothalamo-pituitary axis, so that it can be used to stimulate ovulation in subfertile women (8). In breast cancer, tamoxifen is the most commonly used first-line endocrine therapy in advanced disease, and has recently become established in the adjuvant therapy following removal of the tumour. Overviews of randomized clinical trials world-wide indicate that the 5- and 10-year mortality rates of patients with breast cancer can be reduced by 20–25% (13). Survival advantage appears to be greatest if the primary tumour was oestrogen-receptor positive (ER^+). Patients with metastatic breast cancer also respond to tamoxifen treatment, with an objective response rate of approximately 50% when the tumour is ER^+, 70% when it is ER^+ and progesterone-receptor positive (PR^+), and 10% when it is ER^-. Thus it is assumed that tamoxifen is acting as an antioestrogen by means of its interaction with the ER. Although the molecular basis for the positive response of ER^- tumours is unknown, a number of possibilities have been suggested. The arbitrary division between ER^+ and ER^- evaluation is usually 10 fmol per mg protein. Levels of receptor below this may be sufficient to mediate a response. Alternatively it is possible that tamoxifen blocks ER action not just in the tumour but in other cells that regulate tumour growth by endocrine or paracrine mechanisms. On the other hand, the possibility that tamoxifen might also be capable of inhibiting tumour cell growth by additional ER-independent mechanisms cannot be ruled out (14).

Tamoxifen is a particularly useful drug because it has relatively few deleterious side-effects. In particular, it does not cause osteoporosis, which was a potential risk given the fact that oestrogen is important for the maintenance of bone. It appears that the partial agonist activity of tamoxifen is sufficient to maintain bone density levels (15–18). Tamoxifen also has the beneficial side-effect of reducing blood cholesterol levels, which may lead to reduced mortality from coronary heart disease (9, 19).

Despite the efficacy of tamoxifen treatment, most patients eventually develop drug resistance. The basis for this resistance is unknown but is unlikely simply to reflect loss of receptor. It may result from mutations in the ER (20) or alterations in the metabolism of tamoxifen (21, 22), or alternatively it may be a consequence of additional ER-independent events that result in increased basal proliferation. It would seem likely that tamoxifen-resistant tumours whose growth is still sensitive to oestrogens might nevertheless still respond to an alternative form of endocrine therapy. This was indeed found to be the case when second-line treatment such as progestins (23), androgens (24), or aromatase inhibitors that block oestrogen synthesis (25) was used after tamoxifen failure. Pure antioestrogens are another potential treatment as the appear to function by a different mechanism of action to that of tamoxifen (26). One such compound, ICI 182780, is currently beginning clinical evaluation.

2.1.2 *In vitro* activity

Oestrogens function as mitogens in ER-positive breast cancer cells *in vitro* and stimulate the expression of a number of genes including those for certain growth factors, secretory proteins, and the progesterone receptor (for review see reference 27). Tamoxifen is able to reverse these oestrogen-stimulated responses, albeit to differing extents. For example, the inhibition of growth is accompanied by a decrease in the percentage of S-phase cells and a concomitant increase in the percentage of cells in the G0–G1 phase of the cell cycle, showing that antioestrogens are mainly cytostatic (28–30). However, in addition to its antagonistic behaviour, tamoxifen also behaves as a weak agonist in breast cancer cells. It is able to stimulate slightly both their growth (31) and the expression of the progesterone receptor (32), when used alone.

The expression of a number of growth factors transforming growth factor (TGF)-α, insulin-like growth factor (IGF) 1, and IGF-2 have been shown to be stimulated by oestrogen (27) but it does not appear that any one of these is solely responsible for mediating the mitogenic activity of oestrogen (2). It has been argued that the effects of tamoxifen are not simply to antagonize oestrogen action, because the drug is also able to inhibit the stimulatory effects of growth factors such as epidermal growth factor (EGF) or IGF on breast cancer cell growth in the absence of oestrogen (33, 34). For this reason, it has been suggested that tamoxifen is more likely to result in the synthesis of a growth inhibitory factor. One candidate is TGF-β, which inhibits the growth of a number of cell types, including breast cancer cells (2, 27). In MCF-7 breast cancer cells, tamoxifen was shown to increase the secretion of TGF-β, presumably by means of its interaction with the ER (35, 36). While this may represent the primary mechanism by which tamoxifen inhibits the growth of breast cancer cells, it has also been argued that tamoxifen might stimulate the secretion of TGF-β, not by the tumour itself but by adjacent stromal cells, by means of a novel receptor (14).

In contrast to the action of tamoxifen on breast cancer cells, the compound functions as a strong agonist in endometrial cells, stimulating both cell growth and progesterone receptor (PR) expression to the same extent as oestradiol (37, 38). The molecular basis for this cell-specific difference is not known. General metabolism does not seem to be responsible: breast cancer MCF-7 cells and endometrial cancer EnCa101 cells implanted in opposite flanks of the same mouse retain the difference in their growth response to tamoxifen (39).

3. Antiandrogens

The first synthetic antiandrogen to be described was cyproterone acetate (40). However, in addition to its antiandrogenic properties, cyproterone acetate is a potent progestin and a partial androgen agonist in a number of physiological responses. So-called pure antiandrogens have subsequently been developed that lack additional steroidal activities and agonist activity. The first of these was flutamide (41). In fact, *in vivo*, it is converted into 4-hydroxyflutamide in the liver, resulting in an increase in its affinity for the androgen receptor by approximately 100-fold (42). A disadvantage of this antiandrogen, however, is its short half-life of approximately 8 h. Additional pure antiandrogens with longer half-lives are now available, and these include, Anandron (RU 23908) (43) and Casodex (or ICI 176324) (6) (Fig. 3).

The use of antiandrogens for the treatment of prostatic carcinoma stems from the discovery by Huggins 50 years ago that castration, by reducing circulating androgen levels, is an effective form of therapy. Cyproterone acetate was the first antiandrogen to be used, but treatment leads to a number of side-effects because the drug possesses other steroidal activities. One of these, progestational activity, could be considered advantageous because it results in reduced circulating gonadotrophin and thereby further reduces androgen levels. Nevertheless, because cyproterone acetate is a partial agonist, it has been argued that use or pure antiandrogens is preferable (6). In practice, prostatic cancer is treated by surgical or chemical castration (luteinizing hormone releasing hormone agonists) to block testicular androgen production, in combination with antiandrogens to antagonize the action of androgens produced by the adrenals (44). Unfortunately, as with breast cancer, patients eventually become resistant to treatment irrespective of the antiandrogen used.

Prostate cell lines whose growth is stimulated by androgens have not been isolated. While a number of human prostate tumours retain their androgen responsiveness when transplanted into animals, this characteristic is lost when the tumours

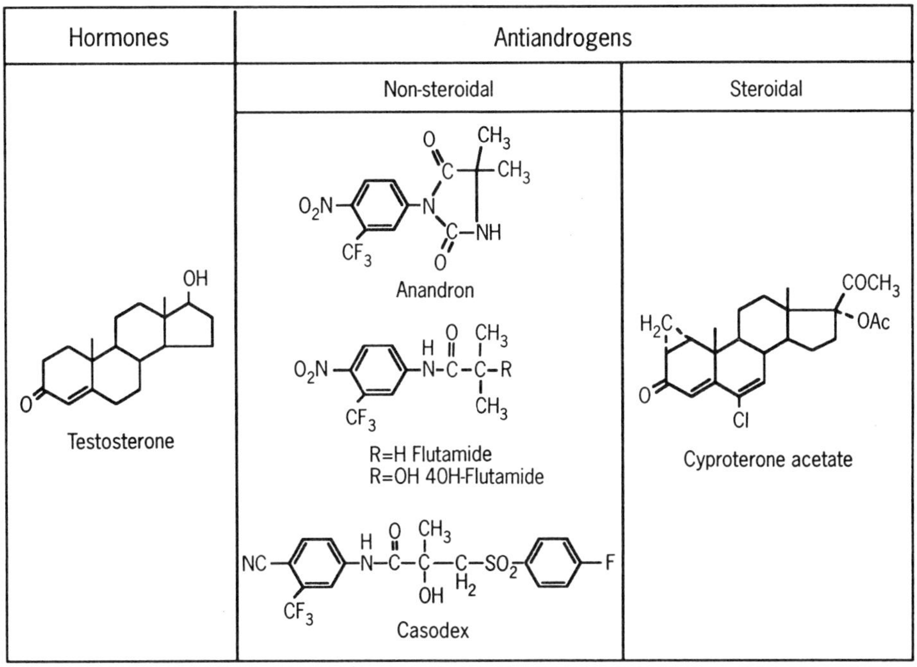

Hormones	Antiandrogens	
	Non-steroidal	Steroidal
Testosterone	Anandron R=H Flutamide R=OH 4OH-Flutamide Casodex	Cyproterone acetate

Fig. 3 Structure of antiandrogens

are cultured as cell lines *in vitro*. Curiously, the growth of one line, LNCaP, is actually stimulated to grow in the presence of antiandrogens (45; see below).

4. Antiprogestins

The most well characterized antiprogestin to date is RU486, which was actually discovered in a search for compounds with antiglucocorticoid activity (5). It binds with high affinity to the receptors for both progestins and glucocorticoids, and antagonizes both their activities. Its structure resembles that of progesterone and cortisol but contains an 11β aryl substitution. The main use of RU486 is for the termination of pregnancy (3). Additional antiprogestins with reduced antiglucocorticoid activity have now been developed (5). The best of these appears to be ZK98299 (46) which differs from RU486 in its side-chain on C17 (Fig. 4). ZK98299 is being examined for its potential to inhibit the growth of hormone-dependent breast cancer.

5. Mechanism of action of steroid hormones

The mechanism by which steroid hormones regulate rates of gene transcription will only be summarized here (Fig. 5) because it has been described in several excellent reviews (47–51) and is dealt with in Chapters 3, 5, and 6.

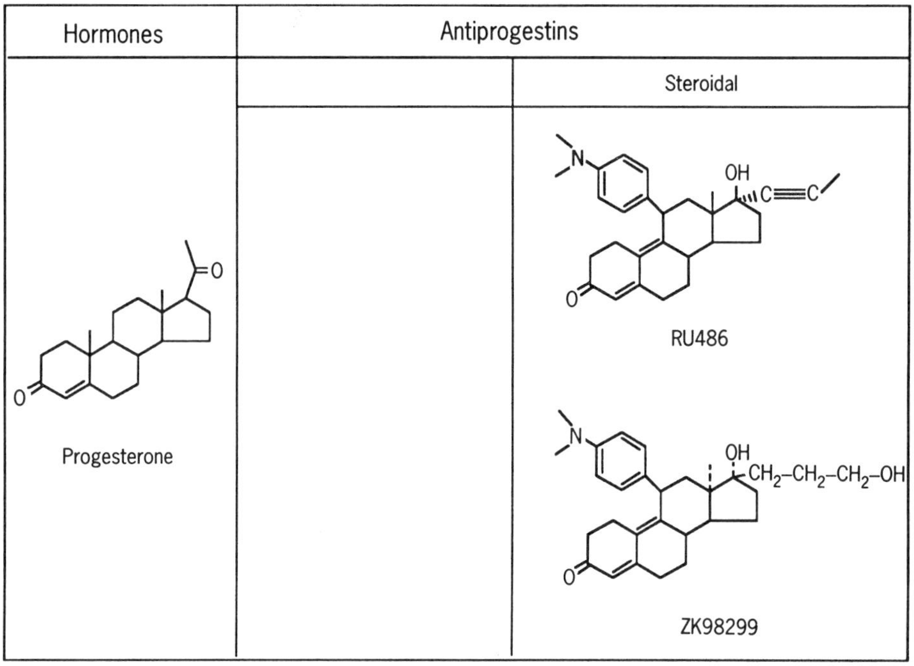

Hormones	Antiprogestins	
		Steroidal
Progesterone		RU486 ZK98299

Fig. 4 Structure of antiprogestins

Briefly, steroid receptors are located predominantly in the nucleus (52) with the exception of the glucocorticoid receptor (GR) which is also found in the cytoplasm (53). Each class of receptor exists in an inactive oligomeric complex with a number of heat-shock proteins including hsp90, hsp70, p56, and p27 (54–56) (Chapter 4). Upon steroid hormone binding, this complex dissociates either to liberate a preformed dimer or to allow dimerization of receptor monomers (57, 58). Receptor dimerization appears to be mediated by a number of different domains of the protein, depending on the receptor. In the ER, the major dimer interface is in the hormone-binding domain (HBD), but there is also another region in the DNA-binding domain that seems to form protein–protein contacts when the reeptor is bound to DNA (57). Some of the residues involved in these interactions have been identified by functional analysis of mutant receptors. Those in the HBD have been shown to overlap with residues that are essential for oestrogen binding (59). It is conceivable, therefore, that the hormone-binding pocket is at or near the dimer interface and that hormone binding may result in stabilized dimerization, perhaps on the basis of hydrophobic shielding. In other receptors the *N*-terminal domain may also play a role in dimerization, although the residues involved have not yet been identified.

Transcriptional activation is brought about when the steroid receptor binds to

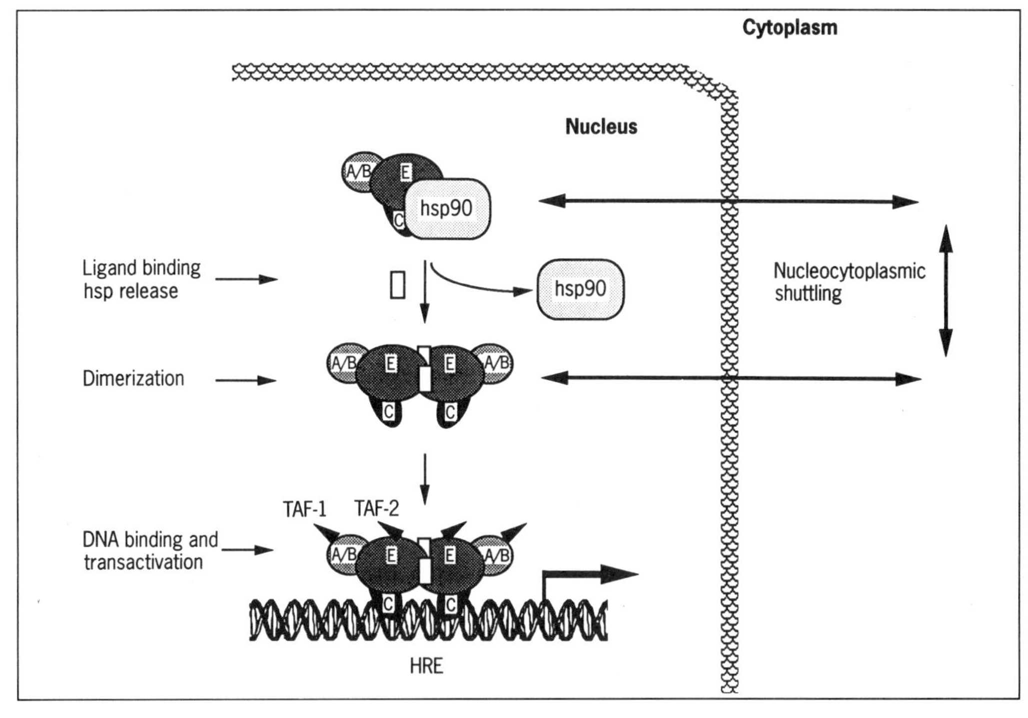

Fig. 5 Model of the action of steroid hormones on receptor activation. The various steps represented are targets of anti-hormone action. hsp, heat-shock protein; HRE, hormone response element; TAF, transcriptional activation function A/B, C, and E represent different regions of the oestrogen receptor

palindromic DNA-binding sites in the vicinity of hormone-responsive genes. The rate at which RNA polymerase initiates gene transcription depends on the formation of a pre-initiation complex that includes a number of basic transcription factors (60–62). It is postulated that transcriptional activators, including steroid receptors, stimulate transcriptional initiation by stabilizing this complex. The binding of a hormonal ligand to the receptor is thought to be required not only to promote DNA binding but also in transcriptional activation, perhaps in the process of stabilization of the pre-initiation transcription complex (Chapter 4).

Transcriptional activation by steroid receptors is mediated by at least two distinct activation regions in the protein, although there is some disagreement regarding their precise location in certain receptors (63–68). One of these, referred to as transcriptional activation function (TAF)- 1, is located in the *N*-terminal domain and another, TAF-2, is in the HBD. In the ER, and probably in other receptors as well, these two activation domains have the potential to act independently, but in the intact receptor they appear to be capable of interacting with one another in some ill-defined way. In addition, their absolute and relative activities vary depending on the cell type (63, 69–72). For example TAF-1 contributes approximately 50% of the transcriptional activity of the receptor in chicken embryo fibroblasts but only 5% in HeLa or 3T3 cells. Although these cells are not usually considered target cells for oestrogens, it is assumed that such differences are likely to mimic the situation in normal target tissues. The most likely explanation to account for the differences is that the activity of different TAFs is mediated by proteins whose expression or activity is cell dependent.

6. Anti-hormone binding

Individual residues throughout the HBD of all the receptors are likely to be required for high-affinity binding of steroid hormone. The first approach in the identification of important residues was to affinity-label the receptor with synthetic ligands. In the human ER, a cysteine at position 530 (equivalent to 534 in the mouse receptor) can be affinity-labelled with both an oestrogen, ketononestrol aziridine, and an antioestrogen, tamoxifen aziridine (73), suggesting that it is in the vicinity of the ligand-binding pocket. This residue is unlikely to be involved in oestrogen binding itself because it can be mutated, at least in the mouse ER, without affecting its ability to bind oestrogen (59). That it is close to the hormone-binding site was supported by examining other mutant mouse receptors which indicated that amino acids between residues 518 and 525 were required for oestrogen-binding activity.

Interestingly, although mutation of a glycine at position 525 and a methionine and/or serine at positions 521/522 essentially abolished oestrogen-binding activity, they did not markedly affect the sensitivity of the receptor for tamoxifen (74). The mutant receptors retained the partial agonist activity exhibited by the wild-type receptor in the presence of 4-hydroxytamoxifen. Analysis of the human receptor has shown that mutation of lysines at positions 529 and 531 (corresponding to residues 533 and 535, respectively, in the mouse protein) to glutamines reduces the affinity of the receptor for oestradiol by 5–10-fold but has no effect on hydroxytamoxifen binding (75). Thus, a number of residues in the C-terminal region of the HBD are able to confer differential sensitivity to oestrogen and 4-hydroxytamoxifen.

In the PR, more is probably known about the residues involved in the binding of antiprogestin RU486 than of progesterone or synthetic progestins. This is because an interesting difference in the binding properties of the receptor in different species has facilitated the identification of critical residues. In contrast to the human PR, the chicken and hamster receptors are incapable of binding RU486 (3). Since there are 31 differences in amino acids between the human and chicken receptors, they were systematically exchanged in order to generate a mutant chicken receptor that was capable of binding the antagonist. In this way it was found that substitution of cysteine 575 with glycine was sufficient to restore RU486 binding. Substitution of the glycine with cysteine in the human PR abrogated binding of RU486 but not that of the progestin R5020. It is conceivable that the glycine is located in the hydrophobic hormone-binding pocket close to the position of the 11β side-chain of RU486, and that the cysteine side-chain present in the chicken and hamster receptors may interfere sterically with binding (76).

In the human androgen receptor (AR) a number of residues have been implicated in androgen binding on the basis of naturally occurring mutations in the receptor that give rise to androgen insensitivity syndromes (77–79; see Chapter 9). To date, approximately a dozen mutations throughout the HBD have been identified that reduce or abolish androgen-binding activity, but it is unclear whether the substituted residue itself is directly involved or whether the reduced binding is caused by the disruptive effects of the substituting amino acid. In addition, a mutation has been identified in the AR found in the human prostate tumour cell line LNCaP which seems to account for the surprising response of this cell line to a variety of hormonal ligands (80). The antiandrogens cyproterone acetate and flutamide, progestins, and oestrogens were able to stimulate LNCaP cell growth. This seems to be accounted for by a mutation at position 868 which results in a threonine → serine substitution. This mutation accounts not only for the altered growth properties of the cell line but also for the altered sensitivity of reporter genes in transient transfection experiments.

7. Mechanism of action of antagonists with partial agonist activity

More is known about the action of the antioestrogen tamoxifen and the antiprogestin/antiglucocorticoid RU486 than any other type of hormone antagonist. Since both of these agents exhibit partial agonist activity, and seem to share many features in common, they will be considered together. In both cases, it appears that the two antagonists bind and induce a different conformation and/or post-translational modification in their respective receptors than that induced by the normal hormonal ligand. Such a difference is supported by the observation that the protease digestion patterns and the mobilities of receptor–DNA complexes after gel shift analysis are different (57, 69, 81–83). These DNA-binding studies also provided the first evidence that the two antagonists did not inhibit DNA-binding activity, at least *in vitro*. This has been confirmed *in vivo* in two types of transient transfection experiment. First, both tamoxifen and RU486 are able to stimulate transcription from a reporter plasmid, the magnitude of the response depending on the gene promoter and the cell type (63, 69, 84). Second, competition assays have demonstrated that the receptor is able to compete for binding to target DNA even when it is complexed with the antagonist (63, 84).

As convincing as the transient transfection experiments appear to be, there is one experiment that seems to indicate that RU486 does not always promote DNA binding. Genomic footprinting has indicated that the GR is able to bind to the promoter of the tyrosine aminotransferase gene in the presence of glucocorticoid but not RU486 (85). Thus, this observation appears to be incompatible with the results of the transfection experiments. It may be a feature only of the GR, or may reflect variation in the ability of different DNA-binding sites to bind receptor complexed with antagonist. Despite this result, it has been argued that tamoxifen and RU486 promote DNA binding of the ER and PR, respectively, and that their inhibitory effects must occur at a subsequent step (63, 84).

The first model proposed by Chambon and co-workers (82, 84) is based on the observation that there are two discrete transcriptional activation regions in both the ER and PR, TAF-1 and TAF-2. The activity of TAF-2 is dependent on the binding of hormonal ligand and is negligible in the presence of the antagonist. Since TAF-1 has the potential to function constitutively, it has been proposed that TAF-1 is likely to be active in the presence not only of agonists but also of antagonists that promote DNA binding. Therefore, in situations where TAF-1 and TAF-2 function independently of one another, tamoxifen and RU486 would act predominantly as antagonists on gene promoters on which TAF-2 contributed most activity, but as agonists when TAF-1 was most active. If both TAF-1 and TAF-2 contributed activity, tamoxifen and RU486 would act as mixed agonists/antagonists (Fig. 6).

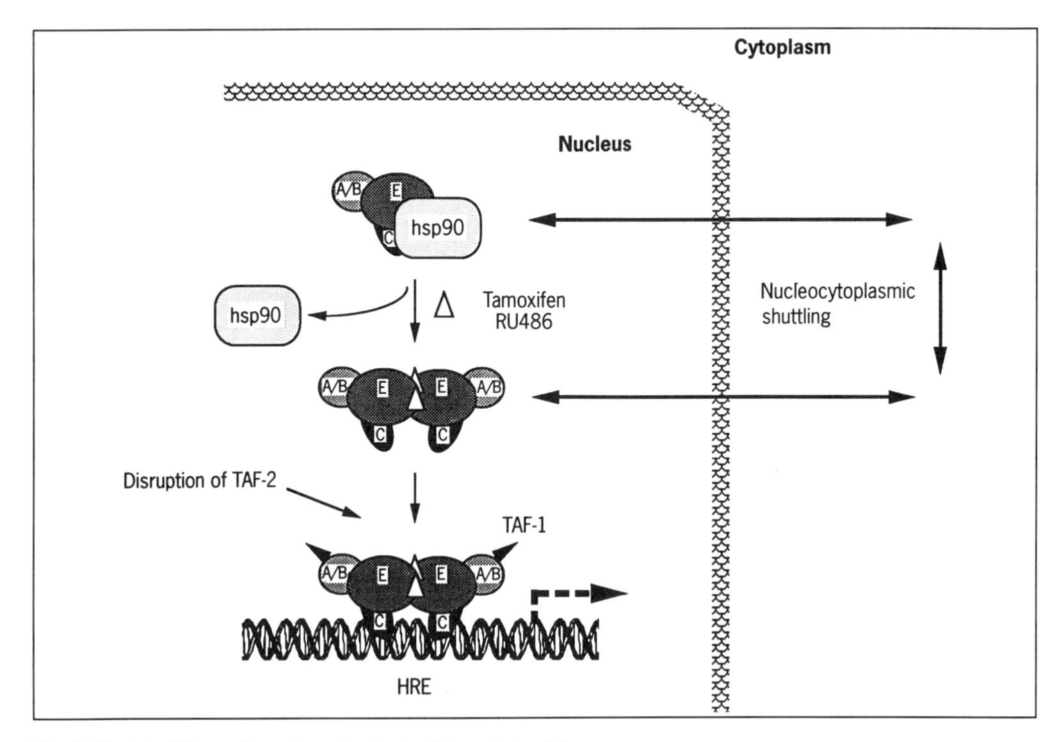

Fig. 6 Model of the action of a mixed agonist–antagonist

Recently, O'Malley and co-workers proposed an alternative mechanism to account for the action of RU486 (86) (Chapter 3). In the PR, they identified a *C*-terminal region whose ability to suppress the transcriptional activity of the receptor is overcome by agonists, but not by antagonists. Consistent with this result was the observation that, when the negative regulatory region was deleted from the PR, RU486 was able to stimulate its transcriptional activity. How this type of mechanism fits in, if at all, with that proposed by Chambon and colleagues is not known; it is quite conceivable that they are not mutually exclusive.

8. Mechanism of action of 'pure' oestrogen antagonists

Because ICI 164384 has been shown to be devoid of agonist activity in the majority of functional assays, particularly those involving cell growth, it has been referred to as a pure antioestrogen (11, 12, 87). The action of ICI 164384 is somewhat controversial, because contradictory results have been obtained in various types of experiments, and these allow completely different interpretations to be made. It has been found, for example, that the DNA-binding activity of the ER from some, but not all, sources is inhibited upon ICI 164384 binding *in vitro* (88–91). In our work, we have found that the DNA-binding activity of receptors overexpressed in insect or mammalian cells, but not that present in human breast cancer cells, is inhibited by the antioestrogen. Moreover, we demonstrated that the ability of analogues of ICI 164384 with different side-chain lengths to inhibit DNA binding correlates with their ability to inhibit the increase in uterine weight *in vivo* (11, 88). We assume that ICI 164384 binds to a similar, if not identical, site to that of oestradiol, which we have shown to overlap with a region involved in receptor dimerization (59). As a consequence we proposed that the antioestrogens, by means of their 7α side-chains (Fig. 1), sterically interfere with dimerization, and as a consequence reduce the affinity with which the ER binds to target DNA sites *in vitro* (88, 89).

It is apparent that the DNA-binding activity of receptors is inhibited significantly only when they are overexpressed (88, 90, 91), possibly reflecting differences in the stability of dimerization. Moreover, ICI 164384 binding to the ER does not seem to prevent DNA binding in transiently transfected cells or when expressed in yeast, and so it has been argued that the antioestrogen must act at a step subsequent to DNA binding (63, 92, 93).

As it may not always be possible to dissociate preformed receptor dimers and inhibit DNA binding in cell-free extracts, we investigated the effects of ICI 164384 and ICI 182780 in intact cells to test whether it might be possible to prevent the formation of dimers. Surprisingly, the antioestrogens were found to cause a decrease in the cellular content of receptor protein by markedly reducing its half-life (94). A similar reduction has been noted in mouse uterine tissue (95), human breast cancer cells (94), and pituitary cells (93) following antioestrogen treatment. It appears that ICI 164384 and ICI 182780 cause an increase in the turnover of ERs in all target cells and tissues, which must contribute to its effectiveness as an anti-oestrogen (Fig. 7).

Recently, the PR and ER have been shown to undergo a process called nucleo-cytoplasmic shuttling in which they exit from the cell nucleus but are rapidly transported back, so that at equilibrium they appear to be predominantly nuclear (96). Immunofluorescence studies have shown that the PR could associate with and co-translocate a nuclear transport defective mutant receptor into the nucleus in a hormone-dependent manner. Corresponding studies with the ER showed that oestrogen and tamoxifen, but not ICI 164384, promoted the co-translocation (97). It is likely that co-translocation is mediated by dimerization through the HBD, but

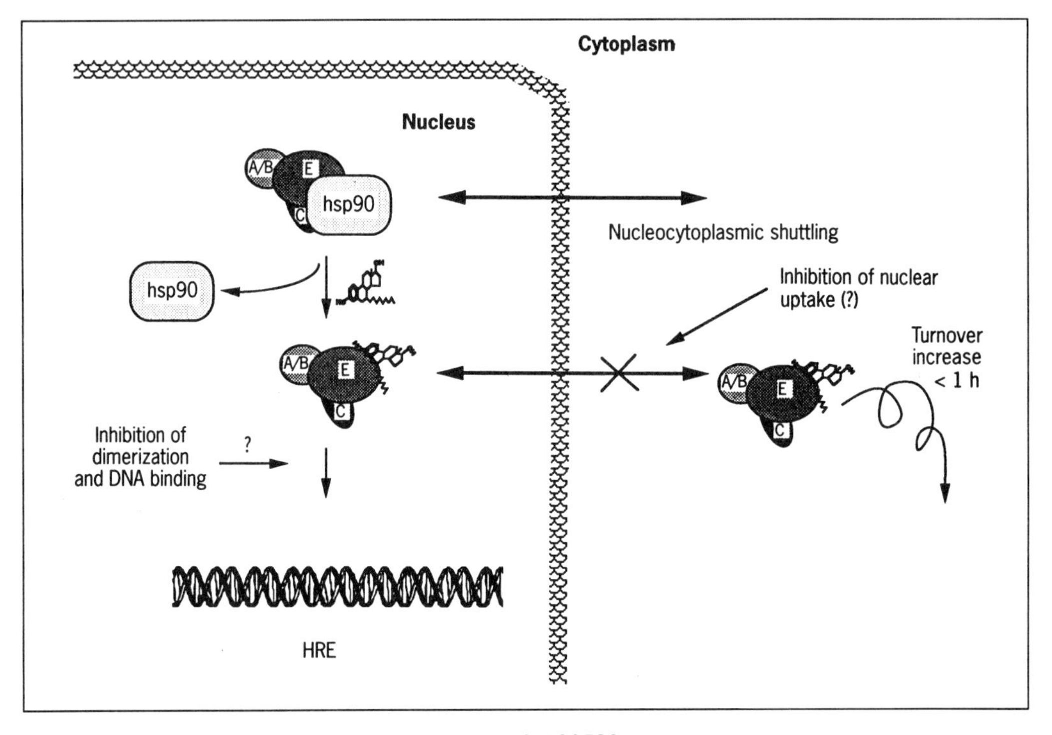

Fig. 7 Model of the action of the pure antioestrogen ICI 182780

this has yet to be shown directly. In our own work we have found that, in contrast to oestrogen and tamoxifen, ICI 164384 and ICI 182780 failed to promote nuclear uptake of the receptor (S. Dauvois, unpublished results). In view of the potential for these antioestrogens to inhibit dimerization, we favour the suggestion that their inability to promote nuclear uptake might result from impaired dimerization. In the absence of nuclear uptake, it appears that the receptor is more rapidly degraded, perhaps in lysosomes.

9. Mechanism of action of other types of antagonist

The antiprogestin ZK98299 appears to act by a different mechanism of action to that of RU486. By using gel shift experiments Klein-Hitpass *et al.* (98) have shown that ZK98299 fails to promote DNA binding, in contrast to progestins and RU486. One possibility they suggest is that ZK98299 might fail to elicit the structural alterations of the steroid receptor necessary for the formation of stable receptor dimers. Further work is required to establish whether this is in fact the case and whether the mechanism of action of ZK98299 in any way resembles that of the antioestrogen ICI 164384.

10. Concluding remarks

There has clearly been tremendous progress over the past 5 years in the analysis of the mechanisms of action of tamoxifen and RU486. Experiments are now being carried out to investigate other types of hormone antagonists, including the anti-androgens. Two features of anti-hormone action remain largely unexplained. These are the wide variation in the ability of antagonists to inhibit different physiological responses *in vivo* and the molecular mechanisms involved when breast and prostate cancers become resistant to hormone therapy.

The effectiveness of hormone antagonists *in vivo* depends on the physiological response being examined, and can vary in different organisms and cell types. It is conceivable that part of this variation is due to the fact that the antagonists can be metabolized in either the liver or the target tissue and that their bioavailability varies in different tissues of the body. These variations, however, are unlikely to explain completely the wide range of effects of, for example, tamoxifen, which is found to be an antagonist in breast cancer cells but an agonist in endometrial cancer cells. When the effect of tamoxifen on uterine or oviduct weight is measured, the drug is found to be a pure antioestrogen in the chicken, a partial agonist in human and rat, and a full agonist in mouse (99). It is extremely doubtful whether the current model, which predicts that tamoxifen inhibits the activity of TAF-2 but not that of TAF-1 (Fig. 6), can account for all these differences. One possible explanation might be that cells could contain specific proteins or co-activators that determine the transcriptional activity of the different activation regions in the receptor; work is in progress to attempt to identify candidate proteins. At present, however, it is not possible to predict the action of a hormone antagonist *in vivo* with any degree of confidence. This is illustrated by the unexpected observation that tamoxifen is able to maintain bone density and lower blood cholesterol levels, and that its success in the treatment of breast cancer might in part be attributed to these beneficial side-effects.

The second area about which little is known is tamoxifen resistance in breast cancer. Alteration in either tamoxifen metabolism or the ER itself remain possibilities, but other explanations are also conceivable. Because 40–50% of tamoxifen-resistant tumours respond to alternative endocrine therapies, it is reasonable to assume that the ER continues to play a role in their growth. The efficacy of ICI 1812780 in the treatment of breast cancer is now being analysed in clinical trials. Since ICI 182780 appears to function by a different mechanism to that described for tamoxifen, there is a strong possibility that it might be effective in the treatment of tamoxifen-resistant breast cancer, provided that the tumour is still hormone dependent. On the other hand, pure antioestrogens might not offer the same beneficial side-effects as tamoxifen; the effect of ICI 182780 on bone density will obviously have to be monitored carefully.

References

1. Jordan, V. C. (1984) Biochemical pharmacology of antiestrogen action. *Pharmacol. Rev.*, **36,** 245.
2. Jordan, V.C. and Murphy, C. S. (1990) Endocrine pharmacology of antiestrogens as antitumor agents. *Endocr. Rev.*, **11,** 578.
3. Baulieu, E.-E. (1989) Contragestion and other clinical applications of RU486, an antiprogesterone at the receptor. *Science,* **245,** 1351.
4. Ojasoo, T. and Raynaud, J.-P. (1990) Steroid hormone receptors. In *Comprehensive Medicinal Chemistry,* Vol. 3, Emmett, J. C. (ed.). Pergamon Press, Oxford, p. 1175.
5. Henderson, D. (1987) Antiprogestational and antiglucocorticoid activities of some novel 11β-aryl substituted steroids. In *Pharmacology and Clinical Uses of Inhibitors of Hormone Secretion and Action*, Furr, B. J. A. and Wakeling, A. E. (eds). Baillière Tindall, London, p. 184.
6. Furr, B. J. A. (1988) The case for pure antiandrogens. *Baillieres Clin. Oncol.*, **2,** 581.
7. Lerner, L. J., Holthaus, F. J., Jr., and Thompson, C. R. (1958) A nonsteroidal estrogen antagonist 1-(*p*-2-diethylaminoethoxyphenyl)-1-phenyl-2-*p*-methoxyphenylethanol. *Endocrinology,* **63,** 295.
8. Huppert, L. C. (1979) Induction of ovulation with clomiphene citrate. *Fertil. Steril.*, **31,** 1.
9. Early Breast Cancer Trialists' Collaborative Group (1988) Effect of adjuvant tamoxifen and cytotoxic therapy on mortality in early breast cancer. *N. Engl. J. Med.*, **319,** 1681.
10. Jordan, V. C., Collins, M. M., Rowsby, L., and Prestwich, G. (1977) A monohydroxylated metabolite of tamoxifen with potent antioestrogenic activity. *J. Endocrinol.*, **75,** 305.
11. Bowler, J., Lilley, T. J., Pittam, J. D., and Wakeling, A. E. (1989) Novel steroidal pure antiestrogens. *Steroids,* **54,** 71.
12. Wakeling, A. E., Dukes, M., and Bowler, J. (1991) A potent specific pure antiestrogen with clinical potential. *Cancer Res.*, **51,** 3867.
13. Bentley, A., Fentiman, I. S., Rubens, R. D., Cuzick, J., Crossley, E., Durrant, K., Harris, A., Clarke, M., Collins, R., Godwin, J., Gray, R., Greaves, E., Harwood, C., Mead, G., Peto, R., and Wheatley, K. (1992) Systemic treatment of early breast cancer by hormonal, cytotoxic, or immune therapy: 133 randomised trials involving 31 000 recurrences and 24 000 deaths among 75 000 women. Part 2. *Lancet,* **339,** 71.
14. Butta, A., MacLennan, K., Flanders, K. C., Sacks, N. P. M., Smith, I., McKinna, A., Dowsett, M., Wakefield, L. M., Sporn, M. B., Baum, M., and Colletta, A. A. (1992) Induction of transforming growth factors β1 in human breast cancer *in vivo* following tamoxifen treatment. *Cancer Res.*, **52,** 4261.
15. Fentiman, I. S., Caleffi, M., and Rodin, A. (1989) Bone mineral content of women receiving tamoxifen for mastalgia. *Br. J. Cancer,* **60,** 262.
16. Love, R. R., Mazess, R. B., Tormey, D. C., Harden, H. S., Newcomb, P. A., and Jordan, V. C. (1988) Some mineral density in women with breast cancer treated for at least two years with tamoxifen. *Breast Cancer Res. Treat.*, **12,** 297.
17. Powles, T. J., Hardy, J. R., Ashley, S. E., Farrington, G. M., Cosgrove, D., Davey, J. B., Dowset, M., McKinna, J. A., Nash, A. G., Sinnett, H. D., Tillyer, C. R., and Treleven, J. G. (1989) A pilot trial to evaluate the acute toxicity and feasibility of tamoxifen for prevention of breast cancer. *Br. J. Cancer,* **60,** 126.
18. Turken, S., Siris, E., Seldin, E., Seldin, D., Flaster, E., Hyman, G., and Linsday, R. (1989) Effects of tamoxifen on spinal bone density in women with breast cancer. *J. Natl Cancer Inst.*, **81,** 1086.

19. Nayfield, S. G., Karp, J. E., Ford, L. G., Dorr, F. A., and Kramer, B. S. (1991) Potential role of tamoxifen in prevention of breast cancer. *J. Natl Cancer Inst.*, **83**, 1450.
20. Fuqua, S. A., Fitzgerald, S. D., Chamness, G. C., Tandon, A. K., McDonnell, D. P., Nawaz, Z., O'Malley, B. W., and McGuire, W. L. (1991) Variant human breast tumor estrogen receptor with constitutive transcriptional activity. *Cancer Res.*, **51**, 105.
21. Osborne, C. K., Coronado, E., Allred, D. C., Wiebe, V., and DeGregorio, M. (1991) Acquired tamoxifen resistance: correlation with reduced breast tumor levels of tamoxifen and isomerization of trans-4-hydroxytamoxifen. *J. Natl Cancer Inst.*, **83**, 1477.
22. Osborne, C. K., Wiebe, V. J., McGuire, W. L., Ciocca, D. R., and DeGregorio, M. W. (1992) Tamoxifen and the isomers of 4-hydroxytamoxifen in tamoxifen-resistant tumors from breast cancer patients. *J. Clin. Oncol.*, **10**, 304.
23. Sedlacek, S. and Horwitz, K. (1984) The role of progestins and progesterone receptors in the treatment of breast cancer. *Steroids*, **44**, 467.
24. Manni, A., Arafah, B., and Pearson, O. H. (1981) Androgen-induced remissions after antiestrogens and hypophysectomy in stage IV. *Breast Cancer Res. Treat.*, **48**, 2507.
25. Santen, R. J., Manni, A., Harvey, H., and Redmond, C. (1990) Endocrine treatment of breast cancer in women. *Endocr. Rev.*, **11**, 221.
26. Gottardis, M. M., Jiang, S. Y., Jeng, M. H., and Jordan, V. C. (1989) Inhibition of tamoxifen-stimulated growth of an MCF-7 tumor variant in athymic mice by novel steroidal antiestrogens. *Cancer Res.*, **49**, 4090.
27. Clarke, R., Dickson, R. B., and Lippman, M. E. (1990) The role of steroid hormones and growth factors in the control of normal and malignant breast. In *Nuclear Hormone Receptors*, Parker, M. (ed.). Academic Press, Lopndon, p. 297.
28. Sutherland, R. L., Green, M. D., Hall, R. E., Reddel, R. R., and Taylor, I. W. (1983) Tamoxifen induces accumulation of MCF-7 human mammary carcinoma cells in the G0/G1 phase of the cell cycle. *Eur. J. Cancer Clin. Oncol.*, **19**, 615.
29. Musgrove, E. A., Wakeling, A. E., and Sutherland, R. L. (1989) Points of action of estrogen antagonists and a calmodulin antagonist within the MCF-7 human breast cancer cell cycle. *Cancer Res.*, **49**, 2398.
30. de Launoit, Y., Dauvois, D., Dufour, M., Simard, J., and Labrie, F. (1991) Inhibition of cell cycle kinetics and proliferation by the androgen 5α-dihydrotestosterone and antiestrogen *N,n*-butyl-*N*-methyl-11-[16′*a*-chloro-3′, 17β-dihydroxy-estra-1′,3′,5′-(10′) triene-7′*a*-yl] undecanamide in human breast cancer ZR-75-1 cells. *Cancer Res.*, **51**, 2797.
31. Poulin, R., Mewrand, Y., Poirier, D., Levesque, C., Dufour, J.-M., and Labrie, F. (1989) Antiestrogenic properties of keoxifene, *trans*-4-hydroxytamoxifen, and ICI 164384, a new steroidal antiestrogen, in ZR-75-1 human breast cancer cells. *Breast Cancer Res. Treawt.*, **14**, 65.
32. Westley, B., May, F. E. B., Brown, A. M. C., Krust, A., Chambon, P., Lippman, M. E., and Rochefort, H. (1984) Effects of antiestrogens on the estrogen-regulated pS2 RNA and the 52- and 160-kilodalton proteins in MCF7 cells and two tamoxifen-resistant sublines. *J. Biol. Chem.*, **259**, 10 030.
33. Freiss, G., Rochefort, H., and Vignon, F. (1990) Mechanisms of 4-hydroxytamoxifen anti-growth factor activity in breast cancer cells: alterations of growth factor receptor binding sites and tyrosine kinase activity. *Biochem. Biophys. Res. Commun.*, **173**, 919.
34. Vignon, F., Bouton, M.-M., and Rochefort, H. (1987) Antiestrogens inhibit the mitogenic effect of growth factors on breast cancer cells in the total absence of estrogens. *Biochem. Biophys. Res. Commun.*, **146**, 1502.

35. Knabbe, C., Lippman, M. E., Wakefield, L. M., Flanders, K. C., Kasid, A., Derynck, R., and Dickson, R. B. (1987) Evidence that transforming growth factor-β is a hormonally regulated negative growth factor in human breast cancer cells. *Cell,* **48,** 417.
36. Knabbe, C., Zugmaier, G., Schmahl, M., Dietel, M., Lippman, M. E., and Dickson, R. B. (1991) Induction of transforming growth factor β by the antiestrogens droloxifene, tamoxifen, and toremifene in MCF-7 cells. *Am. J. Clin. Oncol.,* **14** (Supplement 2), S15.
37. Anzai, Y., Holinka, C. F., Kuramoto, H., and Gurpide, E. (1989) Stimulatory effects of 4-hydroxytamoxifen on proliferation of human endometrial adenocarcinoma cells (Ishikawa line). *Cancer Res.,* **49,** 2362.
38. Jamil, A., Croxtall, J. D., and White, J. O. (1991) The effect of anti-oestrogens on cell growth and progesterone receptor concentration in human endometrial cancer cells (Ishikawa). *J. Mol. Endocrinol.,* **6,** 215.
39. Gottardis, M. M., Robinson, S. P., Satyaswaroop, P. G., and Jordan, V. C. (1988) Contrasting actions of tamoxifen on endometrial and breast tumor growth in the athymic mouse. *Cancer Res.,* **48,** 812.
40. Neumann, F., Berswordt-Wallrabe, R., and Von Elger, W. (1970) Aspects of androgen-dependent events as studied by antiandrogens. *Recent Prog. Horm. Res.,* **26,** 337.
41. Neri, R., Florance, K., Koziol, P., and Van Cleave, S. (1972) A biological profile of a nonsteroidal antiandrogen, SCH 13521 (4′-nitro-3′-trifluoromethyl isobutryanilide). *Endocrinology,* **91,** 427.
42. Simard, J., Luthy, J., Guay, A., Bélanger, A., and Labrie, F. (1986) Characteristics of interaction of the antiandrogen flutamide with the androgen receptor in various target tissues. *Mol. Cell. Endocrinol.,* **44,** 261.
43. Moguilewsky, M., Bertagna, C., and Hucher, M. (1987) Pharmacological and clinical studies of the antiandrogen Anandron. *J. Steroid Biochem.,* **27,** 871.
44. Crawford, E. D., Eisenberger, M. A., McLeod, D. G., Spaulding, J. T., Benson, R., Dorr, A., Blumenstein, B. A., Davis, M. A., and Goodman, P. J. (1989) A controlled trial of leuprolide with and without flutamide in prostatic carcinoma. *N. Engl. J. Med.,* **321,** 419.
45. Wilding, G., Chen, M., and Gelmann, E. P. (1989) Aberrant response *in vitro* of hormone-responsive prostate cancer cells to antiandrogens. *Prostate,* **14,** 103.
46. Neef, G., Beier, S., Elger, W., Henderson, D., and Wiechert, R. (1984) *Steroids,* **44,** 349.
47. Beato, M. (1989) Gene regulation by steroid hormones. *Cell,* **56,** 335.
48. Gronemeyer, H. (1991) Transcription activation by estrogen and progesterone receptors. *Annu. Rev. Genet.,* **25,** 89.
49. Ham, J. and Parker, M. G. (1989) Regulation of gene expression by nuclear hormone receptors. *Curr. Opin. Cell Biol.,* **1,** 503.
50. Evans, R. M. (1988) The steroid and thyroid hormone receptor superfamily. *Science,* **240,** 889.
51. Green, S. and Chambon, P. (1988) Nuclear receptors enhance our understanding of transcription regulation. *Trends Genet.,* **4,** 309.
52. Greene, G. L., Sobel, N. B., King, W. J., and Jensen, E. V. (1984) Immunochemical studies of estrogen receptors. *J. Steroid Biochem.,* **20,** 51.
53. Picard, D. and Yamamoto, K. R. (1987) Two signals mediate hormone-dependent nuclear localization of the glucocorticoid receptor. *EMBO J.,* **6,** 3333.
54. Catelli, M. G., Binart, N., Jung, T. I., Renoir, J. M., Baulieu, E. E., Feramisco, J. R., and Welch, W. J. (1985) The common 90-kD protein component of non-transformed '8S' steroid receptors is a heat-shock protein. *EMBO J.,* **4,** 3131.

55. Denis, M., Poellinger, L., Wikstrom, A. C., and Gustafsson, J. A. (1988) Requirement of hormone for thermal conversion of the glucocorticoid receptor to a DNA binding state. *Nature,* **333,** 686.
56. Sanchez, E. R., Toft, D. O., Schlesinger, M. J., and Pratt, W. B. (1985) Evidence that the 90-kDa phosphoprotein associated with the untransformed L-cell glucocorticoid receptor is a murine heat shock protein. *J. Biol. Chem.,* **260,** 12 398.
57. Kumar, V. and Chambon, P. (1988) The estrogen receptor binds tightly to its responsive element as a ligand-induced homodimer. *Cell,* **55,** 145.
58. Tsai, S. Y., Carlstedt-Duke, J., Weigel, N. L., Dahlman, K., Gustafsson, J.-A., Tsai, M.-J., and O'Malley, B. W. (1988) Molecular interactions of steroid hormone receptor with its enhancer element: evidence for receptor dimer formation. *Cell,* **55,** 361.
59. Fawell, S. E., Lees, J. A., White, R., and Parker, M. G. (1990) Characterization and colocalization of steroid binding and dimerization activities in the mouse estrogen receptor. *Cell,* **60,** 953.
60. Lewin, B. (1990) Commitment and activation at polII promoters: a tail of protein–protein interactions. *Cell,* **61,** 1161.
61. Martin, K. J., Lillie, J. W., and Green, M. R. (1990) Evidence for interaction of different eukaryotic transcriptional activators with distinct cellular targets. *Nature,* **346,** 147.
62. Mitchell, P. J. and Tjian, R. (1989) Transcriptional regulation in mammalian cells by sequence-specific DNA binding proteins. *Science,* **245,** 371.
63. Webster, N. J. G., Green, S., Jin, J. R., and Chambon, P. (1988) The hormone-binding domains of the estrogen and glucocorticoid receptors contain an inducible transcription activation function. *Cell,* **54,** 199.
64. Hollenberg, S. M. and Evans, R. M. (1988) Multiple and co-operative transactivation domains of the human glucocorticoid receptor. *Cell,* **55,** 899.
65. Godowski, P. J., Picard, D., and Yamamoto, K. R. (1988) Signal transduction and transcriptional regulation by glucocorticoid receptor—*lexA* fusion proteins. *Science,* **24,** 812.
66. Danielsen, M., Northrop, J. P., Jonklaas, J., and Ringold, G. M. (1987) Domains of the glucocorticoid receptor involved in specific and nonspecific deoxyribonucleic acid binding, hormone activation, and transcriptional enhancement. *Mol. Endocrinol.,* **1,** 816.
67. Hollenberg, S. M., Giguere, V., Segui, P., and Evans, R. M. (1987) Colocalization of DNA-binding and transcriptional activation functions in the human glucocorticoid receptor. *Cell,* **49,** 39.
68. Miesfeld, R., Godowski, P. J., Maler, B. A., and Yamamoto, K. R. (1987) Glucocorticoid receptor mutants that define a small region sufficient for enhancer activation. *Science,* **236,** 423.
69. Lees, J. A., Fawell, S. E., and Parker, M. G. (1989) Identification of two transactivation domains in the mouse oestrogen receptor. *Nucleic Acids Res.,* **17,** 5477.
70. Webster, N. J., Green, S., Tasset, D., Ponglikitmongkol, M., and Chambon, P. (1989) The transcriptional activation function located in the hormone-binding domain of the human oestrogen receptor is not encoded in a single exon. *EMBO J.,* **8,** 1441.
71. Kumar, V., Green, S., Stack, G., Berry, M., Jin, J. R., and Chambon, P. (1987) Functional domains of the human estrogen receptor. *Cell,* **51,** 941.
72. Tora, L., White, J., Brou, C., Tasset, D., Webster, N., Scheer, E., and Chambon, P. (1989) The human estrogen receptor has two independent nonacidic transcriptional activation functions. *Cell,* **59,** 477.

73. Harlow, K. W., Smith, D. N., Katzenellenbogen, J. A., Green, G. L., and Katzenellenbogen, B. S. (1989) Identification of cysteine 530 as the covalent attachment site of an affinity-labelling estrogen (ketononestrol aziridine) and antiestrogen (tamoxifen aziridine) in the human estrogen receptor. *J. Biol. Chem.*, **164,** 17 476.
74. Danielian, P. S., White, R., Hoare, S. A., Fawell, S. E., and Parker, M. G. (1992) Identification of residues in the estrogen receptor which confer differential sensitivity to estrogen and hydrotamoxifen. *Mol. Endocrinol.*, submitted.
75. Pakdel, F. and Katzenellenbogen, B. S. (1992) Human estrogen receptor mutants with altered estrogen and antiestrogen ligand discrimination. *J. Biol. Chem.*, **267,** 3429.
76. Benhamou, B., Garcia, T., Lerouge, T., Vergezac, A., Gofflo, D., Bigogne, C., Chambon, P., and Gronemeyer, H. (1992) A single amino acid that determines the sensitivity of progesterone receptors to RU486. *Science*, **255,** 206.
77. Brinkmann, A. O. and Trapman, J. (1992) Androgen receptor mutants that affect normal growth and development. In *Cancer Surveys*, Vol. 14, M. G. Parker (ed.). Cold Spring Harbor Laboratory Press, New York, p. 95.
78. Nakao, R., Haji, M., Yanase, T., Ogo, A., Takayanagi, R., Katsube, T., Fukumaki, Y., and Nawata, H. (1992) A single amino acid substitution (Met786–Val) in the steroid-binding domain of a human androgen receptor leads to complete androgen insensitivity syndrome. *J. Clin. Endocrinol. Metab.*, **74,** 1152.
79. Brown, T. R., Lubahn, D. B., Wilson, E. M., French, F. S., Migeon, C. J., and Corden, J. L. (1990) Functional characterization of naturally occurring mutant androgen receptors from subjects with complete androgen insensitivity. *Mol. Endocrinol.*, **4,** 1759.
80. Veldscholte, J., Berrevoets, C. A., Ris-Stalpers, C., Kuiper, G. G. J. M., Jenster, G., Trapman, J., Brinkmann, A. O., and Mulder, E. (1992) The androgen receptor in LNCaP cells contains a mutation in the ligand binding domain which affects steroid binding characteristics and response to antiandrogens. *J. Steroid Biochem. Mol. Biol.*, **41,** 665.
81. Allan, G. F., Leng, X., Tsai, S. Y., Weigel, N. L., Edwards, D. P., Tsai, M.-J., and O'Malley, B. W. (1992) Hormone and anti-hormone induce distinct conformational changes which are central to steroid receptor activation. *J. Biol. Chem.*, **267,** 19 513.
82. Berry, M., Metzger, D., and Chambon, P. (1990) Role of the two activating domains of the oestrogen receptor in the cell-type and promoter-context dependent agonistic activity of the antioestrogen 4-hydroxytamoxifen. *EMBO J.*, **9,** 2811.
83. Brown, M. and Sharp, P. A. (1990) Human estrogen receptor forms multiple protein–DNA complexes. *J. Biol. Chem.*, **265,** 11 238.
84. Meyer, M.-E., Pornon, A., Ji, J., Bocquel, M.-T., Chambon, P., and Gronemeyer, H. (1990) Agonistic and antagonistic activities of RU486 on the functions of the human progesterone receptor. *EMBO J.*, **8,** 3923.
85. Becker, P. B., Gloss, B., Schmid, W., Strähle, U., and Schütz, G. (1986) *In vivo* protein–DNA interactions in a glucocorticoid response element require the presence of hormone. *Nature*, **324,** 686.
86. Vegeto, E., Allan, G. F., Schrader, W. T., Tsai, M.-J., McDonnell, D. P., and O'Malley, B. W. (1992) The mechanism of RU486 antagonism is dependent on the conformation of the carboxy-terminal tail of the human progesterone receptor. *Cell*, **69,** 703.
87. Wakeling, A. E. and Bowler, J. (1988) Biology and mode of action of pure antioestrogens. *J. Steroid Biochem.*, **30,** 141.
88. Arbuckle, N. D., Dauvois, S., and Parker, M. G. (1992) Effects of antioestrogens on the DNA binding activity of oestrogen receptors *in vitro*. *Nucleic Acids Res.*, **20,** 3839.

89. Fawell, S. E., White, R., Hoare, S., Sydenham, M., Page, M., and Parker, M. G. (1990) Inhibition of estrogen receptor–DNA binding by the 'pure' antiestrogen ICI 164 384 appears to be mediated by impaired receptor dimerization. *Proc. Natl Acad. Sci. USA*, **87,** 6883.
90. Sabbah, M., Gouilleux, F., Sola, B., Redeuilh, G., AND Baulieu, E.-E. (1991) Structural differences between the hormone and antihormone estrogen receptor complexes bound to the hormone response element. *Proc. Natl Acad. Sci. USA*, **—8,** 390.
91. Martinez, E. and Wahli, W. (1989) Cooperative binding of estrogen receptor to imperfect estrogen-responsive DNA elements correlates with their synergistic hormone-dependent enhancer activity. *EMBO J.*, **8,** 3781.
92. Pham, T. A., Elliston, J. F., Nawaz, Z., McDonnell, D. P., Tsai, M. J., and O'Malley, B. W. (1991) Antiestrogen can establish nonproductive receptor complexes and alter chromatin structure at target enhancers. *Proc. Natl Acad. Sci. USA*, **88,** 3125.
93. Reese, J. C. and Katzenellenbogen, B. S. (1992) Examination of the DNA-binding ability of estrogen receptor in whole cells: implications for hormone-dependent transactivation and the actions of antiestrogens. *Mol. Cell. Biol.*, **12,** 4531.
94. Dauvois, S., Danielian, P. S., White, R., and Parker, M. G. (1992) Antiestrogen ICI 164 384 reduces cellular estrogen receptor content by increasing its turnover. *Proc. Natl Acad. Sci. USA*, **89,** 4037.
95. Gibson, M. K., Nemmers, L. A., Beckman, W. C., Jr., Davis, V. L., Curtis, S. W., and Korach, K. S. (1991) The mechanism of ICI 164 384 antiestrogenicity involves rapid loss of estrogen receptor in uterine tissue. *Endocrinology*, **129,** 2000.
96. Guiochon-Mantel, A., Lescop P., Christin-Maitre, S., Loosfelt, H., Perrot-Applanat, M., and Milgrom, E. (1991) Nucleocytoplasmic shuttling of the progesterone receptor. *EMBO J.*, **10,** 3851.
97. Ylikomi, T., Bocquel, M. T., Berry, M., Gronemeyer, H., and Chambon, P. (1992) Cooperation of proto-signals for nuclear accumulation of estrogen and progesterone receptors. *EMBO J.*, **11,** 1.
98. Klein-Hitpass, L., Cato, A. C. B., Henderson, D., and Ryffel, G. U. (1991) Two types of antiprogestins identified by their differential action in transcriptionally active extracts from T47D cells. *Nucleic Acids Res.*, **19,** 1227.
99. Furr, B. J. A. and Jordan, V. C. (1984) The pharmacology and clinical uses of tamoxifen. *Pharmacol. Ther.*, **25,** 127.

9 Mutations in steroid hormone receptors causing clinical disease

MICHAEL J. McPHAUL

1. Introduction

The syndromes of hormone resistance have been of interest to scientists for decades. Following the detection of specific receptors for thyroid hormone, vitamin D, and steroid hormones, substantial effort was made to study patients affected with these disorders using assays of hormone binding. Although interesting patterns often emerged from such studies, the pathogenesis of the diseases remained obscure.

In 1984, the nucleotide sequence of the glucocorticoid receptor was reported (1) and was followed in quick succession by the isolation of complementary DNAs (cDNAs) encoding the classic steroid receptors, the thyroid hormone receptors, retinoic acid receptors, and vitamin D receptors (reviewed in reference 2). These molecules were found to comprise a large gene family that share important structural similarities: an amino-terminus of variable length, a highly conserved cysteine-rich DNA-binding domain and a carboxy-terminal hormone-binding domain. The determination of the nucleotide sequences of the genes encoding these molecules has permitted a glimpse of the mechanisms resulting in abnormalities of steroid hormone action. Although patients have been described in whom cellular resistance to actions of mineralocorticoids has been postulated on the basis of clinical phenotype and endocrinological evaluation (3–5), and mutations in steroid receptors have been implicated in the pathogenesis of steroid resistance in neoplasms (6, 7), the present discussion will focus on those clinical resistant states in which causative mutations have been reported, namely those caused by defects in the androgen, thyroid hormone, vitamin D, and glucocorticoid receptors.

2. Androgen resistance

The principal circulating androgens, testosterone and its 5α reduced metabolite, 5α-dihydrotestosterone, mediate a range of biological processes in embryonic development and in adults (8). During embryogenesis, each of these hormones exerts distinctive effects: testosterone is sufficient to effect wolffian duct virilization, while 5α-dihydrotestosterone formation is necessary for normal development of the prostate and external genitalia (9–12).

Resistance to the actions of androgen results in a range of phenotypic abnormalities in 46,XY genotypic men. Severely affected individuals with complete androgen resistance (complete testicular feminization) have a phenotypic appearance of normal women. Although their external genitalia are completely female, testes are present either within the labia or within the abdomen, and the wolffian duct structures (epididymis and vas deferens) drain into a blind-ending vagina. Less severely affected individuals can have predominant female (incomplete testicular feminization) or predominant male (Reifenstein syndrome) phenotypes. In some individuals slight undervirilization or infertility is believed to be due to defects of androgen receptor function (13).

Defects in the androgen receptor are relatively common due to the location of the gene on the X chromosome and the function of the androgen receptor in hemizygous (46,XY) males. Abnormalities of the androgen receptor causing androgen resistance were first recognized in 46,XY women with complete androgen resistance (complete testicular feminization) and associated with absent binding of tritiated androgens in monolayer cultures of fibroblasts. Subsequent studies have demonstrated that clinical phenotypes ranging from those of phenotypic women to those undervirilized but normal men can be due to abnormalities of the androgen receptor (14, 15). Such abnormalities result in alterations in receptor abundance or function, as summarized in Table 1.

Inspection of these binding data leads to several conclusions. First, there is no absolute correlation evident between the type of receptor binding abnormality and the clinical phenotype, although the receptor-binding negative category is most often associated with complete testicular feminization. Second, numerous patients with significant developmental abnormalities can be shown to have only a diminished level of receptor or no identifiable abnormality of the androgen receptor. These results imply a substantial heterogeneity at the genetic level.

Table 1 Phenotype and receptor binding in patients with androgen resistance

Receptor binding abnormality	Complete testicular feminization	Incomplete testicular feminization	Reifenstein syndrome	Undervirilized males	Infertile males
Binding negative	24	6	3	0	0
Reduced	13	9	23	8	3
Qualitative	2	4	7	5	0
No abnormality	4	6	12	2	0

Cultures of skin fibroblasts were established from genital skin biopsies and analysed to assess the level and nature of receptor binding. Reduced receptor binding refers to decreased levels (< 12 fmol/mg protein) of qualitatively normal receptor binding. The category designated qualitative abnormality refers to thermal instability of receptor binding or abnormal dissociation of ligand. No abnormality refers to samples in which normal levels of qualitatively normal receptor binding was detected. Adapted from reference 13.

2.1 Molecular basis of androgen resistance

The definition of the structure of the androgen receptor gene has permitted the identification of mutations causing androgen resistance in a number of families. As will become evident, there is a marked heterogeneity in the pathogenesis of androgen resistance. For the purposes of discussion, the results of the genetic analysis have been arbitrarily divided into those gene alterations that result in large-scale alterations of receptor structure and those that result in small-scale alterations of the primary structure of the androgen receptor.

2.1.1 Large-scale alterations of receptor structure

The first pedigree in which the cause of androgen resistance was defined was that reported by Brown *et al.* (16). Using Southern analysis, these authors demonstrated a deletion of a large portion of the carboxy-terminal segments of the androgen receptor gene encoding the hormone-binding domain in a patient with complete androgen resistance. More recently, two groups have reported the deletion of the entire androgen receptor gene in two unrelated patients with complete testicular feminization (17, 18). In some instances, the boundaries of the deleted segments of the androgen receptor gene have been more localized, as in the patient described by Quigley *et al.* (19) in which genomic DNA encoding only a single exon (exon 3) has been deleted. On the whole, partial or complete deletion of the androgen receptor gene appears to be relatively infrequent and would appear to account for approximately 5% of patients with androgen resistance. To date, these patients have all demonstrated a phenotype of complete androgen resistance.

Interruption of the primary amino acid sequence of the androgen receptor has also been shown to be due to single nucleotide substitutions that result in the introduction of premature termination codons into the coding segment of the androgen receptor gene. Mutations causing premature termination codons that delete the carboxy-terminal segments of the receptor have been detected in seven patients and have been localized to exons 3, 4, 5, 6, and 7 (21–24). In each instance for which the data have been presented, although immunoreactive androgen receptor protein is produced, receptor binding[1] is absent when cDNAs encoding the mutant receptors are introduced into transfected cells. The truncated receptor proteins are also inactive in assays of reporter gene activation.

An interesting variation on this mechanism is that which has been reported to cause androgen resistance in the *Tfm* mouse (25–27). This animal model has been studied by three groups and is caused by a single nucleotide deletion in the amino-terminus of the receptor which results in a shift of the translation reading frame and premature termination of the receptor protein in exon 1. Binding studies and

[1] In this discussion, the term 'receptor binding' is used to refer to specific ligand binding by the androgen receptor in fibroblast cultures or transfected cells.

in vitro translation analysis of the *Tfm* mouse mutant cDNA (27) suggest that a small amount of an abnormal androgen receptor is synthesized. These findings have remarkable similarity to the analysis of the human pedigree analysed by Zoppi *et al.* (28). In this family, premature termination of the androgen receptor also occurred in exon 1 due to a single nucleotide substitution that introduces a premature termination codon into the receptor protein. Cell transfection studies demonstrated initiation of the receptor at an internal methionine and the expression of an androgen receptor protein lacking the first 187 amino acids of the amino-terminus. This study suggests that the androgen resistance was due to a combination of reduced receptor abundance and function.

The results of this class of patients are informative in several respects. First, they demonstrate that the phenotype exhibited by patients with complete or partial deletions of the androgen receptor gene is complete testicular feminization. Second, they demonstrate that the *in vivo* effects of the removal of segments of the hormone-binding domain render the androgen receptor unable to bind hormone and to activate gene transcription. Thus, the *in vivo* data parallel the results of extensive *in vitro* mutagenesis (2). Finally, two mutant receptors, that of the *Tfm* mouse (25–27) and the human mutation described by Zoppi *et al.* (28), suggest that functionally important segments of the androgen receptor reside within the amino-terminus, a finding that has been suggested by studies of androgen receptor mutagenized and studied *in vitro* (29, 30).

2.1.2 Amino acid substitutions causing androgen resistance

Although important information has been derived from patients with androgen resistance caused by large-scale alterations of androgen receptor structure, androgen resistance caused by amino acid substitutions in the androgen receptor have been more informative regarding the structure–function relationships of the androgen receptor protein.

2.1.3 Mutations causing receptor binding-positive androgen resistance

Receptor binding-positive androgen resistance refers to those patients with endocrine profiles and family histories of androgen resistance, but in whom the level of receptor binding is qualitatively and quantitatively abnormal. This entity has been the object of great interest and has been postulated to represent either subtle abnormalities of the androgen receptor that are not evident in assays of receptor binding or defects in other genes required for androgen receptor function.

The question of the aetiology of receptor binding-positive androgen resistance has been at least partially resolved by the analysis of four patients from this category (31, 32). Receptor binding levels in each patient were well within the normal range. In three instances, no qualitative abnormality was evident, and in one there was a very subtle disturbance of ligand dissociation. When the androgen receptor gene of each patient was analysed, an amino acid substitution was identified within the DNA-binding domain of the androgen receptor. In each instance, the effect of the mutation on ligand binding by the mutant receptor was minimal.

Despite normal hormone-binding properties, the mutant receptors were unable to activate a target reporter gene. This inactivity in gene activation assays was found to correlate with an inability of the receptor to bind to target DNA sequences.

2.1.4 Qualitative receptor-binding abnormalities

A number of mutations causing qualitative defects of receptor binding have been identified (33–38). In most instances, these mutations have been shown to be localized to the hormone-binding domain of the receptor. Mutations within this region of the receptor have been shown to result in several types of qualitative defects, including thermolablity, abnormal dissociation of ligand from the receptor, or both. Regardless of the type of defect (thermolability, dissociation rate), receptor function is substantially impaired, although in some instances incompletely.

An intriguing observation has been made in several of these papers. In several instances, the use of supraphysiological doses of androgen results in a demonstrable increase in receptor activity in assays of gene activation (34–36). The use of 20 nM mibolerone resulted in a substantial increase in receptor function compared with 2 nM, although 2 nM was well above the level required to saturate the receptor. In one instance, this effect was demonstrated clinically, when a patient with a qualitative receptor binding abnormality was treated with high-dose testosterone therapy (34). This response to supraphysiological levels of androgen may be a general attribute of mutations causing qualitative abnormalities of receptor binding and may have implications for the treatment of such patients.

2.1.5 Receptor binding-negative androgen resistance

Within the category of patients with receptor binding-negative androgen resistance, mutations resulting in single amino acid substitutions appear to be responsible in over half of such patients (36–40). In each instance, the amino acid substitutions are localized to the hormone-binding domain of the receptor. The detection of these amino acid substitutions has permitted a detailed analysis of the effects of these mutations on receptor function in cells transfected with cDNAs encoding the mutant receptors. Two results were unexpected. First, several of the mutations in this category were found to be capable of binding hormone, albeit with qualitative abnormalities, when expressed in eukaryotic or bacterial systems (36, 39, 40). Second, the mutations causing this form of receptor abnormality localized to regions of the hormone-binding domain that are similar to those causing qualitative receptor-binding abnormalities (37). These findings suggest that the abnormalities identified in the receptor binding-negative category differ from those within the qualitatively abnormal category only in degree.

2.2 Distribution and type of mutations causing androgen resistance

The results of the genetic analysis of 40 patients is shown in Table 2. Inspection of this summary reveals that single nucleotide changes are responsible for causing

Table 2 Summary of published mutations in the androgen receptor gene causing androgen resistance

Type or mutation	Patient phenotype	No. of patients	References
Gene deletion			
Partial	CTF	1	16, 19
Complete	CTF	3	17, 18
Splicing defects	CTF	1	20
Premature termination codons			
Exons 1, 3, 5, 7, 8	CTF	5	21–28
Amino acid substitution			
DNA-binding domain	CTF	4	31–32
Hormone-binding domain	CTF	10	33, 35–40
	ITF	3	37
	Reif	8	34, 37
Glutamine repeat polymorphism			
Amino-terminus	Reif	1	34
	SMA	>20	41

CTF, complete testicular feminization; Reif, Reifenstein phenotype; SMA, X-linked spinomuscular atrophy. The identified lesions, region of the receptor affected, and range of phenotypes in each class is shown. The number of patients reported in each class is indicated. In addition to classic forms of androgen resistance, lengthening of the glutamine polymeric segment has been implicated in the pathogenesis of X-linked spinomuscular atrophy. Patients with this disease have evidence of mild androgen resistance (41). A truncation of the amino-terminal glutamine repeat has been implicated in the pathogenesis of androgen resistance in a patient with the Reifenstein phenotype (34).

most of the androgen resistance. While these substitutions at times act via unusual mechanisms, such as alterations in mRNA splicing or frame shifts, the majority act via the insertion of premature termination codons or the alteration of single amino acid residues. The distribution of mutations is also interesting. Premature termination codons have been identified in many segments of the androgen receptor protein. Mutations causing receptor binding-positive androgen resistance are localized to the DNA-binding domain of the receptor. Mutations causing receptor binding-negative or qualitative abnormalities of receptor binding are localized to the hormone-binding domain of the receptor and are located predominantly in two regions of the molecule, suggesting that these regions are particularly important to the function of the hormone-binding domain (Fig. 1).

2.3 Patient phenotype and genetic mutations

Attempts to correlate patient phenotype and genetic mutations in the androgen receptor gene require information regarding both receptor function and abundance. The former has been obtained for several mutant receptors in assays using co-transfected reporter genes. Recently, a study has been completed using specific antibodies to detect the androgen receptor expressed in genital skin fibroblasts

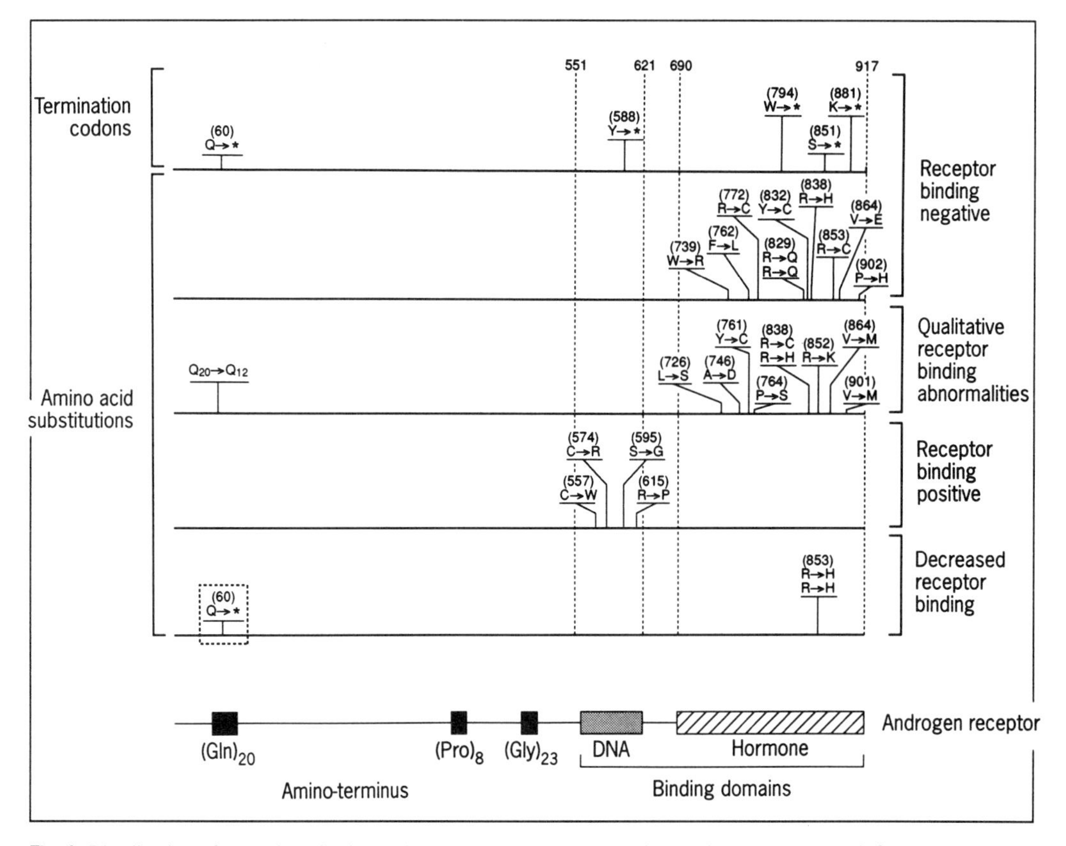

Fig. 1 Distribution of mutations in the androgen receptor gene causing androgen receptor defects. The type of receptor binding defects is indicated on the right, and the nature of the mutation on the left. The position of the mutated amino acid residue is shown using the co-ordinates of Tiley *et al.* (42). The schematic structure of the androgen receptor is shown below. Adapted with permission from reference 43

(44). This study indicated that a good correlation exists between receptor binding and immunoreactive receptor in normal patients and patients with receptor-positive androgen resistance. This correlation was also quite good, although somewhat weaker, for patients with qualitative abnormalities of receptor binding. In the bulk of patients with receptor binding-negative androgen resistance, the effect on receptor expression was minimal, and near normal levels of immunoreactive receptor expression were detected. Thus, in many patients, the phenotype of complete testicular feminization is associated with severe impairments of receptor function, receptor abundance, or both. There are clearly instances in which this correlation is imperfect (37) and a single genetic mutation can give rise to different patient phenotypes. This suggests that genetic determinants outside the androgen receptor gene can influence patient phenotype, presumably by altering receptor expression or function.

3. Thyroid hormone resistance

Resistance to the action of thyroid hormone has been recognized since the initial description in 1967 by Refetoff *et al.* of a pedigree in which affected subjects were found to have clinical signs (growth retardation and bony lesions) suggesting hypothyroidism despite elevated serum levels of thyroid hormone (45). In this initial family, this resistance state was inherited as an autosomal recessive trait. Following this initial report, subsequent studies have established that substantial heterogeneity exists within this syndrome. In most pedigrees, a less severely affected phenotype is observed, with elevated levels of thyroid hormone, elevated levels of thyroid-stimulating hormone (TSH), mild to moderate enlargement of the thyroid gland, and an absence of clinical signs of hyperthyroidism. In contrast to the initial report of Refetoff *et al.* (45), an autosomal dominant inheritance pattern is evident in most pedigrees. Subsequent studies have also suggested that a range of clinical phenotypes is possible (46, 47). Less frequently, patients have been described in whom the thyroid hormone resistance appears to be confined to the pituitary (48) or peripheral tissues alone (49).

Although the molecular basis of this clinical diversity has not been completely defined, molecular studies have established several levels that may ultimately offer some explanation for the heterogeneity. Two distinctive genes (α and β) encode two different high-affinity receptors for thyroid hormone (50, 51). These two receptors are expressed to some level in all tissues, but the ratios of receptor expression vary from tissue to tissue (52). In addition, there are distinctive forms of both the α and β receptors (53–55) generated by alternative slicing of messenger RNA (mRNA), and in some instances these thyroid hormone receptor (TR) isoforms are expressed in a tissue-specific fashion (e.g. TR-β_2, in the rat pituitary). Due to the wealth of information currently available, the subsequent discussion will focus exclusively on the genetic mutations causing the syndrome of generalized thyroid hormone resistance.

3.1 Receptor-binding abnormalities in thyroid hormone resistance

A substantial amount of effort has been expended to examine the binding of thyroid hormone in subjects with syndromes of thyroid hormone resistance (reviewed in reference 56). Despite these efforts, no consistent patterns have emerged. In no pedigree was receptor binding reported to be completely absent, and in families with similar phenotypes thyroid hormone binding has been reported as increased, decreased, or unchanged. These findings have become somewhat easier to interpret with the discovery that both α and β receptor forms are expressed in most tissues (52). Thus, even a complete deficiency of one receptor type would likely be masked by the continued expression of the other.

3.2 Generalized resistance to thyroid hormone

The demonstration by Usala *et al.* in 1988 (57) of a genetic linkage between the TR-β gene and the inheritance of generalized resistance to thyroid hormone (GRTH) focused attention on the structural analysis of the TR-β gene. Following the initial reports by Sakurai *et al.* (58), a number of mutations have been detected within the TR-β gene (59–66). The vast majority of these mutations are amino acid substitutions, although a single amino acid deletion (61) and a frame-shift (47) mutation have been described. An intriguing result of this analysis is the tight clustering of these mutations within specific regions of the TR-β gene. As shown schematically in Figure 2, a large fraction of these mutations are localized to two short segments of the hormone-binding domain of the receptor. Unlike the mutations described for the androgen receptor causing androgen resistance, mutations in the amino-terminus or DNA-binding domains of the TR-β gene causing thyroid hormone resistance have not been described.

The genetic analysis of patients with GRTH indicates that, with only two exceptions (see below), affected individuals are heterozygotes. This result indicates that the expression of a single mutant allele is sufficient to cause a state of partial thyroid hormone resistance. This type of mutation has been termed 'dominant negative'. This inference has been supported by the analysis of the two exceptional cases. Analysis of the original family with GRTH (60) which exhibited an autosomal recessive inheritance pattern indicated that the syndrome of thyroid resist-

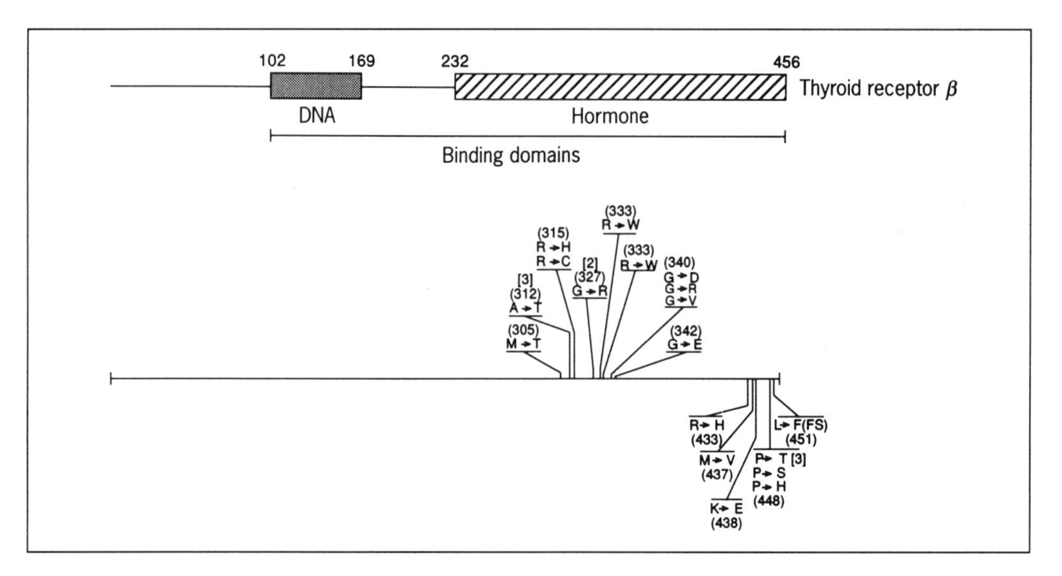

Fig. 2 Distribution of mutations in the thyroid receptor β gene causing thyroid hormone resistance. The positions of mutated amino acids are indicated using the co-ordinates of Weinberger *et al.* (51). In instances of repetitive mutation, the number of times the same mutation has been observed is given in parentheses. The schematic structure of the thyroid receptor β is shown above; note that this diagram is drawn on a scale approximately twice that used for the androgen receptor

ance in this pedigree was due to a partial deletion of the TR-β gene and that affected individuals were homozygous for this deletion. Coupled with the normal phenotype of heterozygous subjects within this pedigree, these findings indicate that the effect of the amino acid substitutions causing GRTH is not simply due to the loss of one TR-β allele but to an interference of the mutant (dominant negative) allele on the function of the remaining normal TR-β allele. This conclusion has been further reinforced by the severely affected phenotype observed in the only patient known to be homozygous for a mutant receptor that can cause GRTH in the heterozygous state (61, 62). The phenotype of this profoundly affected subject was even more pronounced than that observed in patients affected with severe forms of GRTH, and suggests that the two mutant TR-β alleles are able to interfere with the function of the normal TR-α genes.

3.2.1 Mechanism of 'dominant negative' thyroid hormone receptor mutations

The effects of the TR-β amino acid substitutions causing thyroid hormone resistance have been studied using a number of techniques. The effects that these mutations have on hormone binding have been published for 14 different mutations using receptor preparations derived from *in vitro* transcription and translation of mRNA encoding normal and mutant thyroid receptors. As shown in Table 3, these data indicate that amino acid substitutions at residue 340 of the TR-β result in the complete loss of specific hormone binding. This same result is also obtained when TR-β receptor proteins containing a carboxy-terminal frame-shift mutation or an amino acid deletion at residue 332 are analysed. By contrast, TR-β receptors containing amino acid substitutions at either residues 312, 335, 437, or 488 demonstrate appreciable specific binding, but reduced affinity for T_3.

The effects of these mutations on other aspects of receptor function have also been examined. Studies using *in vitro* synthesized receptor suggest that receptors with amino acid substitutions at either residues 448 or 340 retain the ability to bind to DNA, despite the observation that only the 448 substitution mutation was capable of binding hormone. The binding of both receptors could not be differentiated on the basis of the binding to several target DNA sequences.

An important prediction of the molecular and genetic studies is that the 'dominant negative' mutant thyroid receptors should interfere with the action of normal TR-β molecules. This has been examined directly in several studies (66–69). In these experiments mutant receptors containing amino acid substitution mutations were studied in cell transfection assays using reporter genes controlled by thyroid hormone response elements (TREs). One type of reporter gene (e.g. TRE-TKCAT) is regulated in a positive fashion and one type (e.g. TSH-αCAT) is regulated in a negative fashion by the thyroid receptor. The effects of these mutations on receptor function are summarized in Tables 3 and 4. The normal TR-β stimulated the TRE-TK reporter and inhibited the TSH-α reporter gene. Neither mutant receptor was able to alter the activity of either reporter gene in the absence or presence of

Table 3 Summary of published mutations causing generalized resistance to thyroid hormone and the effects on receptor function

Mutation	Mutant[a]	Normal[a]	DNA binding	Transfection	References
G340R	+	8.3×10^9	+		58, 64
—	—	3.4×10^{10}			47
A312T	0.7×10^{10}		NR	NR	
M437V	0.6×10^{10}		NR	NR	
ins443	$<0.17 \times 10^{10}$		NR	NR	
P448T	1.4×10^{10}		NR	NR	
—	—	1.4×10^{10}	+		66
G340S	ND		+	T Dom	
G340R	ND		+	T Dom	
—	—	2.3×10^{10}			64
P448H	4.5×10^9		N or ↑	NR	
—	—	1.5×10^{10}			68
G340R	ND		+	T Dom	
P448H	1.5×10^9		+	T Dom	
Δ448–456	ND		+	T Dom	
—	—	$7 \pm 2 \times 10^{10}$	+		61
G340R	ND			NR	
del332	ND		+	NR	
P448H	4.5×10^9		NR	NR	
	1.1×10^{10}			NR	
—	—	$3.2 \pm 0.5 \times 10^{10}$		8 nM*/—	69
Q335H	$1.6 \pm 0.4 \times 10^{10}$		NR	50 nM*/T Dom	
A312T	$0.7 \pm 0.3 \times 10^{10}$		NR	150 nM*/T Dom	
Frame-shift	ND		NR	Inactive/NR	
del332	ND		NR	Inactive/T Dom	

The locations of mutation are indicated according to the co-ordinates of Weinberger *et al.* (51). In most instances, hormone binding was assessed in binding assays of *in vitro* transcribed and translated receptor proteins. NR, value not reported; ND, not detectable. T Dom indicates that the mutation could be demonstrated to have a 'transdominant' effect on the function of the normal TR-β receptor in transfection assays. Values that are not applicable are indicated by dashes. *Value at which half-maximal reporter gene activation was obtained. [a]indicates the affinity of hormone binding observed for the normal and mutant receptors (expressed as M^{-1}). N, normal.

thyroid hormone. Despite this absence of activity, both the 340 and 448 mutant receptors substantially affected the capacity of the wild-type thyroid receptor to alter gene transcription when both were introduced simultaneously. Interestingly, this effect could be minimized when the assays were performed after incubation of the transfected cell's with a supraphysiological level of thyroid hormone (68, 69). This result is similar to those found in experiments where qualitatively abnormal androgen receptors could be activated by supraphysiological concentrations of androgen (see above).

3.3 Genetic mutation and phenotype

The mechanisms by which the genetic mutations in the TR-β gene give rise to the variable phenotypes of patients with GRTH have been difficult to establish. Not

Table 4 Functional properties of the mutant thyroid hormone receptors change with alterations in thyroid hormone concentrations

Normal TR	Mutant TR	T_3 concentration (nM)	Increase in TRE-TKCAT (%)	Decrease in TSH-αCAT (%)
β	0	5	73	80
β	G340R	5	21	16
β	P448H	5	24	50
α	0	5	118	84
α	G340R	5	36	61
α	P448H	5	43	64
β	0	0.75	25	75
β	0	100	30	70
0	P448H	0.75	4	40
0	P448H	100	55	60
β	P448H	0.75	4	40
β	P448H	100	39	80

The data show the change in the activity of two reporter genes TRE-TKCAT and TSH-αCAT that are regulated by thyroid hormone in a positive or negative fashion, respectively, following transfection with normal or mutant thyroid hormone receptor (TR), or both. Data are derived from reference 68. Similar data were obtained for different mutations by Meier *et al.* (69).

only are the mutations themselves heterogeneous, but the levels of expression of the normal and mutant alleles may vary from individual to individual, or even from tissue to tissue. Although quantitative analyses of mutant mRNA and protein levels have not yet been published, there are suggestions that variable degrees of impairment of receptor function may explain, at least in part, the phenotypic variation observed within the syndrome of GRTH (69). In addition, it is now evident that other auxiliary proteins, such as retinoid X receptor, can further modify the activity of normal (and presumably mutant) thyroid receptor (70–73).

4. Vitamin D resistance

Several kindreds have been reported with hereditary resistance to the action of vitamin D. This syndrome is characterized by manifestations associated with vitamin D deficiency (osteomalacia, hypocalcaemia, secondary hyperparathyroidism, and hypophosphataemia) despite elevated levels of circulating 1,25-dihydroxyvitamin D (74, 75). In addition to these characteristics, several patients have been shown to have complete alopecia (76, 77).

4.1 Receptor binding in fibroblasts in patients with vitamin D resistance

Following the demonstration of specific receptors for vitamin D in skin fibroblasts, several groups analysed the nature of the vitamin D receptors present in cultured fibroblasts from patients with vitamin D resistance (78, 79). These analyses indicated substantial heterogeneity, with diminished or absent receptor binding encountered in several pedigrees. These studies also clearly delineated a group of patients with levels of receptor binding comparable to that of control patients. Chromatography studies suggested that the DNA-binding properties of the mutant vitamin D receptors were abnormal. The lack of induction of 24-hydroxylase activity (a vitamin D-responsive gene) in these same cell strains indicated a true form of hormone resistance (79).

4.2 Genetic defects causing vitamin D resistance

The first patients in whom the molecular basis of vitamin D resistance was determined were those described by Hughes *et al.* (80). The vitamin D receptor genes of these patients were analysed and found to contain amino acid substitutions within the DNA-binding domain of the receptor protein. Amino acid substitutions in the DNA-binding domain were subsequently discovered in patients with similar phenotypes (81, 82). The effects of these mutations have been studied in detail. Although the binding of hormone by the mutant receptors from these pedigrees is indistinguishable from that of the normal vitamin D receptor, the receptor does not bind with normal affinity to target DNA sequences (Chapter 7). This inability to bind to DNA renders the hormone–receptor complexes ineffective and unable to activate gene transcription in a variety of reporter gene assays (82, 83).

Four other mutations have been identified outside the DNA-binding domain of the vitamin D receptor that cause a similar phenotype. In two pedigrees, single nucleotide substitutions were identified that result in the insertion of premature termination codons into the receptor protein in place of residues 150 and 292. When expressed in transfected cells, these receptor proteins are unable to bind ligand and cannot activate a reporter gene (84, 85). Immunoblot and S_1 nuclease assays suggest that the mRNAs containing these mutations are unstable and that little or none of the truncated receptor protein accumulates in fibroblasts or lymphocytes. The authors have suggested that the premature termination codons destabilize the mRNA encoding the vitamin D receptor. The truncated receptor protein may likewise be unstable (84, 85).

Two other pedigrees of vitamin D resistance have been reported in abstract form and are apparently due to amino acid substitutions in the hormone-binding domain of the vitamin D receptor. In one pedigree the defect has been traced to the replacement of a cysteine (residue 187) with a tryptophan (86). In the other family, replacement of arginine 271 by a leucine is the apparent cause of the vitamin D-resistant state (87). Interestingly, the latter mutation alters the affinity of hormone binding but does not decrease the amount of receptor binding.

5. Glucocorticoid resistance

End-organ resistance to the action of glucocorticoids was first reported in 1976 in a patient with no clinical signs or symptoms of hypercortisolism but with abundant biochemical evidence indicating extraordinary production of adrenal steroid (88). Subsequent investigations identified and studied other affected members in the same pedigree (89). In each of these affected members, the degree of excess cortisol production was less than that observed for the index case. Receptor binding studies, performed in both lymphoblasts and fibroblasts, indicated a reduced affinity of the receptor for ligand. The diminished levels of receptor binding detected in broken cell extracts suggested that the glucocorticoid receptor synthesized in these patients had reduced stability. These data suggested either an autosomal recessive form of inheritance or an autosomal dominant inheritance with variable penetrance, with the bulk of data suggesting the latter (90).

The clinical presentation of other patients with glucocorticoid resistance have been variable, presenting with fatigue (91) or precocious sexual maturation (92). Several others have been identified, apparently incidentally. Hypertension has been observed in several families, but is not a consistent feature. A range of abnormalities of receptor binding have been identified in these pedigrees, including decreased receptor number, receptor thermolability, and diminished binding of the receptor to DNA. These data are summarized in Table 5.

Recently, progress has been made in identifying the molecular defects causing these syndromes. The genetic defect in three families have been reported. In the original pedigree of Vingerhoeds *et al.* (88), Hurley *et al.* (99) depicted a single amino acid substitution (aspartate → valine 641) within the hormone-binding domain of the receptor. The effects of this mutation on receptor function have been studied and shown substantially to impair receptor function both in assays of

Table 5 Summary of clinical findings and receptor binding assays in patients with glucocorticoid resistance

Presentation	Index case	Receptor binding	Inheritance	References
A. Hypertension, hypokalemia	Male	Decreased affinity	Recessive	88, 89, **99**
B. Hirsutism, male pattern baldness	Female	Normal affinity, decreasd β_{max}	NR	93, 94, **101**
C. Fatigue	Female	Thermolabile	NR	91
D. Isosexual precocity	Male	Decreased affinity, normal β_{max}	Recessive	92, **100**
E. Moderate hypertension, hirsutism, acne	Female	NR	NR	95
F. Obesity, mild hypertension	Female	Diminished DNA binding, decreased affinity	NR	96
G. Mild hypertension	Male and female	Diminished number	NR	97, 98

Clinical findings, receptor binding properties, and inheritance are indicated for pedigrees with hereditary resistance to glucocorticoids. NR, not reported. Bold references are those that describe the causative genetic mutation.

ligand binding and in assays of reporter gene activation. These studies indicated that the most severely affected individual was homozygous for this mutation and that less affected subjects were heterozygotes.

In a second family, Brufsky *et al.* (100) detected a single nucleotide substitution that results in the replacement of valine 729 by an isoleucine residue. In a third family, a single nucleotide substitution causes an alteration of mRNA splicing which is believed to result in a reduction of receptor number (101). This reduction of receptor number is presumed to cause the glucocorticoid resistance.

Several conclusions can be derived from the genetic studies of the molecular basis of glucocorticoid resistance. First, as in the case of androgen and thyroid hormone resistance, these studies localize critical sites within the hormone-binding domain of the glucocorticoid receptor necessary for normal receptor stability and function. Interestingly, the two amino acid substitution mutations are near homologous clusters in the androgen receptor (37). Second, in the most intensively studied pedigree, the studies indicate that the most severely affected individual is in fact a homozygote and that the less severely affected individuals represent the heterozygous phenotype. In fact, in two of the pedigrees (99, 101) functional characterization has indicated that the loss of one functioning allele results in detectable biochemical abnormalities (101). This finding suggests that there is a delicate balance between receptor number and normal glucocorticoid action.

6. Conclusions

At this juncture a burgeoning number of mutations have been described that cause androgen, thyroid hormone, glucocorticoid, and vitamin D resistance. These results provide some interesting contrasts, both by what has been discovered and by what has not yet been encountered.

The available data suggest that patients harbouring mutations that completely abolish the function of the androgen receptor, TR-β, and vitamin D receptor can be viable. Androgen receptor mutations are more frequently encountered than such mutations in the vitamin D receptor or TR-β because of the hemizygous state of the X-linked androgen receptor in 46,XY males where androgens exert their most profound effects. The fact that mutations affecting the TR-α receptor, oestrogen receptor, or progesterone receptors have not yet been described suggests that either the phenotypes of such patients are unexpected or the effects of such mutations are lethal during development. The lessons gleaned from analysis of families of index cases with thyroid hormone and glucocorticoid resistance suggest that the clinical effects of such mutations in heterozygotes can be quite subtle. The analysis of different families with identical mutations causing androgen or thyroid hormone resistance indicates that factors outside the coding sequence of the receptor can influence patient phenotype. In the case of thyroid hormone resistance, the wide variability of phenotype within a family suggests that these effects are due to separate genetic loci.

The types of mutations identified for the different receptors is interesting. Mutations in the DNA-binding domain of the receptors have been detected in the vitamin D receptor and androgen receptor, the syndromes for which the largest number of pedigrees with complete hormone resistance state have been identified. Although no DNA-binding domain mutation in TR-β has yet been described, the pedigree of Refetoff *et al.* (45, 60), in which partial gene deletion led to a loss of TR-β function, makes it likely that such mutations will eventually be identified.

With only three exceptions (all mutations in the androgen receptor) abnormalities affecting the amino-terminus and causing clinical hormone resistance have not been identified. As *in vitro* mutagenesis studies have suggested that this segment plays a significant functional role, it would seem that the paucity of mutations affecting this region is either because the function of this segment is not discrete (i.e. single amino acid substitutions do not abolish functions) or because the correct patient groups have not been examined. This latter line of reasoning would suggest that as additional, less severely affected patients are analysed, mutations in the amino-terminus may be encountered.

The results of the clinical and genetic analyses of patients with mutations in the androgen receptor, glucocorticoid receptor, vitamin D receptor, and TR-β indicate fundamental differences in the clinical manifestations of these mutations. Defects in the androgen receptor are most evident, due to the hemizygous nature of the 46,XY men in which androgens exert their most profound effects. The varied patient phenotypes and the wide range of levels of androgen receptor expression in fibroblast cultures from normal subjects suggest that androgen receptor may be present in normals well in excess of that required to effect normal male development. This is in contrast to the glucocorticoid receptor, where a decline in functional glucocorticoid receptor by 50% is apparently sufficient to result in discernible biochemical abnormalities (101). No phenotype has been described for patients heterozygous for a null vitamin D receptor allele, suggesting that the remaining functional receptors may be sufficient to perform the physiological role. Finally, analysis of patients with GRTH indicates fundamental differences in the manner by which these mutations ('dominant negative') alter thyroid receptor function compared with mutations in the androgen receptor, glucocorticoid receptor, and vitamin D receptor.

References

1. Hollenberg, S. M., Weinberger, C., Ong, E. S., Cerelli, G., Oro, A., Lebo, R., Thompson, E. B., Rosenfeld, M. G., and Evans, R. M. (1985) Primary structure and expression of a functional human glucocorticoid receptor cDNA. *Nature,* **318,** 635.
2. Evans, R. M. (1988) The steroid and thyroid hormone receptor superfamily. *Science,* **240,** 889.
3. Cheek, D. B. and Perry, J. W. (1958) A salt wasting syndrome in infancy. *Arch. Dis. Child.,* **33,** 252.
4. Speiser, P. W., Stoner, E., and New, M. I. (1986) Pseudohypoaldosteronism: a review

and report of two new cases. In *Steroid hormone resistance*, Chrousos, G. P., Loriaux, D. L., Lipsett, M. B. (eds). Plenum Press, New York, p. 173.

5. Armanini, D., Kuhnle, U., Strasser, T., Dorr, H., Butenandt, I., Weber, P. C., Stockigt, J. R., Pearce, P., and Funder, J. W. (1985) Aldosterone receptor deficiency in pseudohypoaldosteronism. *N. Engl. J. Med.*, **313,** 1178.
6. Fuqua, S. A. W., Chamness, G. C., and McGuire, W. L. (1992) Estrogen receptor mutations in breast cancer. *J. Cell. Biochem.*, **50,** 1.
7. Bourgeois, S. and Gasson, J. D. (1985) Genetic and epigenetic bases of glucocorticoid resistance in lymphoid cell lines. *Biochem. Acta Horm.*, **12,** 311.
8. George, F. W. and Wilson, J. D. (1986) Hormonal control of sexual development. *Vitam. Horm.*, **43,** 145.
9. Walsh, P. C., Madden, J. D., Harrod, M. J., Goldstein, J. L., MacDonald, P. C., and Wilson, J. D. (1974) Familial incomplete male pseudohermaphroditism type 2. Decreased dihydrotestosterone formation in pseudovaginal perineoscrotal hypospadias. *N. Engl. J. Med.*, **291,** 944.
10. Imperato-McGinley, J., Guerrero, L., Gautier, T., and Peterson, R. E. (1974) STeroid 5α-reductase deficiency in man: an inherited form of male pseudohermaphroditism. *Science*, **186,** 1213.
11. Imperato-McGinley, J., Binienda, Z., Arthur, A., Mininberg, D. T., Vaughan, E. D., Jr., and Quimby, F. W. (1985) The development of a male pseudohermaphroditic rat using an inhibitor of the enzyme 5α-reductase. *Endocrinology*, **116,** 807.
12. George, F. W., Johnson, L., and Wilson, J. D. (1989) The effect of a 5α-reductase inhibitor on androgen physiology in the immature male rat. *Endocrinology*, **125,** 2434.
13. Griffin, J. E. and Wilson, J. D. (1989) The androgen resistance syndromes: 5α-reductase deficiency, testicular feminization, and related disorders. In *The Metabolic Basis of Inherited Disease*, 6th edn, Scriver, C. R., Beaudet, A. L., Sly, W. S., and Valle, D. (eds). McGraw-Hill, New york, p. 1919.
14. Keenan, B., Meyer, W. J., III, Hadjian, A. J., Jones, H. W., and Migeon, C. J. (1974) Syndrome of androgen insensitivity in man: absence of 5α-dihydro-testosterone binding protein in skin fibroblasts. *J. Clin. Endocrinol. Metab.*, **38,** 1143.
15. Griffin, J. E. and Durrant, J. L. (1982) Qualitative receptor defects in families with androgen resistance: failure of stabilization of the fibroblast cytosol androgen receptor. *J. Clin. Endocrinol. Metab.*, **55,** 465.
16. Brown, T. R., Lubahn, D. B., Wilson, E. M., Joseph, D. R., French, F. S., and Migeon, C. J. (1988) Deletion of the steroid-binding domain of the human androgen receptor gene in one family with complete androgen insensitivity syndrome: evidence for further genetic heterogeneity in this syndrome. *Proc. Natl Acad. Sci. USA*, **85,** 8151.
17. Quigley, C. A., Friedman, K. J., Johnson, A., Lafreniere, R. G., Silverman, L. M., Lubahn, D. B., Brown, T. R., Wilson, E. M., Willard, H. F., and French, F. S. (1992) Complete deletion of the androgen receptor gene: definition of the null phenotype of the androgen insensitivity syndrome and determination of carrier status. *J. Clin. Endocrinol. Metab.*, **74,** 928.
18. Trifiro, M., Gottlieb, B., Pinsky, L., Kaufman, M., Prior, L., Belsham, D. D. *et al.* (1991) The 56/58KDa androgen-binding protein in male genital skin fibroblasts with a deleted androgen receptor gene. *Mol. Cell. Endo.*, **75,** 37.
19. Quigley, C. A., Evans, B. A. J., Simental, J. A., Marschke, K. B., Sar, M., Lubahn, D. B., Davies, P., Hughes, I. A., Wilson, E. M., and French, F. S. (1992) Complete androgen insensitivity due to deletion of exon C of the androgen receptor gene

highlights the functional importance of the second zinc finger of the androgen receptor *in vivo*. *Mol. Endocrinol.*, **6**, 1992.

20. Ris-Stalpers, C., Kuiper, G. G. J. M., Faber, P. W., Schweikert, H.-U., van Rooij, H. C. J., Zegers, N. D., Hodgins, M. B., deGenhart, H. J., Trapman, J., and Brinkmann, A. O. (1990) Aberrant splicing of androgen receptor mRNA results in the synthesis of a nonfunctional receptor protein in a patient with androgen insensitivity. *Proc. Natl Acad, Sci. USA*, **87**, 7866.
21. Marcelli, M., Tilley, W. D., Wilson, C. M., Griffin, J. E., Wilson, J. D., and McPhaul, M. J. (1990) Definition of the human androgen receptor gene structure permits the identification of mutations that cause androgen resistance: premature termination of the receptor protein at amino acid 588 causes complete androgen resistance. *Mol. Endocrinol.*, **4**, 1105.
22. Marcelli, M., Tilley, W. D., Wilson, C. M., Wilson, J. D., Griffin, J. E., and McPhaul, M. J. (1990) A single nucleotide substitution introduces a premature termination codon into the androgen receptor gene of a patient with receptor negative androgen resistance. *J. Clin. Invest.*, **85**, 1522.
23. Sai, T., Seino, S., Chang, C., Trifiro, M., Pinsky, L., Mhatre, A., Kaufman, M., Lambert, B., Trapman, J., Brinkmann, A. O., Rosenfield, R. L., and Liao, S. (1990) An exonic point mutation of the androgen receptor gene in a patient with complete androgen insensitivity. *Am. J. Hum. Genet.*, **46**, 1095.
24. Trifiro, M., Prior, R. L., Sabbaghian, N., Pinsky, L., Kaufman, M., Nylen, E. G., Belsham, D. D., Greenberg, C. R., and Wrogemen, K. (1991) Amber mutation creates a diagnostic *NaeI* site in the androgen receptor gene of a family with complete androgen insensitivity. *Am. J. Med. Genet.*, **40**, 493.
25. He, W. W., Kumar, M. V., and Tindall, D. J. (1991) A frame-shift mutation in the androgen receptor gene causes complete androgen insensitivity in the testicular-feminized mouse. *Nucleic Acids Res.*, **19**, 2373.
26. Charest, N. J., Zhou, Z.-X., Lubahn, D. B., Olsen, K. L., Wilson, E. M., and French, F. S. (1991) A frame-shift mutation destablizes androgen receptor messenger RNA in the *Tfm* mouse. *Mol. Endocrinol.*, **5**, 573.
27. Gaspar, M. L., Meo, T., Bourgarel, P. Gueret, J. L., and Tosi, M. (1991) A single base deletion in the *Tfm* androgen receptor gene creates a short lived messenger RNA that directs internal translation initiation. *Proc. Natl Acad. Sci. USA*, **88**, 8606.
28. Zoppi, S., Wilson, C. M., Harbison, M. D., Griffin, J. E., Wilson, J. D., McPhaul, M. J., and Marcelli, M. (1992) Complete testicular feminization caused by an amino terminal truncation of the androgen receptor with downstream initiation. *J. Clin. Invest.*, in press.
29. Simental, J. A., Sar, M., Lane, M. V., French, F. S., and Wilson, E. M. (1991) Transcriptional activation and nuclear targeting signals of the human androgen receptor. *J. Biol. Chem.*, **266**, 510.
30. Jentsen, G., vander Korput, H. A. G. M., van Vroonhoven, C., vander Kwast, T., Trapman, J., and Brinkmann, A. O. (1991) Domains of the human androgen receptor involved in steroid binding and transcriptional activation and subcellular localization. *Mol. Endocrinol.*, **5**, 1396.
31. Marcelli, M., Zoppi, S., Grino, P. B., Griffin, J. E., Wilson, J. D., and McPhaul, M. J. (1991) A mutation in the DNA-binding domain of the androgen receptor gene causes complete testicular feminization in a patient with receptor-positive androgen resistance. *J. Clin. Invest.*, **87**, 1123.

32. Zoppi, S., Marcelli, M., Deslypere, J.-P., Griffin, J. E., Wilson, J. D., and McPhaul, M. J. (1992) Amino acid substitutions in the DNA-binding domain of the human androgen receptor are a frequent cause of receptor positive androgen resistance. *Mol. Endocrinol.*, **6,** 409.
33. Lubahn, D. B., Brown, T. R., Simental, J. A., Higgs, H. N., Migeon, C. J., Wilson, E. M., and French, F. S. (1988) Sequence of the intron/exon junctions of the human androgen receptor gene and identification of a point mutation in a family with complete androgen insensitivity. *Proc. Natl Acad. Sci. USA*, **86,** 9534.
34. McPhaul, M. J., Marcelli, M., Tilley, W. D., Griffin, J. E., Isidro-Gutierrez, R. F., and Wilson, J. D. (1991) Molecular basis of androgen resistance in a family with a qualitative abnormality of the androgen receptor and responsive to high-dose androgen therapy. *J. Clin. Invest.*, **87,** 1413.
35. Ris-Stalpers, C., Trifiro, M. A., Kuiper, G. G. J. M., Jenster, G., Romolo, G., Sai, T., vanRooij, H. C. J., Kaufman, M., Rosenfield, R. L., Liao, S., Schweikert, H.-U., Trapman, J., Pinsky, L., and Brinkmann, A. O. (1991) Substitution of aspartic acid 686 by histidine or asparagine in the human androgen receptor leads to a functionally inactive protein with altered hormone-binding characteristics. *Mol. Endocrinol.*, **5,** 1562.
36. Brown, T. R., Lubahn, D. B., Wilson, E. M., French, F. S., Migeon, C. J., and Corden, J. L. (1990) Functional characterization of naturally occurring mutant androgen receptors from patients with complete androgen insensitivity. *Mol. Endocrinol.*, **4,** 1759.
37. McPhaul, M. J., Marcelli, M., Zoppi, S., Griffin, J. E., and Wilson, J. D. (1992) Mutations in the ligand-binding domain of the androgen receptor gene cluster in two regions of the gene. *J. Clin. Invest.*, **90,** 2097.
38. DeBellis, A., Quigley, C. A., Lane, M. V., Wilson, E. M., and French, F. S. (1992) Complete and partial androgen insensitivity syndromes due to point mutations in the androgen receptor gene. *J. Cell. Biochem. Suppl.*, **16C,** 1307 (Abstract).
39. Marcelli, M., Tilley, W. D., Zoppi, S., Griffin, J. E., Wilson, J. D., and McPhaul, M. J. (1991) Androgen resistance associated with a mutation of the androgen receptor at amino acid 772 (Arg → Cys) results from a combination of decreased messenger ribonucleic acid levels and impairment of receptor function. *J. Clin. Endocrinol. Metab.*, **73,** 318.
40. Marcelli, M., Zoppi, S., Wilson, C. M., Griffin, J. F., Wilson, J. D., and McPhaul, M. J. (1992) Amino acid substitutions in a small segment of exon 5 of the human androgen receptor gene cause complete testicular feminization by different mechanisms. *74th Annual Meeting of The Endocrine Society*, San Antonio, Texas, 24–27 June, Abstract 224.
41. LaSpada, A. R., Wilson, E. M., Lubahn, D. B., Harding, A. E., and Fischbeck, K. H. (1991) Androgen receptor gene mutations in X-linked spinal and bulbar muscular atrophy. *Nature*, **352,** 77.
42. Tilley, W. D., Marcelli, M., Wilson, J. D., and McPhaul, M. J. (1989) Characterization and expression of a cDNA encoding the human androgen receptor. *Proc. Natl Acad. Sci. USA*, **86,** 327.
43. McPhaul, M. J., Marcelli, M., Zoppi, S., Griffin, J. E., and Wilson, J. D. (1993) The spectrum of mutations in the androgen receptor gene that cause androgen resistance. *J. Clin. Endocrinol. Metab.*, **76,** 17.
44. Wilson, C. M., Griffin, J. E., Wilson, J. D., Marcelli, M., Zoppi, S., and McPhaul, M. J. (1992) Immunoreactive androgen receptor expression in patients with androgen resistance. *J. Clin. Endocrinol. Metab.*, **75,** 1474.

45. Refetoff, S., DeWind, L. T., and DeGroot, L. J. (1967) Familial syndrome combining deaf-mutism, stippled epiphyses, goiter, and abnormally high PBI: possible target organ refractoriness to thyroid hormone. *J. Clin. Endocrinol. Metab.*, **27**, 279.
46. Weiss, R. E. and Refetoff, S. (1992) Thyroid hormone resistance. *Annu. Rev. Med.*, **43**, 363.
47. Parilla, R., Mixson, A. J., McPherson, J. A., McClaskey, J. H., and Weintraub, B. D. (1991) Characterization of seven novel mutations of the c-*erb*Aβ gene in unrelated kindreds with generalized thyroid hormone resistance. *J. Clin. Invest.*, **88**, 2123.
48. Gershengon, M. C. and Weintraub, B. D. (1975) Thyrotropin-induced hyperthyroidism caused by selective pituitary resistance to thyroid hormone. *J. Clin. Invest.*, **56**, 633.
49. Kaplan, M. M., Swartz, S. L., and Larsen, P. R. (1981) Partial peripheral resistance to thyroid hormone. *Am. J. Med.*, **70**, 1115.
50. Sap, J., Munnoz, A., Damm, K., Goldberg, Y., Ghysdael, J., Leutz, A., Beug, H., and Vennström, B. (1986) The c-*erb*A protein is a high-affinity receptor for thyroid hormone. *Nature*, **324**, 635.
51. Weinberger, C., Thompson, C. C., Ong, E. S., Lebo, R., Gruol, D. J., and Evans, R. M. (1986) The c-*erb*A gene encodes a thyroid hormone receptor. *Nature*, **324**, 641.
52. Sakurai, A., Nakai, A., and DeGroot, L. J. (1989) Expression of three forms of thyroid hormone receptor in human tissues. *Mol. Endocrinol.*, **3**, 392.
53. Benbrook, D. and Pfahl, M. (1987) A novel thyroid hormone receptor encoded by a cDNA clone from a human testis library. *Science*, **238**, 788.
54. Mitsuhashi, T., Tennyson, G. E., and Nikodem, V. M. (1988) Alternative splicing generates messages encoding rat c-*erb*A proteins that do not bind thyroid hormone. *Proc. Natl Acad. Sci. USA*, **85**, 5804.
55. Hodin, R. A., Lazar, M. A., Wintman, B. I., Darling, D. S., Koenig, R. J., Larsen, P. R., Moore, D. D., and Chin, W. W. (1989) Identification of a thyroid hormone receptor that is pituitary-specific. *Science*, **244**, 76.
56. Refetoff, S. (1989) The syndrome of generalized resistance to thyroid hormone (GRTH). *Endocr. Res.*, **15**, 717.
57. Usala, S. J., Bale, A. E., Gesundleit, N., Weinberger, C., Lash, R. W., Wondisford, F. E., McBride, O. W., and Weintraub, B. D. (1988) Tight linkage between the syndrome of generalized thyroid hormone resistance and the human c-*erb*Aβ gene. *Mol. Endocrinol.*, **2**, 1217.
58. Sakurai, A., Takeda, K., Ain, K., Ceccarelli, P., Nakai, A., Seino, S., Bell, G. I., Refetoff, S., and DeGroot, L. J. (1989) Generalized resistance to thyroid hormone associated with a mutation in the ligand-binding domain of the human thyroid hormone receptor β. *Proc. Natl Acad. Sci. USA*, **86**, 8977.
59. Usala, S. J., Tennyson, G. E., Bale, A. E., Sash, R. W., Gesunheit, N., Wondisford, F. E., Accili, D., Hauser, P., and Weintraub, B. D. (1990) A base mutation of the c-*erb*Aβ thyroid hormone receptor in a kindred with generalized thyroid hormone resistance. *J. Clin. Invest.*, **85**, 93.
60. Takeda, K., Balzano, S., Sakurai, A., DeGroot, L. J., and Refetoff, S. (1991) Screening of nineteen unrelated families with generalized resistance to thyroid hormone for know point mutations in the thyroid hormone receptor β gene and the detection of a new mutation. *J. Clin. Invest.*, **87**, 496.
61. Usala, S. J., Menke, J. B., Watson, T. L., Wondisford, F. E., Weintraub, B. D., Bérard, J., Bradley, W. E. C., Ono, S., Mueller, O. T., and Bercu, B. B. (1991) A homozygous deletion on c-*erb*Aβ thyroid receptor gene in a patient with generalized thyroid hormone

resistance: isolation and characterization of the mutant receptor. *Mol. Endocrinol.*, **5,** 327.

62. Ono, S., Schwartz, I. D., Mueller, O. T., Root, A. W., Usala, S. J., and Bercu, B. B. (1991) Homozygosity for a dominant negative thyroid hormone receptor gene responsible for generalized resistance to thyroid hormone. *J. Clin. Endocrinol. Metab.*, **73,** 990.
63. Usala, S. J., Menke, J. B., Watson, T. L., Bérard, J., Bradley, W. E. C., Bale, A. E., Lash, R. W., and Weintraub, B. D. (1991) A new point mutation in the 3,5,3',-triidodthyronine-binding domain of the c-*erb*Aβ thyroid hormone receptor is tightly linked to generalized thyroid hormone resistance. *J. Clin. Endocrinol. Metab.*, **73,** 32.
64. Usala, S. J., Wondisford, F. E., Watson, T. L., Menke, J. B., and Weintraub, B. D. (1990) Thyroid hormone and DNA binding properties of a mutant C-*erb*Aβ receptor associated with generalized thyroid hormone resistance. *Biophys. Biochem. Res. Commun.*, **171,** 575.
65. Takeda, K., Weiss, R. E., and Refetoff, S. (1992) Rapid localization of mutations in the thyroid hormone receptor-β gene by denaturing gradient gel electrophoresis in 18 families with thyroid hormone resistance. *J. Clin. Endocrinol. Metab.*, **74,** 712.
66. Adams, M., Nagaya, T., Tone, Y., Jameson, J. L., and Chatterjee, V. K. K. (1992) Functional properties of a novel mutant thyroid hormone receptor in a family with generalized thyroid hormone resistance syndrome. *Clin. Endocrinol.*, **36,** 281.
67. Sakurai, A., Miyamoto, T., Refetoff, S., and DeGroot, L. J. (1990) Dominant negative transcriptional regulation by a mutant thyroid hormone receptor-β in a family with generalized resistance to thyroid hormone. *Mol. Endocrinol.*, **4,** 1988.
68. Chatterjee, V. K. K., Nagaya, T., Madison, L. D., Datta, S., Rentoumis, A., and Jameson, J. L. (1991) Thyroid hormone resistance syndrome. *J. Clin. Invest.*, **87,** 1977.
69. Meier, C. A., Dickstein, B. M., Ashizawa, K., McClaskey, J. H., Muchmore, P., Ransom, S. C., Menke, J. B., Hao, E.-H., Usola, S. J., Bercu, B. B., Cheng, S.-Y., and Weintraub, B. D. (1992) Variable transcriptional activity and ligand binding of mutant β_1 3,5,3'-triiodothyronine receptors from 4 families with generalized resistance to thyroid hormone. *Mol. Endocrinol.*, **6,** 248.
70. Kliewer, S. A., Umesono, K., Mangelsdorf, D. J., and Evans, R. M. (1992) Retinoid X receptor interacts with nuclear receptors in retinoic acid, thyroid hormone and vitamin D_3 signalling. *Nature*, **355,** 446.
71. Zhang, X.-kun, Hoffmann, B., Tran, P. B-V., Graupner, G., and Pfahl, J. (1991) Retinoid X receptor is an auxiliary protein for thyroid hormone and retinoic acid receptors. *Nature*, **355,** 441.
72. Yu, V. C., Delsert, C., Andersen, B., Holloway, J. M., Devary, O. V., Näär, A. M., Kim, S. Y., Boutin, J.-M., Glass, C. K., and Rosenfeld, M. G. (1991) RXRβ: a coregulator that enhances binding of retinoic acid, thyroid hormone, and vitamin D receptors to their cognate response elements. *Cell*, **67,** 1251.
73. Leid, M., Kastner, P., Lyons, R., Nakshatri, H., Saunders, M., Zacharewski, T., Chen, J.-Y., Staub, A., Garnier, J.-M., Mader, S., and Chambon, P. (1992) Purification, cloning, and RXR identity of the HeLa cell factor with which RAR or TR heterodimerizes to bind target sequences efficiently. *Cell*, **68,** 377.
74. Brooks, M. H., Bell, N. H., Love, L., Stern, P. H., Orfei, E., Queener, S. F., Hamstra, A. J., and DeLuca, H. F. (1978) Vitamin D resistant rickets type II. *N. Engl. J. Med.*, **298,** 996.
75. Rogers, J. F., Fleischman, A. R., Finberg, L., Hamstra, A., and DeLuca, H. F. (1979) Rickets with alopecia: an inborn error of vitamin D metabolism. *J. Pediatr.*, **94,** 729.

76. Liberman, U. A., Eil, C., and Marx, S. J. (1983) Resistance to 1,25-dihydroxy-vitamin D. *J. Clin. Invest.*, **71,** 192.
77. Hirst, M. A., Hockman, H. I., and Feldman, D. (1985) Vitamin D resistance and alopecia: a kindred with normal, 1,25-dihydroxyvitamin D binding, but decreased receptor affinity for deoxyribonucleic acid. *J. Clin. Endocrinol. Metab.*, **60,** 490.
78. Malloy, P. J., Hochberg, Z., Pike, J. W., and Feldman, D. (1989) Abnormal binding of vitamin D receptors to deoxyribonucleic acid in a kindred with vitamin D-dependent rickets, type II. *J. Clin. Endocrinol. Metab.*, **68,** 263.
79. Gamblin, G. T., Liberman, U. A., Eil, C., Downs, R. W., DeGrange, D. A., and Marx, S. J. (1985) Vitamin D-dependent rickets, type II. *J. Clin. Invest.*, **75,** 954.
80. Hughes, M. R., Malloy, P. J., Kieback, D. G., Kesterson, R. A., Pike, J. W., Feldman, D., and O'Malley, B. W. (1988) Point mutations in the human vitamin D receptor gene associated with hypocalcemic rickets. *Science*, **242,** 1702.
81. Sone, T., Marx, S. J., Liberman, U. A., and Pike, J. W. (1990) A unique point mutation in the human vitamin D receptor chromosomal gene confers hereditary resistance to 1,25-dihydroxyvitamin D_3. *Mol. Endocrinol.*, **4,** 623.
82. Saijo, T., Ito, M., Takeda, E., Mahbubul Huq, A. H. M., Naito, E., Yokota, I., Sone, T., Pike, J. W., and Kuroda, Y. (1991) A unique mutation in the vitamin D receptor gene in three Japanese patients with vitamin D-dependent rickets type II: utility of single-strand conformation polymorphism analysis for heterozygous carrier determination. *Am. J. Hum. Genet.*, **49,** 668.
83. Sone, T., Scott, R. A., Hughes, M. R., Malloy, P. J., Feldman, D., O'Malley, B. W., and Pike, J. W. (1989) Mutant vitamin D receptors which confer resistance to 1,25-dihydroxyvitamin D_3 in humans are transcriptionally inactive *in vitro*. *J. Biol. Chem.*, **264,** 20 230.
84. Ritchie, H. H., Hughes, M. R., Thompson, E. T., Malloy, P. J., Hochberg, Z., Feldman, D., Pike, J. W., and O'Malley, B. W. (1989) An ochre mutation in the vitamin D receptor gene causes hereditary 1,25-dihydroxyvitamin D_3-resistant rickets in three families. *Proc. Natl Acad. Sci. USA*, **86,** 9783.
85. Malloy, P. J., Hochberg, Z., Tiosano, D., Pike, J. W., Hughes, M. R., and Feldman, D. (1990) The molecular basis of hereditary 1,25-dihydroxyvitamin D_3 resistant rickets in seven related families. *J. Clin. Invest.*, **86,** 2071.
86. Malloy, P. J., Hughes, M. R., Pike, J. W., and Feldman, D. (1991) Vitamin D receptor mutations and hereditary 1,25-dihydroxyvitamin D resistant rickets. In *Vitamin D: Gene Regulation, Structure-Function Analysis and Clinical Application*, Norman, A. W., Bouillon, R., and Thomasset, M. (eds). Walter deGruyter, New York, p. 116.
87. Kristjansson, K., Rut, A. R., Hewison, M., O'Riordan, J. L. H., and Hughes, M. R. (1993) Two mutations in the hormone binding domain of the vitamin D receptor cause tissue resistance to 1,25 dihydroxyvitamin-D_3. *J. Clin. Invest.*, in press.
88. Vingerhoeds, A. C. M., Thijssen, J. H. H., and Schwarz, F. (1976) Spontaneous hypercortisolism without Cushing's syndrome. *J. Clin. Endocrinol. Metab.*, **43,** 1128.
89. Chrousos, G. P., Vingerhoeds, A., Brandon, D., Eil, C., Pugeat, M., DeVroede, M., Loriaux, D. L., and Lipsett, M. B. (1982) Primary cortisol resistance in man. *J. Clin. Invest.*, **69,** 1261.
90. Chrousos, G. P., Vingerhoeds, A. C. M., Loriaux, D. L., and Lipsett, M. B. (1983) Primary cortisol resistance: a family study. *J. Clin. Endocrinol. Metab.*, **56,** 1243.
91. Brönnegard, M., Werner, S., and Gustafsson, J.-A. (1986) Primary cortisol resistance

associated with a thermolabile glucocorticoid receptor in a patient with fatigue as the only symptom. *J. Clin. Invest.*, **78,** 1270.

92. Malchoff, C. D., Javier, E. C., Malchoff, D. M., Martin, T., Rogol, A., Brandon, D., Loriaux, D. L., and Reardon, G. E. (1990) Primary cortisol resistance presenting as isosexual precocity. *J. Clin. Endocrinol. Metab.*, **70,** 503.
93. Lamberts, S. W. J., Koper, J. W., Biemond, P., denHolder, F. H., and deJong, F. H. (1989) Familial and iatrogenic cortisol receptor resistance. *Cancer Res.*, **49,** 2217s.
94. Lamberts, S. W. J., Koper, J. W., Biewond, P., den Holder, F. H., and de Jong, F. H. (1992) Cortisol receptor resistance: the variability of its clinical presentation and response to treatment. *J. Clin. Endocrinol. Metab.*, **74,** 313.
95. Vecsei, P., Frank, K., Haack, D., Heinze, V., Ho, A.-D., Honour, J. W., Lewicka, S., Schoosch, M., and Ziegler, R. (1991) Primary glucocorticoid receptor defect with likely familial involvement. *Cancer Res.*, **49,** 2220s.
96. Nawata, H., Sekiya, K., Higuchi, K., Kato, K.-I., and Ibayashi, H. (1987) Decreased deoxyribonucleic acid binding of glucocorticoid–receptor complex in cultured skin fibroblasts from a patient with the glucocorticoid resistance syndrome. *J. Clin. Endocrinol. Metab.*, **65,** 219.
97. Iida, S., Gomi, M., Moriwaki, K., Itoh, Y., Hirobe, K., Matsuzawa, Y., Katagiri, S., Yonezawa, T., and Tarui, S. (1985) Primary cortisol resistance accompanied by a reduction in glucocorticoid receptors in two members of the same family. *J. Clin. Endocrinol. Metab.*, **60,** 967.
98. Iida, S., Gomi, M., Moriwaki, K., Fujii, H., Tsugawa, M., and Tarui, S. (1989) Properties of glucocorticoid receptors in Epstein–Barr virus-transformed lymphocytes from patients with familial cortisol resistance. *Cancer Res.*, **49,** 2214s.
99. Hurley, D. M., Accili, D., Stratakis, C. A., Karl, M., Vamvakopoulos, N., Rorer, E., Constantine, K., Taylor, S. I., and Chrousos, G. P. (1991) Point mutation causing a single amino acid substitution in the hormone-binding domain of the glucocorticoid receptor in familial glucocorticoid resistance. *J. Clin. Invest.*, **87,** 680.
100. Brufsky, A. M., Malchoff, D. M., Javier, E. C., Reardon, G., Rowe, D., and Malchoff, C. D. (1990) A glucocorticoid receptor mutation in a subject with primary cortisol resistance. *Trans. Assoc. Am. Physicians*, **103,** 53.
101. Karl, M., Lamberts, S. W. J., Detera-Wadleigh, S. D., Stratakis, C. A., Hurley, D., Accili, D., and Chrousos, G. P. (1992) Splice site deletion in the glucocorticoid receptor gene causes familial glucocorticoid resistance. *74th Annual Meeting of The Endocrine Society*, San Antonio, Texas, 24–27 June, Abstract 564.

Index

androgen binding protein 1–10
androgen insensitivity 186–92
androgen receptor 70, 188–96
androgen resistance 186–92
androgens 39
antiandrogens 170–1
α-antitrypsin, AAT 12–14
antioestrogens 166–70
antiprogestins 171–2
AP-1 134
  fos 135
  jun 134

breast cancer
  cell lines 169–70
  human MCF-7 cells 169–70

cancer
  breast 168–70
  prostate 170
chicken ovalbumin upstream promoter (COUP) transcription factor 50–1
clomiphene 168
composite response elements 38–9
conservation of nuclear receptors 45–6, 55, 96, 141–2
co-operative interaction between steroid receptors 50, 102
co-operative interaction between transcription factors 50
corticosteroid-binding globulin 10–17
cortisol 26–34
COUP-TF 50–1
crystallographic analysis 153–8

dimerization 49–50, 98–105, 148–50
  heterodimers 50, 128
DNA binding 98–105
  binding domain 103–5, 145–8
DNA binding sites
  competition for 129–30
  composite response elements 38–9
  negative hormone response elements 118, 122–6
  negative glucocorticoid response elements 122–4
  oestrogen response elements 98, 126, 143–5
  steroid response elements 48–50
  thyroid hormone response elements 60, 102–3, 143–5
  vitamin D response elements 127–8, 195–6
dominant negative effects of receptors 127–8, 195–6

evolutionary conversion of receptors 45–6, 55, 96, 141–2

α-fetoprotein 132
flutamide 170–1
follicle stimulating hormone, FSH 8
fos 135

glucocorticoid receptor 27, 66–7, 199–200
glucocorticoid resistance 198–200
glucocorticoid response elements 98, 143–5
glucocorticoids 15–16
glycoprotein hormones 124, 131

heat-shock proteins 64–85
  hsp56 53, 73–5
  hsp70 53, 71–4, 81–2
  hsp90 48, 53, 66–71, 74, 84
heterodimers 128
hirsutism 199
hormone antagonists 166–78
  antiandrogens 170–1
  antioestrogens 166–70
  antiprogestins 171–2
  clomiphene 168
  flutamide 170–1
  4-hydroxytamoxifen 166–70, 175–6
  ICI 164384 166–9, 177–8
  ICI 182780 166–9, 177–8
  RU 486 171, 174–6
  tamoxifen 166–70, 175–6
hormone resistance disorders 186–201
hormone response elements 35, 143–5
hsp56 53, 73–5
hsp70 53, 71–4, 81–2
hsp90 48, 53, 66–71, 74, 84
human MCF-7 cells 169–70
hypertension 199
hypocalcaemic vitamin D resistant rickets 197
11β-hydroxysteroid dehydrogenase 29–37
4-hydroxytamoxifen 166–70, 175–6

ICI 164384 166–9, 177–8
ICI 182780 166–9, 177–8

jun 134

mineralocorticoid receptors 28–32, 70
mineralocorticoid specificity 32–4
mouse mammary tumour virus 99

negative glucocorticoid response elements 122–4
negative hormone response elements 118, 122–6
nuclear localization 97–8
nuclear magnetic resonance 151–3
nucleocytoplasmic shuttling 177–8
Nur 77 55–7

oestrogen receptor 18, 70, 99–100
oestrogen response elements 98, 126, 143–5
oestrogens 39
orphan receptors 45–8, 54–6
osteocalcin 123, 132

partial agonists 175–6
phosphorylation 56–7
phylogenetic tree of receptors 55
progesterone receptor 27, 51, 68, 101–2
prolactin 123–4, 131
proliferin 38–9, 123–4
pro-opiomelanocortin gene 122–3, 131
prostate cancer 170
pure antioestrogens 177–8

receptor heterocomplexes 78–82
receptor isoforms 108
receptor transformation 83–4
receptors
  androgen receptor 70, 188–96
  glucocorticoid receptor 27, 66–7, 199–200
  mineralocorticoid receptors 28–32, 70
  oestrogen receptor 18, 70, 99–100
  orphan receptors 45–8, 54–6
  progesterone receptor 27, 51, 68, 101–2
  retinoic acid receptors 70, 102–3
  thyroid hormone receptors 45–6, 70, 194–7
  vitamin D receptor 27, 197–8

Reifenstein syndrome 187, 191
repression
  by interference 119–21
  by quenching 119–21, 133–4
  by squelching 119–21
  by tethering 119–21
response elements 98–105
  composite response elements 38–9
  negative hormone response elements 118, 122–6
  negative glucocorticoid response elements 122–4
  oestrogen response elements 98, 126, 143–5
  steroid response elements 48–50
  thyroid hormone response elements 60, 102–3, 143–5
  vitamin D response elements 143–4
retinoic acid receptors 70, 102–3
retinoid X receptors 50, 171–2
RU 486 171, 174–6

serine proteinase inhibitor, SERPIN 10–14
sex steroid-binding globulin 1–11
silencer 129
squelching 107
steroid hormone antagonists 166–78
  antiandrogens 170–1
  antioestrogens 166–70
  antiprogestins 171–2
  clomiphene 168
  flutamide 170–1
  ICI 164, 384 166–9, 177–8
  ICI 182, 780 166–9, 177–8
  4-hydroxytamoxifen 166–70, 175–6
  RU 486 171, 174–6
  tamoxifen 166–70, 175–6
steroid hormone receptors
  androgen receptor 70, 188–96
  glucocorticoid receptor 27, 66–7, 199–200
  mineralocorticoid receptors 28–32, 70
  oestrogen receptor 18, 70, 99–100
  progesterone receptor 27, 51, 101–2
steroid hormones
  androgens 39
  glucocorticoids 15–16
  oestrogens 39
steroid receptor heterocomplexes 78–82
steroid response elements 48–50

tamoxifen 166–70, 175–6
testicular feminization 186–92
TFIID 52–2, 106–7
thyroid hormone receptors 45–6, 70, 194–7
thyroid hormone resistance 193–7
thyroid hormone response elements 60, 102–3, 143–5
transcriptional activation 51–3, 56, 105–8
transcriptional repression 118–34
transcription factors 51–3
transcription intermediary factors 106–8
transformation 83
thyroid stimulating hormone, TSH 125–6, 132–3

v-erb A 45
vitamin D 39, 70, 197–8
vitamin D receptor 27, 197–8
vitamin D resistance 158–9, 197–8
vitamin D response elements 143–4

X-ray crystallographic analysis 153–8

zinc finger co-ordination 146–7, 150–1